Applied Mathematical Sciences
Volume 47

T0192167

Springer
New York
Berlin
Heidelberg
Barcelona
Hong Kong
London
Milan
Paris
Singapore
Tokyo

Applied Mathematical Sciences

(continued following index)

Jack K. Hale   Luis T. Magalhães
Waldyr M. Oliva

Dynamics in
Infinite Dimensions

Second Edition

Appendix by Krzysztof P. Rybakowski

With 15 Figures

Springer

Jack K. Hale
School of Mathematics
Georgia Tech
Atlanta, GA 30332-0001
USA
hale@math.gatech.edu

Luis T. Magalhães
Departamento de Matemática
Instituto Superior Técnico
Av. Rovisco Pais
Lisbon 1049-001
Portugal
lmagal@math.ist.utl.pt

Waldyr M. Oliva
Departamento de Matemática
Instituto Superior Técnico
Av. Rovisco Pais
Lisbon 1049-001
Portugal
wamoliva@math.ist.utl.pt

Editors

S.S. Antman
Department of Mathematics
and
Institute for Physical Science
  and Technology
University of Maryland
College Park, MD 20742-4015, USA
ssa@math.umd.edu

J.E. Marsden
Control and Dynamical Systems, 107-81
California Institute of Technology
Pasadena, CA 91125, USA
marsden@cds.caltech.edu

L. Sirovich
Division of Applied Mathematics
Brown University
Providence, RI 02912, USA
chico@camelot.mssm.edu

Mathematics Subject Classification (2000): 37Lxx, 37Kxx

Library of Congress Cataloging-in-Publication Data
Hale, Jack K.
  Dynamics in infinite dimensions / Jack K. Hale, Luis T. Magalhaes, Waldyr M. Oliva.—2nd ed.
    p.   cm. — (Applied mathematical sciences ; 47)
  Rev. ed. of: Introduction to infinite dimensional dynamical system—geometri theory. c1984.
  Includes bilbliographical references and index.
    1. Differentiable dynamical system.   I. Magalhaes, Luis T.   II. Oliva, Waldyr M.   III.
  Hale, Jack K. Introduction to infinite dimensional dynamical system—geometric theory.
  IV. Title.   V. Applied mathematical sciences (Springer-Verlag New York Inc.) ; v. 47.
  QA1 .A647 vol. 47 2002
  [QA614.8]
  510 s—dc21
  [514´.74]   2002024179

ISBN 978-1-4419-3012-5   e-ISBN 978-0-387-22896-9

Printed in the United States of America.

9 8 7 6 5 4 3 2 1

www.springer-ny.com

Springer-Verlag   New York Berlin Heidelberg
A member of BertelsmannSpringer Science+Business Media GmbH

Preface

In our book published in 1984 *An Introduction to Infinite Dimensional Dynamical Systems-Geometric Theory*, we presented some aspects of a geometric theory of infinite dimensional spaces with major emphasis on retarded functional differential equations. In this book, the intent is the same. There are new results on Morse–Smale systems for semiflows, persistence of hyperbolicity under perturbations, nonuniform hyperbolicity, monotone dynamical systems, realization of vector fields on center manifolds and normal forms. In addition, more attention is devoted to neutral functional differential equations although the theory is much less developed. Some parts of the theory also will apply to many other types of equations and applications.

Jack K. Hale
Luis T. Magalhães
Waldyr M. Oliva

Contents

1 Introduction

There is an extensive theory for the flow defined by dynamical systems generated by continuous semigroups $T : \mathbb{R}^+ \times \mathcal{M} \to \mathcal{M}$, $T(t, x) := T(t)x$, where $T(t) : \mathcal{M} \to \mathcal{M}$, $\mathbb{R}^+ = [0, \infty)$, and $\mathcal{M}$ is either a finite dimensional compact manifold without boundary or a compact manifold with boundary provided that the flow is differentiable and transversal to the boundary. The basic problem is to compare the flows defined by different dynamical systems. This comparison is made most often through the notion of topological equivalence. Two semigroups T and S defined on $\mathcal{M}$ are *topologically equivalent* if there is a homeomorphism from $\mathcal{M}$ to $\mathcal{M}$ which takes the orbits of T onto the orbits of S and preserves the sense of direction in time.

If the semigroups are defined on a finite dimensional Banach space X, then extreme care must be exercised in order to compare the orbits with large initial data and only very special cases have been considered. One way to avoid the consideration of large initial data in the comparison of semigroups is to consider only those semigroups for which infinity is unstable; that is, there is a bounded set which attracts the *positive orbit* of each point in X. In this case, there is a *compact global attractor* $\mathcal{A}(T)$ of the semigroup T; that is, $\mathcal{A}(T)$ is *compact invariant* ($T(t)\mathcal{A}(T) = \mathcal{A}(T)$ for all $t \geq 0$) and, in addition, for any bounded set $B \subset X$, dist $_X(T(t)B, \mathcal{A}(T)) \to 0$ as $t \to \infty$. In such situations, it is often possible to find a neighborhood $\mathcal{M}$ of $\mathcal{A}(T)$ for which the closure is a compact manifold with boundary and the boundary is transversal to the flow. Therefore, the global theory of finite dimensional dynamical systems can be applied.

We remark that the invariance of $\mathcal{A}(T)$ implies that, for each $x \in \mathcal{A}(T)$, we can define a bounded *negative orbit* (or a bounded *backward extension*) through x; that is, a function $\varphi : (-\infty, 0] \to X$ such that $\varphi(0) = x$ and, for any $\tau \leq 0$, $T(t)\varphi(\tau) = \varphi(t + \tau)$ for $0 \leq t \leq -\tau$. If the compact global attractor $\mathcal{A}(T)$ exists, then it is given by

$$\mathcal{A}(T) = \{x \in X : T(t)x \text{ is defined and bounded for } t \in \mathbb{R}\}. \qquad (1.1)$$

Many applications involve semigroups T on a non-locally compact space X; for example, semigroups generated by partial differential equations and delay differential or functional differential equations (see for example [8], [78], [86], [198] and the references therein). The first difficulty in the non-locally compact case is to decide how to compare two semigroups. It seems to be almost impossible to make a comparison of all or even an arbitrary bounded set of the space X. On the other hand, we can define in the non-locally compact case the set $\mathcal{A}(T)$ as in (1.1). This set will contain all of the bounded invariant sets of T and, under some reasonable conditions, should contain all of the information about the limiting behavior of solutions. For this reason, we make comparisons of semigroups only on $\mathcal{A}(T)$. This does not mean that the transient behavior is unimportant, but only that our emphasis here is on $\mathcal{A}(T)$. The following definition first appeared in a paper by Hale (see [73]) in 1981.

Definition 1.0.1. *We say that a semigroup T on X is equivalent to a semigroup S on X, $T \sim S$, if there is a homeomorphism $h : \mathcal{A}(T) \to \mathcal{A}(S)$ which preserves orbits and the sense of direction in time.*

We reemphasize that, in the definition of equivalence, we restrict to the set $\mathcal{A}(T)$ and not to a neighborhood of $\mathcal{A}(T)$. Due to the fact that we are not able to take this full neighborhood, adaptation of the finite dimension theory of dynamical systems to our setting is nontrivial. Also, we will need to impose further restrictions on the classes of semigroups that will be considered.

If $\mathcal{A}(T)$ is not compact, there is very little known about general flows. If $\mathcal{A}(T)$ is compact, then we can easily verify the following result.

Proposition 1.0.2. *If $\mathcal{A}(T)$ is compact, then $\mathcal{A}(T)$ is the maximal compact invariant set. If, in addition, for each $t \geq 0$, $T(t)$ is one-to-one on $\mathcal{A}(T)$, then T is a continuous group on $\mathcal{A}(T)$.*

In a particular application, the semigroup defining the dynamical system depends upon parameters. In the case of ordinary differential equations or functional differential equations, the parameter could be a particular class of vector fields. If the semigroup is generated by partial differential equations, the parameters could be a class of vector fields or the boundary of the region of definition or the boundary conditions or all of these. A basic problem is to know if the flow defined by the a semigroup is preserved under the above equivalence relation when one allows variations in the parameters. More precisely, we make the following definitions.

Definition 1.0.3. *Suppose that X is a complete metric space, Λ is a metric space, $T : \Lambda \times \mathbb{R}^+ \times X \to X$ is continuous and, for each $\lambda \in \Lambda$, let $T_\lambda : \mathbb{R}^+ \times X \to X$ be defined by $T_\lambda(t)x = T(\lambda, t, x)$ and suppose that T_λ is a continuous semigroup on X for each $\lambda \in \Lambda$. Define $\mathcal{A}(T_\lambda)$ as above. The semigroup T_λ is said to be $\mathcal{A}$-stable if there is a neighborhood $U \subset \Lambda$ of λ such that $T_\lambda \sim T_\mu$ for each $\mu \in U$. We say that T_λ is a bifurcation point if T_λ is not $\mathcal{A}$-stable.*

The basic problem is to discuss detailed properties of the set $\mathcal{A}(T_\lambda)$, the structure of the flow on $\mathcal{A}(T_\lambda)$ and the manner in which $\mathcal{A}(T_\lambda)$ changes with λ.

Some basic questions that should be discussed are the following:

1. Is T_λ generically one-to-one on $\mathcal{A}(T_\lambda)$?
2. If T_λ is $\mathcal{A}$-stable, is T_λ one-to-one on $\mathcal{A}(T_\lambda)$?
3. For each $x \in \mathcal{A}(T_\lambda)$, what are the smoothness properties of $T_\lambda(t)x$ in t? For example, does it possess the same smoothness properties as the semigroup has in x and or λ?
4. Is the Hausdorff dimension and capacity of $\mathcal{A}(T_\lambda)$ finite?
5. When is $\mathcal{A}(T_\lambda)$ a manifold or the union of a finite number of manifolds?
6. Can $\mathcal{A}(T_\lambda)$ be embedded in a finite dimensional manifold generically in λ?
7. Can $\mathcal{A}(T_\lambda)$ be embedded in a finite dimensional invariant manifold generically in λ?
8. Are Morse–Smale systems open and $\mathcal{A}$-stable?
9. Are Kupka–Smale semigroups generic in the class $\{T_\lambda, \lambda \in \Lambda\}$?

In these notes, we attempt to discuss these questions in some detail in order to indicate how one can begin to obtain a geometric theory for dynamical systems in infinite dimensions. We present some results which apply to many types of situations including functional differential equations of retarded and neutral type, quasilinear parabolic partial differential equations and dissipative hyperbolic partial differential equations. Some of the more detailed results are for a class of semigroups satisfying compactness and smoothness hypotheses and are directly applicable to retarded functional differential equations with finite delay, quasilinear parabolic equations and more general situations. There are many important applications which do not satisfy the compactness and smoothness hypotheses; for example, retarded equations with infinite delay, neutral functional differential equations, the linearly damped nonlinear wave equation as well as other equations of hyperbolic type. Throughout, we will note the difficulties involved in the extensions to more general semigroups. As will be clear, the theory is still in its infancy.

Our theory is presented for the case in which $\mathcal{A}(T)$ is compact. In Chapter 2, we present conditions on the semigroup T and dissipative properties of the flow which will imply that $\mathcal{A}(T)$ is compact and, therefore, is the maximal compact invariant set. Also, we give conditions which are necessary and sufficient for $\mathcal{A}(T)$ to be the compact global attractor.

In Chapter 3, we give the definitions and examples of retarded and neutral functional differential equations on manifolds, discuss the basic properties of the semigroups defined by these equations, the existence of compact global attractors and the differences between these two types of equations.

In Chapter 4, we show that the compact global attractor has finite capacity for a class of mappings which includes the time-one maps of retarded and neutral functional differential equations, linearly damped hyperbolic equations as well as many other types of equations.

In Chapter 5, we give some examples illustrating the importance of discussing the manner in which the flow on the compact global attractor depends upon parameters. A rather complete investigation is made for a retarded functional differential equation serving as a model in viscoelasticity and known as the Levin-Nohel equation with the parameter being the relaxation function. Also, a complete description is given for the flow on the attractor for a scalar parabolic equation in one dimension. Some details in the proof are referred to Chapter 10. A counter-example for the Hartman-Grobman theorem in the setting of Hadamard derivatives is also described.

In Chapter 6, the definitions of Morse–Smale maps and flows are given. The stability of Morse–Smale maps was proved in [87] and is reproduced here. The stability for semiflows is a recent result appearing in [154] and, as will be seen, there are some conditions imposed on the flow which involve smoothness. These smoothness conditions are satisfied for retarded functional differential equations and parabolic equations, but are not satisfied for neutral functional differential equations and partial differential equations for which the solutions do not smooth in time. It would be very interesting to extend the results in this chapter to more general situations.

Chapter 7 is devoted to the persistence under perturbations for uniformly hyperbolic invariant sets of semiflows, assuming the smoothness condition mentioned above. The hypothesis that the flow is one-to-one on a compact invariant set implies the existence of a conjugacy between perturbed and unperturbed semiflows; if the flow on the invariant set is not one-to-one, one obtains only a semi-conjugacy. Hyperbolic measures and nonuniform hyperbolicity together with the corresponding concepts of invariant manifolds are discussed in the finite dimensional case with some remarks on perspectives for the infinite dimensional setting.

Even though the flow defined by evolutionary equations defined by functional differential equations and partial differential equations are defined on an infinite dimensional space, the particular type of equation considered may impose restrictions on the flow. This can play a very important role in the development of the geometric theory and have an important impact on the types of bifurcations that may occur. In Chapter 8, we characterize the flows that can occur on center manifolds for retarded and neutral functional differential equations. This chapter also contains a complete theory of normal forms for these equations as well as abstract evolutionary equations with delays with applications.

In Chapter 9, we give conditions under which the compact global attractor will be a smooth manifold taking into account the recent literature on this subject.

In the new Chapter 10 on monotonicity, we present a general class of monotone operators for which it is possible to show that the stable and unstable manifolds of hyperbolic critical elements are transversal. Applications are given to ordinary and parabolic partial differential equations as well as their time and space discretizations. This chapter also contains a presentation of the Morse decomposition of the flow on the compact global attractor for a differential delay equation with negative feedback. The proofs of all results depend in a significant way upon a discrete Lyapunov function.

Chapter 11 on the Kupka–Smale Theorem as well as the Appendix on Homotopy Index Theorem are essentially the same as in the 1984 book [87].

To assist the reader, we sometimes repeat concepts and theorems in various chapters.

We would like to thank a number of colleagues from several Institutions who motivated and helped us with comments and suggestions when we were writing and typing this text. Among them we mention Carlos Rocha, Luis Barreira, Rui Loja Fernandes, João Palhoto de Matos, Pedro G. Henriques, Pedro Girão, Esmeralda Dias, Luiz Fichmann, Antonio Luis Pereira, Sérgio Oliva, Dan Henry and Maria do Carmo Carbinatto. In a special acknowledgment we want to say thanks to Teresa Faria who developed a complete theory of normal forms in Chapter 8. Thanks also to the members of CAMGSD and ISR of Instituto Superior Técnico (UTL), and for the partial support by FCT (Portugal) through the program POCTI.

2 Invariant Sets and Attractors

In this chapter, the basic theory of invariant sets and attractors is summarized and many examples are given. Complete proofs may be found in [78] and the references cited in the text.

Suppose that X is a complete metric space with metric d and let $\mathbb{R}^+ = [0, \infty)$, $\mathbb{R}^- = (-\infty, 0]$. A mapping $T : \mathbb{R}^+ \times X \to X$, $(t, x) \mapsto T(t)x$, is said to be a C^0-*semigroup* (or a *continuous semiflow*) (or a C^0-*dynamical system*) if

(i) $T(0) = I$,
(ii) $T(t + s) = T(t)T(s), t, s \in \mathbb{R}^+$,
(iii) The map $(t, x) \mapsto T(t)x$ is continuous in t, x for $(t, x) \in \mathbb{R}^+ \times X$.

For any $x \in X$, the *positive orbit* $\gamma^+(x)$ through x is defined as $\gamma^+(x) = \cup_{t \geq 0} T(t)x$. A *negative orbit* $\gamma^-(x)$ through x is the image $y(\mathbb{R}^-)$ of a continuous function $y : \mathbb{R}^- \to X$ such that, for any $t \leq s \leq 0$, $T(s - t)y(t) = y(s)$. A *complete orbit* $\gamma(x)$ through x is the union of $\gamma^+(x)$ and a negative orbit through x.

Since the range of $T(t)$ need not be the whole space, to say that there is a negative orbit through x may impose restrictions on x. Since $T(t)$ may not be one-to-one, there may be more than one negative orbit through x if one exists. We define *the negative orbit* $\Gamma^-(x)$ through x as the union of all negative orbits through x. *The complete orbit* $\Gamma(x)$ through x is $\Gamma(x) = \gamma^+(x) \cup \Gamma^-(x)$.

For any subset B of X, we let $\gamma^+(B) = \cup_{x \in B} \gamma^+(x)$, $\Gamma^-(B) = \cup_{x \in B} \Gamma^-(x)$, $\Gamma(B) = \cup_{x \in B} \Gamma(x)$ be respectively the positive orbit, negative orbit, complete orbit through B.

The limiting behavior of $T(t)$ as $t \to \infty$ is of fundamental importance. For this reason, for $x \in X$, we define $\omega(x)$, the ω-*limit set of x* or the ω-*limit set of the positive orbit through x*, as

$$\omega(x) = \cap_{\tau \geq 0} \mathrm{Cl}\gamma^+(T(\tau)x).$$

This is equivalent to saying that $y \in \omega(x)$ if and only if there is a sequence $t_k \to \infty$ as $k \to \infty$ such that $T(t_k)x \to y$ as $k \to \infty$. In the same way, for any set $B \subset X$, we define $\omega(B)$, the ω-*limit set of B* or the ω-*limit set of the positive orbit through B*, as

$$\omega(B) = \cap_{\tau \geq 0} \mathrm{Cl}\gamma^+(T(\tau)B).$$

This is the same as saying that $y \in \omega(B)$ if and only if there are sequences $t_k \to \infty$ as $k \to \infty$, $x_k \in B$, such that $T(t_k)x_k \to y$ as $k \to \infty$.

Analogously, we can define the α-*limit set of a negative orbit* $\gamma^-(x)$ *or of the negative orbit* $\Gamma^-(x)$ *of a point* x as well as the same concepts for a set $B \subset X$.

We remark that $\omega(B) \supset \cup_{x \in B}\omega(x)$, but equality may not hold. In fact, suppose that $f : \mathbb{R} \to \mathbb{R}$ is a C^1-function for which there is a constant M such that $xf(x) < 0$ for $|x| > M$ and consider the scalar ODE $\dot{x} = f(x)$. For each $x \in \mathbb{R}$, $\omega(x)$ is an equilibrium point. If the zeros of f are simple, then, for any interval B containing at least two equilibrium points, the set $\cup_{x \in B}\omega(x)$ is disconnected, whereas $\omega(B)$ is an interval. For $f(x) = x - x^3$, $B = [-2, 2]$, we have $\cup_{x \in B}\omega(x) = \{0, \pm 1\}$, whereas $\omega(B) = [-1, 1]$.

To state a result about ω-limit sets, we need some additional notation. A set $A \subset X$ is said to be *invariant* (under the semigroup T) if $T(t)A = A$ for $t \geq 0$. We say that a set A *attracts* a set B under the semigroup T if $\lim_{t \to \infty} \mathrm{dist}_X (T(t)B, A) = 0$, where

$$\mathrm{dist}_X (B, A) = \sup_{x \in B} \mathrm{dist}_X (x, A) = \sup_{x \in B} \inf_{y \in A} \mathrm{dist}_X (x, y).$$

Lemma 2.0.1. *If $B \subset X$ is a nonempty bounded set for which there is a compact set J which attracts B, then $\omega(B)$ is nonempty, compact, invariant and attracts B. In addition, if $\omega(B) \subset B$, then*

$$\omega(B) = \cap_{t \geq 0} T(t)B.$$

In particular, if $B \subset X$ is a nonempty subset of X and there is a $t_0 > 0$ such that $\mathrm{Cl}\,\gamma^+(T(t_0)B)$ is compact, then $\omega(B)$ is nonempty, compact, invariant and $\omega(B)$ attracts B. If B is connected, then $\omega(B)$ is connected.

A compact invariant set A is said to be the *maximal compact invariant set* if every compact invariant set of T is contained in A. An invariant set A is said to be a *compact global attractor* if A is a maximal compact invariant set which attracts each bounded set of X. Notice that this implies that $\omega(B) \subset A$ for each bounded set B. It is easy to verify the following result.

Lemma 2.0.2. *If $\mathcal{A}(T)$ is compact, then $\mathcal{A}(T)$ is the maximal compact invariant set. If, for each $x \in X$, $\gamma^+(x)$ has compact closure, then $\mathcal{A}(T)$ attracts points of X. If, for any bounded set $B \subset X$, $\omega(B)$ is compact and attracts B, then $\mathcal{A}(T)$ is the compact global attractor. If $T(t)$ is one-to-one on $\mathcal{A}(T)$ for each $t \geq 0$, then T is a continuous group on $\mathcal{A}(T)$.*

To proceed further, we need some concepts of stability of an invariant set J of a continuous semigroup T. The set J is *stable* if, for any neighborhood V of J, there is a neighborhood U of J such that $T(t)U \subset V$ for all $t \geq 0$. The set J *attracts points locally* if there is a neighborhood W of J such that J attracts points of W. The set J is *asymptotically stable* if it is stable and attracts points locally. The set J is a *local attractor* or, equivalently, *uniformly asymptotically stable* if it is stable and attracts a neighborhood of J.

Lemma 2.0.3. *An invariant set J is stable if, and only if, for any neighborhood V of J, there is a neighborhood $V' \subset V$ such that $T(t)V' \subset V'$ for all $t \geq 0$. A compact invariant set J is a local attractor if and only if there is a neighborhood V of J with $T(t)V \subset V$ for all $t \geq 0$ and J attracts V.*

The following basic result on the existence of the maximal compact invariant set is due to Hale, LaSalle and Slemrod (see [83]).

Theorem 2.0.4. *If the semigroup T on X is continuous and there is a nonempty compact set K that attracts compact sets of X and $A = \cap_{t \geq 0} T(t)K$, then A is independent of K and*

(i) A is the maximal compact invariant set,
(ii) A is connected if X is connected,
(iii) A is stable and attracts compact sets of X.

It is possible to have a semigroup satisfying the conditions of Theorem 2.0.4 and yet the set $A(T)$ is not a compact global attractor even though it is the maximal compact stable invariant set. As noted by Hale [80], this can be seen for linear semigroups on a Banach space X. In the statement of the result, we let $r(E\sigma(A))$ denote the radius of the essential spectrum of a linear operator A on a Banach space.

Theorem 2.0.5. *If T is a linear C^0-semigroup on a Banach space X and the origin $\{0\}$ attracts each point of X, then*

(i) $\{0\}$ is stable, attracts compact sets and is the maximal compact invariant set.
(ii) $\gamma^+(B)$ is bounded if B is bounded.
(iii) $\{0\}$ is the compact global attractor if, and only if, there is a $t_1 > 0$ such that $r(E\sigma(T(t_1))) < 1$.
(iv) If there is a t_1 such that $r(E\sigma(T(t_1))) = 1$, then the origin attracts compact sets, but is not a compact global attractor.

If X is a Banach space and T is a continuous linear semigroup for which there is a $t_1 > 0$ such that $T(t_1)$ is a completely continuous operator, then $r(E\sigma(T(t_1))) = 0$. Property (iii) of Theorem 2.0.5 implies that $\{0\}$ is the compact global attractor if it attracts each point of X.

We give two examples of interesting evolutionary equations which satisfy the conditions in (iv) of Theorem 2.0.5.

Example 2.0.6. (Neutral delay differential equation). Consider the neutral delay differential equation

$$\frac{d}{dt}[x(t) - ax(t-1)] + cx(t) = 0, \quad t \geq 0, \tag{2.1}$$

where c, a are constants. For any $\varphi \in X = C([-1, 0], \mathbb{R})$, X with the sup norm, we can use this equation to define a function $x(t, \varphi)$, $t \geq -1$, with

$x(\theta, \varphi) = \varphi(\theta)$, $-1 \le \theta \le 0$, the function $x(t, \varphi) - ax(t - 1, \varphi)$ is C^1 on $(0, \infty)$ with a continuous right hand derivative on $[0, \infty)$ and (2.1) is satisfied for $t \ge 0$. If we let $T_{(2.1)}(t)\varphi$ be defined by $(T_{(2.1)}(t)\varphi)(\theta) = x(t + \theta, \varphi)$, $-1 \le \theta \le 0$, $t \ge 0$, then $T_{(2.1)} : \mathbb{R}^+ \times X \to X$ is a linear C^0-semigroup on X.

It can be shown that $r(E\sigma(T_{(2.1)}(t))) = e^{t\ln|a|}$ for all $t \ge 0$. Therefore, for $|a| < 1$, we have $r(E\sigma(T_{(2.1)}(t))) < 1$ for $t > 0$. If $\{0\}$ attracts each point of X, then $\{0\}$ is the compact global attractor by (iii) of Theorem 2.0.5. If $a = -1$, then $r(E\sigma(T_{(2.1)}(t))) = 1$. Furthermore, it can be shown (but it is not easy) that $\{0\}$ attracts points of X if $c > 0$. From Theorem 2.0.5 part (iv), it follows that $\{0\}$ is the maximal compact invariant set but is not the compact global attractor.

Example 2.0.7. (Locally damped wave equation). Another interesting example of a linear map that satisfies the conditions of Theorem 2.0.5 is the linearly locally damped wave equation

$$\partial_t^2 u + \beta(x)\partial_t u - \Delta u = 0, \quad x \in \Omega, \tag{2.2}$$

with the Dirichlet boundary conditions $u = 0$ in $\partial\Omega$. The domain $\Omega \subset \mathbb{R}^n$ is assumed to be bounded and have a smooth boundary. The function β is continuous and nonnegative on the closure of Ω. If $X = H_0^1(\Omega) \times L^2(\Omega)$ and $(\varphi, \psi) \in X$, then there is a unique solution $(u, \partial_t u)$ defined and bounded for $t \ge 0$ and coincides with (φ, ψ) at $t = 0$. Define the semigroup $T : \mathbb{R}^+ \times X \to X$, by $(t, \varphi, \psi) \mapsto T(t)(\varphi, \psi)$, the solution of (2.2) through (φ, ψ) at $t = 0$.

If there is an $x_0 \in \Omega$ such that $\beta(x_0) > 0$, then Iwasaki ([104]) has shown that $T(t)(\varphi, \psi) \to (0,0)$ as $t \to \infty$. Dafermos ([39]) gave a simpler proof using generalized Fourier series, the energy function and a density argument. Bardos, Lebeau and Rauch ([11]) showed that the origin is the compact global attractor if and only if there is a $T > 0$ such that every ray of geometric optics intersects $\omega \times (0, T)$, where ω is the support of β. If $n = 1$, this implies that the origin is the compact global attractor. If the support of β contains the boundary of Ω, then the origin is the compact global attractor. On the other hand, if $n = 2$, there is a function β with nonempty support such that $r(E\sigma(T(t))) = 1$ for all t. Therefore, Theorem 2.0.5 implies that $(0,0)$ is not the compact global attractor even though it is the maximal compact invariant set and attracts compact sets, and positive orbits of bounded sets are bounded.

Our next goal is to determine conditions which imply that the maximal compact invariant set is the compact global attractor.

The semigroup T or the map $T(t), t \ge 0$, acting on X is said to be *conditionally completely continuous for $t \ge t_1$* if, for each $t \ge t_1$ and each bounded set $B \subset X$ for which $T(t)B$ is bounded, we have $\operatorname{Cl} T(t)B$ compact. The semigroup T or the map $T(t), t \ge 0$, is said to be *completely continuous for $t \ge t_1$* if it is conditionally completely continuous for $t \ge t_1$ and $T(t)$ takes bounded sets of X to bounded sets for each $t \ge t_1$.

We say that the semigroup T is *point (resp. compact) dissipative* if there is a bounded set $B \subset X$ such that B attracts each point (resp. compact set) of X. The following result is an easy consequence of results by Billotti and LaSalle [17].

Theorem 2.0.8. *If the semigroup T on X is completely continuous for $t \geq t_1 > 0$ and point dissipative, then there is a compact global attractor.*

Theorem 2.0.8 applies to several types of evolutionary equations that occur in the applications. We give two illustrative examples.

Example 2.0.9. (Quasilinear parabolic PDE). Suppose that $u \in \mathbb{R}^k$, Ω is a bounded open set in $\mathbb{R}^n$ with smooth boundary, D is a $k \times k$ diagonal positive matrix, Δ is the Laplacian operator, and consider the quasilinear parabolic partial differential equation

$$\partial_t u - D\Delta u = f(x, u) \text{ in } \Omega, \tag{2.3}$$

$$\frac{\partial u}{\partial n} = 0 \text{ on } \partial\Omega. \tag{2.4}$$

Let $H^1(\Omega)$ be the Sobolev space of functions which together with their first derivatives belong to $L^2(\Omega)$ and let $X = H^1(\Omega)^k$. Under appropriate conditions on the function $f(x, u)$, for any $u_0 \in X$, there is a unique solution $u(t, x, u_0)$ of (2.3) with $u(0, x, u_0) = u_0(x)$ for $x \in \Omega$. If we define $T_f(t)u_0 = u(t, \cdot, u_0)$ and assume that all solutions are defined for all $t \geq 0$, then T_f is a C^0-semigroup on X (see the book by Henry [99]). It can be shown that $T_f(t)$ is a completely continuous map for each $t > 0$. Therefore, there will be a compact global attractor if T is point dissipative.

The same remarks apply to situations where there are different boundary conditions as well as to vector fields $f(x, u, \nabla u)$ if the space X is chosen appropriately.

Example 2.0.10. (Retarded functional differential equation (RFDE) on $\mathbb{R}^n$). Suppose that $r > 0$, $C = C([-r, 0], \mathbb{R}^n)$ with the sup norm, $f \in C^1(C, \mathbb{R}^n)$ and consider the equation

$$\dot{x}(t) = f(x_t), \tag{2.5}$$

where, for each fixed t, x_t designates the function in C given by $x_t(\theta) = x(t + \theta)$, $\theta \in [-r, 0]$.

For any $\varphi \in C$, by a solution of (2.5) with initial value $\varphi \in C$ at $t = 0$, we mean a function $x(t, \varphi)$ defined on an interval $[-r, \alpha)$, $\alpha > 0$, such that $x_0(\cdot, \varphi) = \varphi$, $x(t, \varphi)$ is continuously differentiable on $(0, \alpha)$ with a right hand derivative approaching a limit as $t \to 0$ and the function $x(t, \varphi)$ satisfies (2.5) on $[0, \alpha)$. It can be shown that there is a unique solution through φ. If we let $T_f(t)\varphi = x_t(\cdot, \varphi)$ and assume that all solutions are defined for all $t \geq 0$, then $T_f : \mathbb{R}^+ \times C \to C$ is a continuous semigroup on C. Furthermore, the Arzela-Ascoli theorem implies that $T_f(t)$ is completely continuous for $t \geq r$

if f takes bounded sets into bounded sets. As a consequence, if T_f is point dissipative, then there is a compact global attractor.

Example 2.0.11. (RFDE on submanifolds of $\mathbb{R}^n$). Given a RFDE f on $\mathbb{R}^n$ and an embedded submanifold $M \subset \mathbb{R}^n$ it may happen that for any $\varphi \in C([-r, 0], M)$ the vector $f(\varphi)$ is tangent to M at the point $\varphi(0) \in M$. In this case the solution $x(t, \varphi)$ remains in M for all $t \geq 0$ and we say that it defines a RFDE on the manifold M. For instance, Oliva (see [150]) considered the RFDE on $\mathbb{R}^3$ given by the equations below:

$$\dot{x}(t) = -x(t-1)y(t) - z(t)$$
$$\dot{y}(t) = x(t-1)x(t) - z(t)$$
$$\dot{z}(t) = x(t) + y(t).$$

They satisfy $x\dot{x} + y\dot{y} + z\dot{z} = 0$ (or $x^2 + y^2 + z^2 = constant$) for any $t \geq 0$. So, it defines a RFDE on $M := S^2$. More generally, if M is a separable C^∞ finite dimensional connected manifold (compact or not) and $I = [-r, 0]$ then the set $C^0(I, M) := C([-r, 0], M)$ is a Banach manifold and a RFDE on M is defined by a smooth function $F : C^0(I, M) \to TM$ (TM is the tangent bundle of M),such that $F(\varphi) \in T_{\varphi(0)}M$, for all $\varphi \in C^0(I, M)$. We will describe this situation with more details in Chapter 3.

For other types of applications, the semigroup does not satisfy the smoothing or compactification properties as in Theorem 2.0.8. For example, Theorem 2.0.8 will not apply to neutral functional differential equations, the linearly damped wave equation, as well as many other types of equations. However, there is a class of semigroups which have a type of smoothing or compactification which occurs at $t = \infty$. We now make this precise.

Following Hale, LaSalle and Slemrod (see [83]) we say that the semigroup T on X is *asymptotically smooth* if, for any bounded set B in X with $T(t)B \subset B$, $t \geq 0$, there exists a compact set J in the closure of B such that J attracts B. The definition in [83] is given in a different equivalent form.

Notice that, in the definition of asymptotically smooth, there is the conditional hypothesis: 'for any bounded set B for which $T(t)B \subset B$, $t \geq 0$.' For those B satisfying this conditional hypothesis, the set $\gamma^+(B)$ is bounded. However, it is important to notice that we do not require that $\gamma^+(B)$ be bounded for every, or even any, bounded set B. A semigroup T can be asymptotically smooth without the positive orbits of each bounded set being bounded.

We have the following basic result on ω-limit sets of bounded sets for asymptotically smooth semigroups.

Lemma 2.0.12. *If the semigroup T on X is asymptotically smooth and B is a nonempty subset of X such that there is a t_1 such that $\gamma^+(T(t_1)B))$ is bounded, then $\omega(B)$ is a nonempty, compact invariant set which attracts B. If B is connected, then so is $\omega(B)$.*

The following equivalent characterizations of asymptotically smooth maps are interesting.

Lemma 2.0.13. *The semigroup T on X is asymptotically smooth if and only if either of the following conditions are satisfied:*

(i) *For any nonempty closed bounded set $B \subset X$, there is a nonempty compact set $J = J(B) \subset X$ such that $J(B)$ attracts the set $L(B) = \{x \in B : T(t)x \in B \text{ for } t \geq 0\}$.*

(ii) *For any bounded set B in X for which there is a t_0 such that $\gamma^+(T(t_0)B)$ is bounded, any sequence of the form $T^{t_j}x_j$, $x_j \in B$, $t_j \geq t_0$, $j \geq 1$, is relatively compact.*

Semigroups satisfying (ii) of Lemma 2.0.13 were referred by Ladyzenskaya (see [120]) as *asymptotically compact*. The later term also has been used by Ball (see [9] and [10]) and Sell and You (see [188]) for semigroups with a condition stronger than (ii) which implies that positive orbits of bounded sets are bounded. Among the examples of asymptotically smooth semigroups T are those for which either

(1) There is a $t_1 > 0$ such that $T(t_1)$ is conditionally completely continuous or

(2) T has the representation

$$T(t) = S(t) + U(t),$$

where $S(t)$ is Lipschitzian with Lipschitz constant $k < 1$ for all $t > 0$ and there is a $t_1 > 0$ such that $U(t_1)$ is conditionally completely continuous.

If X is a Banach space and S in (2) is a linear semigroup on some closed subspace of X with spectral radius less than one for $t > 0$, then there is no loss in generality in assuming that the Lipschitz constant for $S(t)$ is < 1 for each $t > 0$.

There are many other applications for which the corresponding semigroups are asymptotically smooth. We remark that, if T is a linear semigroup acting on a Banach space X, then it is asymptotically smooth if, and only if, for each $t > 0$, the essential spectral radius of $T(t)$ is < 1.

The following result is a necessary and sufficient condition for the existence of a compact global attractor.

Theorem 2.0.14. *A continuous semigroup T on X has a compact global attractor if and only if*

(i) *T is asymptotically smooth.*

(ii) *T is point dissipative.*

(iii) *For any bounded set B in X, there is a $t_0 = t_0(B)$ such that $\gamma^+(T(t_0)B)$ is bounded.*

The "if" part of Theorem 2.0.14, with (iii) replaced by $\gamma^+(B)$ bounded for B bounded, is due to Hale, LaSalle and Slemrod (see [83]) with modifications by Cooperman (see [38]) and Massatt (see [138]). The "only if" part requires (iii) to be stated as in the theorem. In fact, it is possible to have (iii) satisfied for a dynamical system and yet there is a bounded set B for which $\gamma^+(B)$ is unbounded (see [29]).

We now give some specific examples of asymptotically smooth semigroups which are generated by evolutionary equations.

Example 2.0.15. (RFDE with infinite delay). Let X be a Banach space of functions taking $(-\infty, 0]$ into $\mathbb{R}^n$, let $f : X \to \mathbb{R}^n$ and consider the equation

$$\dot{x}(t) = f(x_t), \tag{2.6}$$

under the hypotheses $f \in C^1(X, \mathbb{R}^n)$, where, as in Example 2.0.10, $x_t(\theta) = x(t + \theta)$, $\theta \in (-\infty, 0]$. A solution of (2.6) is defined the same as in Example 2.0.10 and we define the semigroup T_f on X as before.

Let $X_0 = \{\psi \in X : \psi(0) = 0\}$ and consider the equation

$$y(t) = 0, t \geq 0, \tag{2.7}$$

with initial data in X_0. If $y(t, \psi)$, $\psi \in X_0$, is the solution of (2.7) and $S(t)\psi = y_t(\cdot, \psi)$, then S is a continuous semigroup on X_0. It is a rather trivial semigroup since $(S(t)\psi)(\theta) = 0$ for $t + \theta \geq 0$ and equal to $\psi(t + \theta)$ for $t + \theta < 0$. On the other hand, we obtain the following interesting representation for the semigroup T_f on $X \cap C((-\infty, 0], \mathbb{R}^n)$:

$$T_f(t)\varphi = S(t)(\varphi - \varphi(0)) + U_f(t)\varphi \tag{2.8}$$

where $U_f(t)$ is completely continuous if f is a bounded map.

As a consequence of this representation, the determination of whether or not T_f is asymptotically smooth is independent of f and depends only upon the topology of the space X. If $\gamma > 0$ is a given constant and we choose X as

$$C_\gamma := \{\varphi \in C((-\infty, 0], \mathbb{R}^n) : \lim_{\theta \to -\infty} e^{\gamma\theta}\varphi(\theta) \text{ exists}\}$$

then it is easy to verify that $S(t)$ is a strict contraction for each $t > 0$. Therefore, T_f is asymptotically smooth.

If we choose the space X to be the same as above except with $\gamma = 0$, then the radius of the essential spectrum of $S(t)$ is one and T_f is not asymptotically smooth.

We remark that the retarded equation with finite delay can be discussed as above to see that it has the above representation for all $t \geq 0$, a fact that is important for some applications.

There is a general theory for RFDE with infinite delay in abstract spaces in a book by Hino, Murakami and Naito (see [101]) including conditions on the underlying space which imply that the semigroup is asymptotically smooth. This book is also a good source for the literature on the historical development of the subject.

Example 2.0.16. (A neutral functional differential equation on $\mathbb{R}^n$). We keep the notation of Example 2.0.10. Suppose that $D : C \to \mathbb{R}^n$ is a continuous linear operator given as

$$D\varphi = \varphi(0) - \int_{-r}^{0} d\mu(\theta)\, \varphi(\theta), \tag{2.9}$$

where $\mu(\theta)$ is an $n \times n$ matrix of bounded variation and there is a continuous nondecreasing function $\gamma(s)$, $s \in [-r, 0]$, $\gamma(0) = 0$, such that, for any $s > 0$,

$$Var_{[-s,0]}\mu \leq \gamma(s). \tag{2.10}$$

Suppose that $f : C \to \mathbb{R}^n$ is a C^k-function, $k \geq 1$, and is a bounded map. A neutral functional differential equation (NFDE) on $\mathbb{R}^n$ is a relation

$$\frac{d}{dt}Dx_t = f(x_t) \tag{2.11}$$

where $\frac{d}{dt}$ is the right hand derivative of Dx_t at t.

For a given function $\varphi \in C$, we say that $x(t, \varphi)$ is a *solution* of (2.11) on the interval $[0, \alpha_\varphi)$, $\alpha_\varphi > 0$, with *initial value* φ at $t = 0$ if $Dx_t(\cdot, \varphi)$ is defined on $[-r, \alpha_\varphi)$, is continuously differentiable on $(0, \alpha_\varphi)$ with a continuous right hand derivative at zero, $x_t(\cdot, \varphi) \in C$ for $t \in [0, \alpha_\varphi)$, $x_0(\cdot, \varphi) = \varphi$ and $x(t, \varphi)$ satisfies (2.11) on the interval $[0, \alpha_\varphi)$.

For any $\varphi \in C$, it is possible to show that there is a unique solution $x(t, \varphi)$ through φ of (2.11) defined on an interval $[-r, \alpha)$ with $\alpha > 0$. Let $T_{D,f}(t) = x_t(\cdot, \varphi)$ for $t \in [0, \alpha)$. If we assume that all solutions are defined for all $t \geq 0$, then $T_{D,f}$ is a continuous semigroup on C.

To obtain a representation of $T_{D,f}$ as in (2) above, we assume that

$$D\varphi = D_0\varphi - \int_{-r}^{0} B(\theta)\varphi(\theta)d\theta \tag{2.12}$$

where

$$D_0\varphi = \varphi(0) - g(\varphi) := \varphi(0) - \Sigma_{j=1}^{\infty} B_j\varphi(-r_j),$$

$$\Sigma_{j=1}^{\infty}|B_j| < \infty, \quad \int_{-r}^{0} |B(\theta)|d\theta < \infty,$$

and each $r_j > 0$.

If we let $C_{D_0} = \{\varphi \in C : D_0\varphi = 0\}$, then the difference equation

$$D_0 y_t = 0, \quad y_t \in C_{D_0} \tag{2.13}$$

defines a C^0-semigroup T_{D_0} on C_{D_0}.

We can now state an important representation formula for the semigroup $T_{D,f}$ of (2.11) with the operator D satisfying (2.12).In fact, there exists a matrix $\Phi_{D_0} = (\varphi_1^{D_0}, \ldots, \varphi_n^{D_0})$ such that $D\Phi_{D_0} = I$. If $\Psi_{D_0} = I - \Phi_{D_0}D_0 : C \to C_{D_0}$,then

$$T_{D,f}(t) = T_{D_0}(t)\Psi_{D_0} + U_{D,f}(t), \qquad (2.14)$$

with $U_{D,f}(t)$ conditionally completely continuous for $t \geq 0$.

From the two last representations, we see that the semigroup $T_{D,f}$ will be asymptotically smooth if the semigroup T_{D_0} has the property that the radius of the spectrum of $T_{D_0,f}(t)$ is < 1 for every $t > 0$. This is equivalent to saying that the zero solution of the difference equation (2.13) is exponentially asymptotically stable. We refer to this by saying that the operator D_0 is *exponentially stable* (see [86] for details).

Example 2.0.17. (Linearly damped wave equation). Let Ω be a bounded domain in $\mathbb{R}^n$, $n \leq 3$, with a boundary which is Lipschitzian, and consider the equation

$$\partial_t^2 u + \beta\partial_t u - \Delta u = f(u) \quad \text{in } \Omega \qquad (2.15)$$

under the condition

$$u = 0 \quad \text{in } \partial\Omega.$$

Here $\beta > 0$ is a constant and $f \in C^2(\mathbb{R}, \mathbb{R})$. Equation (2.15) can be written as a system

$$\partial_t U = CU + F(U), \qquad (2.16)$$

where

$$U = \begin{bmatrix} u \\ v \end{bmatrix} \qquad C = \begin{bmatrix} 0 & I \\ \Delta & -\beta \end{bmatrix} \qquad F(U) = \begin{bmatrix} 0 \\ f(u) \end{bmatrix}.$$

We consider equation (2.15) in $X = H_0^1(\Omega) \times L^2(\Omega)$. It is not difficult to prove that C generates a continuous group e^{Ct} on X. A mild (or weak) solution with initial data $U_0 = (\varphi, \psi) \in X$ is defined to be a function that satisfies the variation of constants formula

$$U(t) = e^{Ct}U_0 + \int_0^t e^{C(t-\tau)}F(U(\tau))d\tau. \qquad (2.17)$$

If we assume the growth conditions

$$|f'(u)| \leq ce^{|u|^\gamma}, \quad \gamma < 2 \text{ if } n = 2, \qquad (2.18)$$

$$|f''(u)| \leq c(1 + |u|^\gamma), \quad \gamma < 1 \text{ if } n = 3,$$

then one can easily prove the local existence and uniqueness of a solution of (2.15) which we denote by $T(t)(\varphi, \psi)$.

Define the energy function

$$V(\varphi, \psi) = \int_\Omega [\frac{1}{2}|\nabla\varphi|^2 + \frac{1}{2}|\psi|^2 - \int_0^\varphi f(s)ds].$$

For a dense set of initial data, one can differentiate V along the solutions $T(t)(\varphi, \psi) = (u(t), v(t))$ of (2.15) and obtain, after an integration by parts,

$$\dot{V}(u(t), v(t)) = -\int_\Omega \beta|\partial_t u|^2 \le 0. \tag{2.19}$$

This implies that $V(T(t)(\varphi, \psi)) \le V(\varphi, \psi)$ for all $t \ge 0$ in the domain of existence of the solution. In order to assert that $T(t)(\varphi, \psi)$ is bounded in its domain of existence (and thus $T(t)(\varphi, \psi)$ exists for all $t \ge 0$), we need some additional conditions on the function f. A sufficient condition is that f satisfies the dissipative condition

$$\limsup_{|u|\to\infty} \frac{f(u)}{u} < \lambda_1, \tag{2.20}$$

where λ_1 is the first eigenvalue of $-\Delta$ with homogeneous Dirichlet boundary conditions. The condition (2.20) also implies that $\gamma^+(B)$ is bounded if B is bounded in X and condition (iii) of Theorem 2.0.14 is satisfied.

The growth condition on f in (2.18) implies that the map $f : H_0^1(\Omega) \to L^2(\Omega)$ is compact. From this fact, we deduce that the integral term in (2.17) is a compact operator for each $t \ge 0$.

The assumption that β is positive implies that there are positive constants K, α, such that

$$|e^{CT}|_{L(X,X)} \le Ke^{-\alpha t}, \quad t \ge 0.$$

As a consequence, the semigroup T is asymptotically smooth and condition (i) of Theorem 2.0.14 is satisfied.

Since T is asymptotically smooth and $\gamma^+(\varphi, \psi)$ is bounded for each $(\varphi, \psi) \in X$, we know that $\omega(\varphi, \psi)$ is compact and invariant. From (2.19), it follows that $\omega(\varphi, \psi) \subset E$, where E is the set of equilibrium points of (2.16). The dissipative condition and a variational argument shows that the set E is bounded. Therefore, (ii) of Theorem 2.0.14 is satisfied and the compact global attractor for (2.15) exists.

The existence of the compact global attractor for (2.15) was first proved by Hale [75] using the above approach and independently by Haraux [96] using a different method and assuming a smoother boundary Ω.

If $\gamma = 3$ and $n = 3$, the mapping $f : H_0^1(\Omega) \to L^2(\Omega)$ is not compact and the above method cannot be used to prove the asymptotic smoothness. However, assuming more smoothness on Ω, Arrieta, Carvalho and Hale (see [7]) used another method to prove that the semigroup was asymptotically smooth and, therefore, the compact global attractor exists. Other proofs now have been given and we refer to Raugel (see[172]) for a survey of recent results. This paper also contains other interesting results and applications as well as an extensive bibliography.

It has been useful in the geometric theory of dynamical systems, to consider sets of recurrent motions, in particular, sets of nonwandering points. For this, assume that $\mathcal{A}(T)$ is compact. An element $\psi \in \mathcal{A}(T)$ is called a *nonwandering point* of T if, for any neighborhood U of ψ in $\mathcal{A}(T)$ and any $\tau > 0$, there exists $t = t(U, \tau) > \tau$ and $\phi \in U$ such that $T(t)\phi \in U$. The set of all nonwandering points of T is called the *nonwandering set* and is denoted by $\Omega(T)$.

Proposition 2.0.18. *If a semigroup T has $\mathcal{A}(T)$ compact, then $\Omega(T)$ is closed, contains all of the ω-limit sets of precompact positive orbits and all of the α-limit sets of precompact negative orbits. Moreover, if $T(t)$ is one-to-one on $\mathcal{A}(T)$, then $\Omega(T)$ is invariant.*

Remark 2.0.19. We have presented the above theory for continuous semigroups. However, the same results are valid for dynamical systems defined by the iterates of a continuous map $T : X \to X$. An exception is that ω-limit sets may not be connected. Gobbino and Sardella (see [68]) have shown that even the compact global attractor may not be connected. However, it is always invariantly connected in the sense of LaSalle (see [122]).

Remark 2.0.20. As the reader will observe in many of the following sections, the principal results will be stated for continuous semigroups T for which there is a $t_1 > 0$ such that $T(t_1)$ is completely continuous. It is interesting and important to characterize the set of asymptotically smooth semigroups for which these results remain valid. In some places in the text, we indicate where this can be done and, in others, where it appears to require new ideas to obtain the appropriate extensions to asymptotically smooth semigroups.

3 Functional Differential Equations on Manifolds

In this chapter we deal, mainly, with the current status of the global and geometric theory of functional differential equations (FDE) on a finite dimensional manifold. Retarded functional differential equations ($RFDE$) and neutral functional differential equations ($NFDE$) (in particular retarded differential delay equations and neutral differential delay equations) will be considered. As we will show, for a generic initial condition, the dependence of solutions with time differs, enormously, from one case to the other. In $RFDE$ with finite delays the solution (as a curve in the phase space) starts continuous and its smoothness increases with time; on the other hand, in $NFDE$, in general, the solution stays only continuous for all time. Also, the semiflow operator of a smooth $RFDE$ is not necessarily one to one even when we restrict its action to the set of all (smooth) global bounded solutions; this corresponds to the existence of collisions, a phenomenon which never occurs in the category of smooth finite dimensional vector fields (see Example 3.2.18 and Chapter 7). A vector field on a finite dimensional manifold is a special case of a $RFDE$ and any $RFDE$ is a particular case of a $NFDE$. The global and geometric theory of flows of vector fields acting on a (finite dimensional) manifold is very well developed, much more than the corresponding theory of semiflows acting on infinite dimensional Banach manifolds, even when the semiflows are defined by $RFDE$. Analogously, the global and geometric theory for $RFDE$ is much more developed than the corresponding one for $NFDE$ and so, many interesting open questions appear naturally.

3.1 RFDE on manifolds

Let M be a separable C^∞ finite dimensional connected manifold, I the closed interval $[-r, 0]$, $r > 0$, and $C^0(I, M)$ the totality of continuous maps φ of I into M. Let TM be the tangent bundle of M and $\tau_M : TM \to M$ its C^∞-canonical projection. Assume there is given on M a complete Riemannian structure (it exists because M is separable) with δ_M the associated complete metric. This metric on M induces an admissible metric on $C^0(I, M)$ by

$$\delta(\varphi, \bar{\varphi}) = \sup \left\{ \delta_M \big(\varphi(\theta), \bar{\varphi}(\theta) \big) : \theta \in I \right\}.$$

The space $C^0(I, M)$ is complete and separable, because M is complete and separable. The function space $C^0(I, M)$ is a C^∞-manifold modeled on a separable Banach space. If M is embedded as a closed submanifold of an Euclidean space V, then $C^0(I, M)$ is a closed C^∞-submanifold of the Banach space $C^0(I, V)$.

If $\rho : C^0(I, M) \to M$ is the evaluation map, $\rho(\varphi) = \varphi(0)$, then ρ is C^∞ and, for each $a \in M$, $\rho^{-1}(a)$ is a closed submanifold of $C^0(I, M)$ of co-dimension $n = \dim M$. A *retarded functional differential equation (RFDE)* on M is a continuous function $F : C^0(I, M) \to TM$, such that $\tau_M F = \rho$. Roughly speaking, an RFDE on M (see Oliva [147]) is a function mapping each continuous path φ lying on M, $\varphi \in C^0(I, M)$, into a vector tangent to M at the point $\varphi(0)$. The notation RFDE(F) is used as short for "retarded functional differential equation F". Nonautonomous RFDE on manifolds could be similarly defined, but we restrict the definition to the autonomous case as these are the only equations discussed in the present notes.

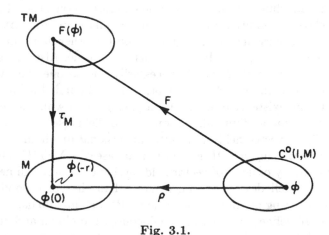

Fig. 3.1.

Given a function x of a real variable and with values in the manifold N, we denote $x_t(\theta) = x(t + \theta)$, $\theta \in I$, whenever the right-hand side is defined for all $\theta \in I$. A *solution of an RFDE(F) on M* with initial condition $\varphi \in C^0(I, M)$ at t_0 is a continuous function $x(t)$ with values on M and defined on $t_0 - r \le t < t_0 + A$, for some $0 < A < \infty$, such that:

(i) $x_{t_0} = \varphi$,

(ii) $x(t)$ is a C^1-function of $t \in [t_0, t_0 + A)$,

(iii) $\big(x(t), (d/dt)x(t)\big) = F(x_t)$, $t \in [t_0, t_0 + A)$, where $\big(x(t), (d/dt)(x(t))\big)$ represents the tangent vector to the curve x at the point t. The number $r > 0$ is called *delay* or *lag* of F.

One can write locally, in natural coordinates of TM,

$$\big(x(t), \dot{x}(t)\big) = F(x_t) = \big(x(t), f(x_t)\big)$$

or simply

$$\dot{x}(t) = f(x_t),$$

for an appropriate function f.

An existence and uniqueness theorem for initial value problems can be established as for the corresponding result for $M = \mathbb{R}^n$; see Hale and Verduyn-Lunel [86]. A function G between two Banach manifolds is said to be *locally Lipschitzian* at a certain point φ of its domain, if there exist coordinate neighborhoods of φ and of $G(\varphi)$, in the domain and in the range of G, respectively, and the representation of G defined through the associated charts is Lipschitz, as a mapping between subsets of Banach spaces.

Theorem 3.1.1. *If F is an RFDE on M which is locally Lipschitzian, then for each $\varphi \in C^0(I, M)$, $t_0 \in \mathbb{R}$, there exists a unique solution $x(t)$ of F with initial condition $x_{t_0} = \varphi$.*

Proof: By Whitney's embedding theorem, M can be considered as a submanifold of $\mathbb{R}^N$ for an appropriate integer N. Accordingly, TM can be considered a submanifold of $\mathbb{R}^N \times \mathbb{R}^N$. We will construct an extension $\bar{F}$ of F which defines an RFDE $\bar{F} : C^0(I, \mathbb{R}^N) \to \mathbb{R}^N \times \mathbb{R}^N$ such that $\bar{F}(\varphi) = F(\varphi)$ if $\varphi \in C^0(I, \mathbb{R}^N)$ and $\bar{F}(\varphi) = 0$ outside a certain neighborhood of $C^0(I, M)$ in $C^0(I, \mathbb{R}^N)$. Let U be a tubular neighborhood of M in $\mathbb{R}^N$ and α the C^∞ projection. Let W be the open set $W = \{\varphi \in C^0(I, \mathbb{R}^N) \mid \varphi(I) \subset U\}$. Define $F_1 : W \to \mathbb{R}$ by $F_1(\varphi) = 1 - \int_{-r}^0 |\alpha(\varphi(s))|^2 ds$ where $|\cdot|$ is the Euclidean norm in $\mathbb{R}^N$. Then $F_1(\varphi) = 1$ if and only if $\varphi \in C^0(I, M)$ and $F_1(\varphi) < 1$ if $\varphi \notin C^0(I, M)$. For every $0 < \varepsilon < 1$ let $W_\varepsilon = F_1^{-1}([1 - \varepsilon, \infty))$. Fix some $0 < \varepsilon < 1$ and take a C^∞ $\psi : \mathbb{R} \to \mathbb{R}$ satisfying $\psi(t) = 1$ for $t \geq 1$ and $\psi(t) = 0$ for $t \leq 1 - \varepsilon/2$. Define $F_2 : C^0(I, \mathbb{R}^N) \to \mathbb{R}$ as $F_2(\varphi) = \psi\big(F_1(\varphi)\big)$ if $\varphi \in W_\varepsilon$ and $F_2(\varphi) = 0$ if $\varphi \notin W_\varepsilon$. Then F_2 is a C^∞-function and satisfies $F_2(W) \leq 1$ for all φ and $F_2(\varphi) = 1$ if and only if $\varphi \in C^0(I, M)$. Finally define $\bar{F}$ as $\bar{F}(\varphi) = 0$ when $\varphi \notin W$ and $\bar{F}(\varphi) = F_2(\varphi)F(\alpha \circ \varphi)$ when $\varphi \in W$. The standard results on existence and uniqueness of solutions of RFDE on $\mathbb{R}^N$ can be applied to finish the proof of the theorem (a previous proof can be seen in Oliva [147]). ∎

Using similar ideas to the ones that appear in the proof of Theorem 3.1.1, it is possible to establish, for RFDE on manifolds, results on continuation of solutions to maximal intervals of existence, and on continuous dependence relative to changes in initial data and in the RFDE itself, which are analogous to the corresponding results in $\mathbb{R}^n$.

Given a locally Lipschitzian RFDE(F) on M, its maximal solution $x(t)$, satisfying the initial condition φ at t_0 is sometimes denoted by $x(t; t_0, \varphi, F)$, and x_t is denoted by $x_t(t_0, \varphi, F)$. The arguments φ and F will be dropped whenever confusion may not arise, and t_0 will be dropped if $t_0 = 0$.

The *semiflow* of an RFDE(F) is defined by $\Phi(t, \varphi, F) = x_t(\varphi, F)$, whenever the right-hand side makes sense. It will be written as $\Phi(t, \varphi)$, whenever confusion is not possible. The notation $\Phi_t \varphi = \Phi(t, \varphi)$ is also used.

The following theorem shows important properties of the semiflow Φ. For the statement of differentiability properties of Φ, it is convenient to introduce the notation $\mathcal{X}^k = \mathcal{X}^k(I, M)$, $k \geq 1$, for the Banach space of all C^k-RFDE defined on the manifold M, which are bounded and have bounded derivatives up to order k, taken with the C^k-uniform norm.

Theorem 3.1.2. *If F is an RFDE on M in $\mathcal{X}^k$, $k \geq 1$ then the family of mappings $\{\Phi_t, 0 < t < \infty\}$ is a strongly continuous semigroup of operators on $C^0 = C^0(I, M)$ i.e.,*

(i) Φ_0 is the identity on C^0,
(ii) $\Phi_{t+s} = \Phi_t \Phi_s$, $t, s \geq 0$, $t + s < T$,
(iii) the map from $[0, \infty) \times C^0$ into C^0 given by $(t, \varphi) \rightarrow \Phi_t(\varphi)$ is continuous.

Furthermore, the solution map, as a function $\Phi : [0, \infty) \times C^0 \times \mathcal{X}^k \rightarrow C^0$, $k \geq 1$, has the following regularity properties:

1. *Φ is continuous on $[0, \infty) \times C^0 \times \mathcal{X}^k$,*
2. *for each fixed $t \geq 0$, the map $\Phi(t, \cdot, \cdot) : C^0 \times \mathcal{X}^k \rightarrow C^0$ is C^k,*
3. *the map $\Phi : (sr, \infty) \times C^0 \times \mathcal{X}^k \rightarrow C^0$ is C^s, for all $0 \leq s \leq k$.*

Proof: As in the proof of Theorem 3.1.1, these properties can be reduced to the analogous properties for RFDE in $\mathbb{R}^n$.   ∎

The solution map $\Phi_t : C^0 \rightarrow C^0$ needs not be one-to-one, but, if there exists $\varphi, \psi \in C^0$ and $t, s \geq 0$ such that $\Phi_t \varphi = \Phi_s \psi$, then $\Phi_{t+\sigma}(\psi) = \Phi_{s+\sigma}(\psi)$ for all $\sigma \geq 0$ for which these terms are defined.

The following property of the solution map is also useful.

Theorem 3.1.3. *If F is an RFDE, $F \in \mathcal{X}^1$ and the corresponding solution map $\Phi_t : C^0([-r, 0], M) \rightarrow C^0([-r, 0], M)$ is uniformly bounded on compact subsets of $[0, \infty)$, then, for $t \geq r$, Φ_t is a compact map, i.e., it maps bounded sets of C^0 into relatively compact subsets of C^0.*

Proof: Again, this property can be reduced to the analogous property for FDE in $\mathbb{R}^n$. Actually, the proof is an application of the Ascoli-Arzela theorem.   ∎

A consequence of this result is that, for an RFDE(F) satisfying the hypothesis in the theorem with $r > 0$, Φ_t can never be a homeomorphism because the unit ball in $C([-r, 0), \mathbb{R}^n)$ is not compact. The hypothesis of Theorem 3.1.3 are satisfied if $F \in \mathcal{X}^k$ and M is compact.

The double tangent space, $T^2 M$, of a manifold M, admits a canonical involution $w : T^2 M \rightarrow T^2 M$, w^2 equal to the identity on $T^2 M$, and w is a C^∞-diffeomorphism on $T^2 M$ which satisfies $\tau_{TM} \cdot w = T\tau_M$ and $T\tau_M \cdot w =$

$\tau_{TM'}$ where $\tau_M : TM \rightarrow M$ and $\tau_{TM} : T^2M \rightarrow TM$ are the corresponding canonical projections. If F is a C^k RFDE on a manifold M, $k \geq 1$, and TF is its derivative, it follows that $w \cdot TF$ is a C^{k-1} RFDE on TM, which is called the *first variational equation* of F. The map w is norm preserving on T^2M and the solution map Ψ_t of $w \cdot TF$ is the derivative of the solution map Φ_t of F, i.e., $\Psi_t = T\Phi_t$.

Let us apply the general notions and results introduced in the previous Chapter 2 for the case of RFDE.

A function $y(t)$ is said to be a *global solution* of an RFDE(F) on M, if it is defined for $t \in (-\infty, +\infty)$ and, for every $\sigma \in (-\infty, +\infty)$, $x_t(\sigma, y_\sigma, F) = y_t$, $t \geq \sigma$. The constant and the periodic solutions are particular cases of global solutions. The solutions with initial data in unstable manifolds of equilibrium points or periodic orbits are often global solutions, for example, when M is compact. An *invariant* set of an RFDE(F) on a manifold M is a subset S of $C^0 = C^0(I, M)$ such that for every $\varphi \in S$ there exists a global solution x of the RFDE, satisfying $x_0 = \varphi$ and $x_t \in S$ for all $t \in \mathbb{R}$. The ω-*limit set* $\omega(\varphi)$ of an orbit $\gamma^+(\varphi) = \{\Phi_t\varphi, t \geq 0\}$ through φ is the set

$$\omega(\varphi) = \bigcap_{\tau \geq 0} C\ell \bigcup_{t \geq \tau} \Phi_t\varphi. \tag{3.1}$$

This is equivalent to saying that $\psi \in \omega(\varphi)$ if and only if there is a sequence $t_n \rightarrow \infty$ as $n \rightarrow \infty$ such that $\Phi_{t_n}\varphi \rightarrow \psi$ as $n \rightarrow \infty$. For any set $S \subset C^0$, one can define

$$\omega(S) = \bigcap_{\tau \geq 0} C\ell \bigcup_{\substack{t \geq \tau \\ \varphi \in S}} \Phi_t\varphi.$$

In a similar way, if $x(t, \varphi)$ is a solution of the RFDE(F) for $t \in (-\infty, 0]$, $x_0(\cdot, \varphi) = \varphi$, one can define the α-*limit set of the negative orbit* $\{x_t(\cdot, \varphi), -\infty < t \leq 0\}$. Since the map Φ_t may not be one-to-one, there may be other negative orbits through φ and, thus, other α-limit points. To take into account this possibility, we define the α-limit set of φ in the following way. For any $\varphi \in C^0$ and any $t \geq 0$, let

$$H(t, \varphi) = \{\psi \in C^0 : \text{there is a solution } x(t, \varphi) \text{ of the RFDE}(F)$$
$$\text{on } (-\infty, 0], \; x_0(\cdot, \varphi) = \varphi, \; x_{-t}(\cdot, \varphi) = \psi\}$$

and define the α-limit set $\alpha(\varphi)$ of φ as

$$\alpha(\varphi) = \bigcap_{\tau \geq 0} C\ell \bigcup_{t \geq \tau} H(t, \varphi). \tag{3.2}$$

Lemma 3.1.4. *Let $F \in \mathcal{X}^k$, $k \geq 1$, be a RFDE on a connected manifold M. Then the ω-limit set $\omega(\varphi)$ of any bounded orbit $\gamma^+(\varphi)$, $\varphi \in M$ is nonempty, compact, connected and invariant. The same conclusion is valid for $\omega(S)$ for any connected set $S \subset C^0$ for which $\gamma^+(S)$ is bounded.*

If $\bigcup_{t\geq 0} H(t,\varphi)$ is non-empty and bounded, then the α-limit set $\alpha(\varphi)$ is nonempty, compact and invariant. If, in addition, $H(t,\varphi)$ is connected, then $\alpha(\varphi)$ is connected.

Remark 3.1.5. It seems plausible that $H(t,\varphi)$ is always connected, but it is not known if this is the case.

Remark 3.1.6. If M is a compact manifold, then $\gamma^+(\varphi)$, $\bigcup_{t\geq 0} H(t,\varphi)$ are bounded sets and, thus, the ω-limit set is nonempty, compact, connected and invariant. The α-limit set is compact and invariant, being connected if $H(t,\varphi)$ is connected and nonempty if $\bigcup_{t\geq 0} H(t,\varphi)$ is nonempty.

Remark 3.1.7. If Φ_t is one-to-one, then $H(t,\varphi)$ is empty or a singleton for each $t \geq 0$ and, thus, the boundedness of the negative orbit of φ implies $\alpha(\varphi)$ is a nonempty, compact, connected invariant set.

Proof: (of Lemma 3.1.4) The proof given here follows the proof of the analogous statement for dynamical systems defined on a Banach space. However, in order to emphasize the ideas behind the result, a direct proof is given.

Let $\gamma^+(\varphi) = \{\Phi_t\varphi, \ t \geq 0\}$ be bounded. Since $F \in \mathcal{X}^1$, Ascoli's Theorem can be used to show that $\gamma^+(\varphi)$ is precompact. It follows now directly from the definition of $\omega(\varphi)$ in (3.1) that it is nonempty and compact.

Assume now that $\text{dist}\big(\Phi_t\varphi, \omega(\varphi)\big) \not\to 0$ as $t \to \infty$, where dist stands for the admissible metric in $C^0(I, M)$. Then there exist $\varepsilon > 0$ and a sequence $t_k \to \infty$ as $k \to \infty$ such that $\text{dist}\big(\Phi_{t_k}\varphi, \omega(\varphi)\big) > \varepsilon$ for $k = 1, 2, \dots$ Since the sequence $\{\Phi_{t_k}\varphi\}$ is in a compact set, it has a convergent subsequence. The limit necessarily belongs to $\omega(\varphi)$, contradicting $\text{dist}\big(\Phi_{t_k}\varphi, \omega(\varphi)\big) > \varepsilon$. Thus, $\text{dist}\big(\Phi_t\varphi, \omega(\varphi)\big) \to 0$ as $t \to \infty$. If $\omega(\varphi)$ were not connected, it would be a union of two disjoint compact sets which would be a distance $\sigma > 0$ apart. This contradicts $\text{dist}\big(\Phi_t\varphi, \omega(\varphi)\big) \to 0$ as $t \to \infty$, and so $\omega(\varphi)$ is connected.

Suppose $\psi \in \omega(\varphi)$. There exists a sequence $t_k \to \infty$ as $k \to \infty$ such that $\Phi_{t_k}\varphi \to \psi$. For any integer $N \geq 0$, there exists an integer $k_0(N)$ such that $\Phi_{t_k+t}\varphi$ is defined for $-N \leq t < +\infty$ if $k \geq k_0(N)$. Since $\gamma^+(\varphi)$ is precompact, one can find a subsequence $\{t_{k,N}\}$ of $\{t_k\}$ and a continuous function $y : [-N, N] \to \omega(\varphi)$ such that $\Phi_{t_{k,N}+t}\varphi \to y(t)$ as $k \to \infty$ uniformly for $t \in [-N, N]$. By the diagonalization procedure, there exists a subsequence, denoted also by $\{t_k\}$, and a continuous function $y : (-\infty, \infty) \to \omega(\varphi)$, such that $\Phi_{t_k+t}\varphi \to y(t)$ as $k \to \infty$, uniformly on compact sets of $(-\infty, +\infty)$. Clearly, $y(t)$, $t \geq \sigma$ is the solution of the RFDE(F) with initial condition y_σ at $t = \sigma$, i.e., $y(t) = x(t; \sigma, y_\sigma, F)$, $t \geq \sigma$. Thus, y is a global solution of the RFDE(F). On the other hand $y(0) = \psi$. Consequently, $\omega(\varphi)$ is invariant.

The assertions for $\omega(S)$, $S \subset M$, which are contained in the statement, can now be easily proved and the assertions relative to $\alpha(\varphi)$, $\varphi \in M$, are proved in an analogous way.   ∎

Given an RFDE(F) on M, we denote by $A(F)$ the set of all initial data of global bounded solutions of F. The set $A(F)$ is clearly an invariant set of F. If $F \in \mathcal{X}^1$ and $\gamma^+(\varphi)$ (or $\bigcup_{t \geq 0} H(t, \varphi)$) is bounded, then Lemma 3.1.4 implies that $\omega(\varphi)$ (or $\alpha(\varphi)$) is contained in $A(F)$. Consequently, if $F \in \mathcal{X}^1$, the set $A(F)$ contains all the information about the limiting behavior of the bounded orbits of the RFDE(F). It is important to know when the set $A(F)$ is compact for, in this case, it is the *maximal compact invariant set* of F. Also, it is important to know if $A(F)$ is the *compact global attractor*, that is, $A(F)$ is the maximal compact invariant set and attracts bounded sets.

In the following, we say that the solution map Φ_t is a *bounded map uniformly on compact subsets* of $[0, \infty)$ if, for any bounded set $B \subset C^0$ and any compact set $K \subset [0, \infty)$, the set $\bigcup_{t \in K} \Phi_t B$ is bounded. An RFDE(F) is said to be *point dissipative* if its semigroup is point dissipative.

Sometimes we deal with discrete dynamical systems, that is, iterates of a map. In this case, the above concepts are defined in the same way.

Theorem 3.1.8. *If $F \in \mathcal{X}^1$ is a point dissipative RFDE on M and the corresponding solution map, Φ_t, is a bounded map uniformly on compact subsets of $[0, \infty)$, then there is a compact set $K \subset C^0$ which attracts all compact sets of C^0. The set $\mathcal{J} = \bigcap_{n \geq 0} \Phi_{nr} K$ is the same for all compact sets K which attract compact sets of C^0, it is the nonempty, connected compact global attractor.*

Proof: Although this theorem is a consequence of Theorem 2.0.8, we present a complete proof for the sake of completeness. Assume the hypotheses in the statement hold and fix $\varepsilon > 0$. Since F is point dissipative, there exists a bounded set B such that, for each $\varphi \in C^0$, there is a $t_0 = t_0(\varphi)$ such that $\Phi_t \varphi \subset \mathcal{B}(B, \varepsilon)$ for $t \geq t_0(\varphi)$. By continuity, for each $\varphi \in C^0$ there is a neighborhood O_φ of φ in M such that $\Phi_t O_\varphi \subset \mathcal{B}(B, \varepsilon)$ for $t_0(\varphi) \leq t \leq t_0(\varphi) + r$. Since, by Theorem 3.1.3, Φ_r is a compact map, it follows that $B^* = \Phi_r \mathcal{B}(B, \varepsilon)$ is a precompact set and $\Phi_{t+r} O_\varphi \subset B^*$ for $t_0(\varphi) \leq t \leq t_0(\varphi) + r$. If H is an arbitrary compact set of C^0, one can form a finite covering $\{O_{\varphi_i}(H)\}$ with $\varphi \in H$ and define $N(H)$ to be the smallest integer greater or equal than $\max_i \{1 + t_0(\varphi_i)/r\}$. Let $H_0 = \bigcup_i O_{\varphi_i}(H)$ and let $K = \bigcup_{i=0}^{N(B^*)} \Phi_{ir} B^*$. The set K is compact. It is then easy to show that $\Phi_{nr} B^* \subset K$ for $n \geq N(B^*)$ and $\Phi_t H \subset \Phi_t H_0 \subset K$ for $t \geq (N(B^*) + N(H))r$. Consequently, the compact set K attracts all compact sets of C^0.

Applying the above argument to the compact set K itself, we get $\Phi_t K \subset K$ for $t \geq (N(K) + N(B^*))r$. Therefore $\omega(K) \subset K$. Let $\mathcal{J} = \bigcap_{n \geq 0} \Phi_{nr} K$. Clearly, $\mathcal{J}$ is compact and $\mathcal{J} \subset \omega(K)$. On the other hand, if $\psi \in \omega(K)$ there are sequences $t_j \to \infty$ as $j \to \infty$ and $\varphi \in K$ such that $\Phi_{t_j} \varphi_j \to \psi$ as $j \to \infty$. Since $\{\Phi_t K, \ t \geq (N(K) + N(B^*))r\}$ is precompact, for any integer i one can find a subsequence of $\{\Phi_{t_j - ir} \varphi_j\}$ which converges to some $\psi_i \in \omega(K) \subset K$, and then $\Phi_{ir} \psi_i = \psi$ for all integer i, implying that $\psi \in \mathcal{J}$. This proves $\omega(K) \subset \mathcal{J}$ and, consequently, $\omega(K) = \mathcal{J}$. From Lemma 3.1.4, $\mathcal{J}$ is nonempty, compact, connected and invariant.

To prove that $\mathcal{J}$ is the maximal compact invariant set, suppose H is any compact invariant set. Since K attracts H and H is invariant, it follows that $H \subset \Phi_{nr}K$ and, therefore, $H \subset \mathcal{J}$.

It remains to prove that $\mathcal{J}$ is independent of the choice of the compact set K which attracts all compact sets of C^0. For this, denote $\mathcal{J} = \mathcal{J}(K)$ and $\mathcal{J}(K_1) = \bigcap_{n \geq 0} \Phi_{nr}K_1$ where K_1 is a compact set which attracts all compact sets of C^0. Both $\mathcal{J}(K)$ and $\mathcal{J}(K_1)$ are invariant and compact, and they are attracted by both K and K_1. Therefore $\mathcal{J}(K) \subset K_1$, $\mathcal{J}(K_1) \subset K$ and $\mathcal{J}(K) \subset \Phi_{nr}K_1$, $\mathcal{J}(K_1) \subset \Phi_{nr}K$ for all $n \geq 0$. Consequently, $\mathcal{J}(K) = \mathcal{J}(K_1)$. For any bounded set $B \subset C^0$, the set $\Phi_r B$ is relatively compact and, thus, $\mathcal{J}$ attracts B and is the compact global attractor.   ∎

Theorem 3.1.9. *If $F \in \mathcal{X}^1$ is a point dissipative RFDE on a connected manifold M and the corresponding solution map, Φ_t, is uniformly bounded on compact subsets of $[0,\infty)$, then $A(F)$ is connected and is the compact global attractor.*

Corollary 3.1.10. *If $F \in \mathcal{X}^1$ is an RFDE on a connected compact manifold M, then $A(F)$ is connected, is the compact global attractor and*

$$A(F) = \bigcap_{n \geq 0} \Phi_{nr}(C^0).$$

Proof: Noting that $K = C\ell\Phi_r(C^0)$ is a compact set (attracting C^0), the corollary is an obvious consequence of Theorem 3.1.9.   ∎

The set $A(F)$ has certain continuity properties in relation to the dependence on F. If M is compact, we have the following theorem, and if M is not compact, some additional hypotheses are needed to obtain a similar result.

Theorem 3.1.11. *If $F \in \mathcal{X}^1$ is an RFDE on a compact manifold M, then the attractor set $A(F)$ is upper semicontinuous in F; that is, for any neighborhood U of $A(F)$ in M, there is a neighborhood V of F in $\mathcal{X}^1$ such that $A(G) \subset U$ if $G \in V$.*

Proof: By Corollary 3.1.10, the set $A(F)$ is the compact global attractor. General results in the theory of stability, based on the construction of "Lyapunov functions", guarantee that, for any neighborhood U of $A(F)$ in C^0, there is a neighborhood V of F in $\mathcal{X}^1$ and a $T > 0$ such that the solution map associated with the RFDE $G \in V$, Φ_t^G, satisfies $\Phi_t^G C^0 \subset U$, for all $G \in V$, $t \geq T$. Since, from Corollary 3.1.10, $A(G) = \bigcap_{n \geq 0} \Phi_{nr}^G(C^0)$, it follows that $A(G) \subset U$.   ∎

The preceding argument requires the use of converse theorems on asymptotic stability, establishing the existence of "Lyapunov functions". An alternative proof can be given as follows. By Corollary 3.1.10, the set $A(F)$ is compact and attracts C^0. Let U denote an arbitrarily small neighborhood of $A(F)$, say consisting of all points at a distance from $A(F)$ smaller than a certain $\varepsilon > 0$. Based on Gronwall's inequality one can show that $\Phi_t^F(\varphi)$ and $\Phi_t^G(\varphi)$ can be made as close as desired, uniformly in $\varphi \in C^0$ and $G \in V \subset \mathcal{X}^1$, by choosing V to be a sufficiently small neighborhood of F in $\mathcal{X}^1$. Since $A(F)$ attracts C^0, denoting by W the neighborhood of $A(F)$ consisting of points at a distance from $A(F)$ smaller than $\varepsilon/2$, it follows that there is an integer $N > 0$ such that $\Phi_{nr}^F(C^0) \subset W$ for $n \geq N$. By choosing V sufficiently small we have $\Phi_{nr}^G(C^0) \subset U$ for all $G \in V$. Since, by Corollary 3.1.10, $A(G) = \bigcap_{n \geq 0} \Phi_{nr}^G(C^0)$, it follows that $A(G) \subset U$.

Remark 3.1.12. The second proof given for the preceding theorem does not generalize for manifolds M which are not compact. However, the first proof can be used, together with some additional hypothesis, to establish a similar result for M not compact.

Recall that for an RFDE(F) on a manifold M, an element $\psi \in A(F)$ is called a *nonwandering point* of F if, for any neighborhood U of ψ in $A(F)$ and any $T > 0$, there exists $t = t(U,T) > T$ and $\tilde{\psi} \in U$ such that $\Phi_t(\tilde{\psi}) \in U$. The set of all nonwandering points of F, that is, its *nonwandering set* of F, is denoted by $\Omega(F)$.

Proposition 3.1.13. *If $F \in \mathcal{X}^1$ is a point dissipative RFDE on a manifold M, then $\Omega(F)$ is closed and, moreover, if Φ_r is one-to-one on $A(F)$, then $\Omega(F)$ is invariant.*

Proof: The proof follows ideas similar to the ones used in the proof of Lemma 3.1.4. ∎

Corollary 3.1.14. *If $F \in \mathcal{X}^1$ is an RFDE on a compact manifold M, then $\Omega(F)$ is closed and, moreover, if Φ_r is one-to-one on $A(F)$, then $\Omega(F)$ is invariant.*

Most of the results in this section are valid in a more abstract setting. We state the results without proof, for maps, and the extension to flows is easy to accomplish.

Throughout the discussion X is a complete metric space and $T : X \to X$ is continuous. The map T is said to be asymptotically smooth if for some bounded set $B \subseteq X$, there is a compact set $J \subseteq X$ such that, for any $\varepsilon > 0$, there is an integer $n_0(\varepsilon, B) > 0$ such that, if $T^n x \in B$ for $n > 0$, then $T^n x \in (J, \varepsilon)$ for $n \geq n_0(\varepsilon, B)$ where (J, ε) is the ε-neighborhood of J.

Theorem 3.1.15. *If $T : X \to X$ is continuous and there is a compact set K which attracts compact sets of X and $J = \bigcap_n T^n K$, then*

(i) J is independent of K;

(ii) J is maximal, compact, invariant;
(iii) J is stable and attracts compact sets of X.

If, in addition, T is asymptotically smooth, then

(iv) for any compact set $H \subseteq X$, there is a neighborhood H_1 of H such that $\bigcup_{n \geq 0} T^n H_1$ is bounded and J attracts H_1. In particular, J is uniformly asymptotically stable.

The following result is useful in the verification of the hypotheses of Theorem 3.1.15 and, in addition, gives more information about the strong attractivity properties of the set J.

Theorem 3.1.16. *If T is asymptotically smooth and T is compact dissipative, then there exists a compact invariant set which attracts compact sets and the conclusions of Theorem 3.1.15 hold. In addition, if $\bigcup_{n \geq 0} T^n B$ is bounded for every bounded set B in X, then J is the compact global attractor.*

We now define a more specific class of mappings which are asymptotically smooth.

A *measure of noncompactness* β on a metric space X is a function β from the bounded sets of X to the nonnegative real numbers satisfying

(i) $\beta(A) = 0$ for $A \subseteq X$ if and only if A is precompact,
(ii) $\beta(A \cup B) = \max[\beta(A), \beta(B)]$.

A classical measure of noncompactness is the *Kuratowskii measure of noncompactness* α defined by

$$\alpha(A) = \inf\{d : A \text{ has a finite cover of diameter } < d\}.$$

A continuous map $T : X \to X$ is a *β-contraction of order $k < 1$* with respect to the measure of noncompactness β if $\beta(TA) \leq k\beta(A)$ for all bounded sets $A \subseteq X$.

Theorem 3.1.17. *β-contractions are asymptotically smooth.*

From Theorem 3.1.17 and Theorem 3.1.16, it follows that T being a β-contraction which is compact dissipative with positive orbits of bounded sets bounded implies there exists a maximal compact invariant set J which attracts bounded sets of X.

It is also very important to know how the set J depends on the map T; that is, a generalization of Theorem 3.1.11. To state the result, we need another definition.

Suppose $T : \Lambda \times X \to X$ is continuous. Λ and X are complete metric spaces. Also suppose $T(\lambda, \cdot) : X \to X$ has a maximal compact invariant set $J(\lambda)$ for each $\lambda \in \Lambda$. We say $T : \Lambda \times X \to X$ is collectively β-contracting if, for all bounded sets B, $\beta(B) > 0$, one has $\beta\left(\bigcup_{\lambda \in \Lambda} T(\lambda, B)\right) < \beta(B)$.

Theorem 3.1.18. *Let X, Λ be complete metric spaces, $T : \Lambda \times X \to X$ continuous and suppose there is a bounded set B independent of $\lambda \in \Lambda$ such that B is compact dissipative under $T(\lambda, \cdot)$ for every $\lambda \in \Lambda$. If T is collectively β-contracting, then the maximal compact invariant set $J(\lambda)$ of $T(\lambda, \cdot)$ is upper semicontinuous in λ.*

For an historical discussion of the existence of maximal compact invariant sets, see Hale [72], [73]. The proofs of all results also can be found there. We remark that more sophisticated results on dissipative systems have been obtained by Massat [138].

3.2 Examples of RFDE on manifolds

Example 3.2.1. RFDE on $\mathbb{R}^n$
    Autonomous retarded functional differential equations on $\mathbb{R}^n$ are usually defined as equations of the form

$$\dot{x}(t) = f(x_t)$$

where f maps $C^0(I, \mathbb{R}^n)$ into $\mathbb{R}^n$. Taking $M = \mathbb{R}^n$ and identifying TM with $\mathbb{R}^n \times \mathbb{R}^n$, one can define the function $F : C^0(I, M) \to TM$ such that $F(\varphi) = (\varphi(0), f(\varphi))$. If f is continuous, then F is an RFDE on $M = \mathbb{R}^n$ which can be identified with the above equation.

Example 3.2.2. Ordinary Differential Equations as RFDE
    Any continuous vector field X on a manifold M defines an RFDE on M by $F = X\rho$ where $\rho : C^0 \to M$ is, as before, the evaluation map $\rho(\varphi) = \varphi(0)$.

Example 3.2.3. Ordinary Differential Equations on $C^0(I, M)$
    Any continuous vector field Z on $C^0 = C^0(I, M)$ for $I = [-r, 0]$, $r > 0$ and M a manifold, defines an RFDE on M by $F = T\rho \circ Z$, where $T\rho$ denotes the derivative of the evaluation map ρ.

Example 3.2.4. Products of Real Functions on $C^0(I, M)$ by RFDE on M
    If $g : C \to \mathbb{R}$ is continuous and F is an RFDE on M, then the map $G : C^0 \to TM$ given by $G(\varphi) = g(\varphi) \cdot F(\varphi)$ is also an RFDE on M.

Example 3.2.5. RFDE on TM
    Retarded functional differential equations on TM are continuous maps $\bar{F} : C^0(I, TM) \to T^2 M$ satisfying $\tau_{TM} \cdot \bar{F} = T\rho$.
    Recall that one can write locally $\bar{F}(\varphi, \psi) = (\varphi(0), \psi(0), f_1(\varphi, \psi), f_2(\varphi, \psi))$. Consequently, for the solutions $(x(t), y(t))$ on TM we have

$$\dot{x}(t) = f_1(x_t, y_t)$$
$$\dot{y}(t) = f_2(x_t, y_t).$$

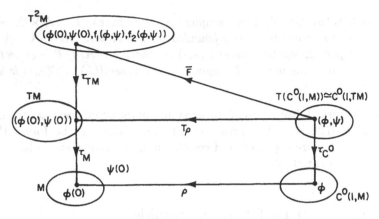

Fig. 3.2.

Given a C^1 RFDE(F), its first variational equation $\bar{F}$ is a special case of an RFDE on TM. Denoting locally, $F(\varphi) = (\varphi(0), f(\varphi))$, we have $\bar{F}(\varphi, \psi) = (\varphi(0), \psi(0), f(\varphi), df(\varphi)\psi)$ where df denotes the derivative of f. The solutions $(x(t), y(t))$ on TM, of the first variational equation $\bar{F}$ must satisfy

$$\dot{x}(t) = f(x_t)$$
$$\dot{y}(t) = df(x_t)y_t.$$

Example 3.2.6. Second order RFDE on M

Another special case of RFDE on TM is associated with second order RFDE on M.

Let $\bar{F} : C^0(I, TM) \to T^2M$ a continuous function such that, locally,

$$\bar{F}(\varphi, \psi) = (\varphi(0), \psi(0), \psi(0), f(\varphi, \psi)).$$

The solutions $(x(t), y(t))$ of the RFDE($\bar{F}$) on TM satisfy

$$\dot{x}(t) = y(t)$$
$$\dot{y}(t) = f(x_t, y_t)$$

or

$$\ddot{x}(t) = f(x_t, \dot{x}_t)$$

where $x(t)$ assumes values in M. We are therefore justified in calling second order RFDE on M to the functions $\bar{F} : C^0(I, TM) \to T^2M$ of the form described above.

Example 3.2.7. Retarded Differential Delay Equations on M

Let $g : M \times M \to TM$ be a continuous function such that $\tau_M \cdot g = \pi_1$ is the first projection of $M \times M$ onto M, and let $d : C^0(I, M) \to M \times M$ be such that $d(\varphi) = (\varphi(0), \varphi(-r))$.

The function $F = g \cdot d$ is an RFDE on M, and for its solutions $x(t)$ one can write, locally,

$$\bigl(x(t), \dot{x}(t)\bigr) = g\bigl(x(t), x(t-r)\bigr) = \Bigl(x(t), \bar{g}\bigl(x(t), x(t-r)\bigr)\Bigr)$$

or simply

$$\dot{x}(t) = \bar{g}\bigl(x(t), x(t-r)\bigr).$$

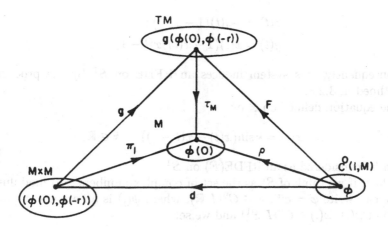

Fig. 3.3.

Example 3.2.8. RFDE on Embedded Submanifolds of $\mathbb{R}^n$

Let S be an embedded submanifold of $\mathbb{R}^n$ which is positively invariant under the RFDE on $\mathbb{R}^n$ given by

$$\dot{x}(t) = f(x_t),$$

i.e., solutions with initial condition φ at $t = 0$ such that $\varphi(0) \in S$, assume values in S for all $t \geq 0$ in their interval of existence.

The function $F : C^0(I, S) \to TS$ such that $F(\varphi) = (\varphi(0), f(\varphi))$ is an RFDE on S.

Example 3.2.9. An RFDE on S^2

Let us come back to the system of differential delay equations on $\mathbb{R}^3$ already considered in Example 2.0.11

$$\dot{x}(t) = -x(t-1)y(t) - z(t)$$
$$\dot{y}(t) = x(t-1)x(t) - z(t)$$
$$\dot{z}(t) = x(t) + y(t).$$

Its solutions satisfy $x\dot{x} + y\dot{y} + z\dot{z} = 0$, or $x^2 + y^2 + z^2 = $ constant, $t \geq 0$. Consequently, if $\varphi \in C^0([-1,0];\mathbb{R}^3)$ and $\varphi(0) \in S^2 = \{(x,y,z) \in \mathbb{R}^3 : x^2 + y^2 + z^2 = 1\}$, S^2 is positively invariant and, therefore, the given system induces an RFDE on S^2 by the construction given in the preceding example.

Example 3.2.10. RFDE on S^1

a) The set $S^1 = \{(x,y) \in \mathbb{R}^2 : x^2 + y^2 = 1\}$ is positively invariant under the system

$$\dot{x}(t) = -y(t)(1 - x(t))x(t - 1)$$
$$\dot{y}(t) = x(t)(1 - x(t))x(t - 1).$$

Consequently, this system induces an RFDE on S^1 by the procedure outlined in 3.2.8.

b) The equation defined on $\mathbb{R}$ by

$$\dot{x}(t) = k\sin(x(t) - x(t - 1)), \quad k \in \mathbb{R},$$

can be considered as an RFDE(F) on S^1.
In fact, if we think of S^1 as the set of complex numbers φ of modulus 1, one can write $\varphi = e^{i\psi}$, $\psi \in C^0(I, \mathbb{R})$, where $\psi(.)$ is an argument (up to 2π) of $\varphi(.)$, $\varphi(.) \in C^0(I, S^1)$ and we set:

$$F(\varphi) = \left(\varphi(0), k\sin(\psi(0) - \psi(-1))u_{\varphi(0)}\right)$$

where $u_{\varphi(0)}$ is the positively oriented unit vector tangent to S^1 at $\varphi(0)$. The determination of ψ up to 2π shows that F is well defined due to the 2π-periodicity of the sinus function.

c) The equation defined on $\mathbb{R}$ by

$$\dot{x}(t) = \frac{\pi}{2}(1 - \cos x(t)) + \frac{\pi}{2}(1 - \cos x(t - 1))$$

is another example of an equation that can be considered as an RFDE on S^1 by the same procedure used in b).

Example 3.2.11. A Second Order Equation on S^1
The second order scalar equation

$$\ddot{x}(t) = A\dot{x}(t) + B\sin x(t - r)$$

can be written as a system

$$\dot{x}(t) = y(t)$$
$$\dot{y}(t) = Ay(t) + B\sin x(t - r)$$

where $A, B \in \mathbb{R}$. This system defines a second order RFDE on S^1 given by map $F : C^0(I, TS^1) \to T^2 S^1$ such that

$$F(\varphi, \psi) = \big(\varphi(0), \psi(0), \psi(0), A\psi(0) + B \sin \bar{\varphi}(-r)\big),$$

where $\bar{\varphi}(.)$ is any argument of $\varphi(.)$.

As a matter of fact, this equation is an RFDE on the cylinder $S^1 \times \mathbb{R} = TS^1$.

This equation has been studied in connection with the circumutation of plants and is sometimes called the sunflower equation.

Example 3.2.12. The Levin-Nohel Equation on $\mathbb{R}$ and on S^1

Let $G : \mathbb{R} \to \mathbb{R}$ and $a : [0, r] \to \mathbb{R}$ be C^1 functions and denote by g the derivative of G. The scalar equation

$$\dot{x}(t) = -\int_{-r}^{0} a(-\theta) g\big(x(t+\theta)\big) d\theta$$

is known as the Levin-Nohel equation on $\mathbb{R}$. It has been studied in connection with nuclear reactor dynamics and also as a viscoelastic model (see [123], [72] and [78]).

A special case of this equation is obtained with $G(x) = 1 - \cos x$. This equation can be considered as an RFDE on S^1 by the same procedure as used for the example in 3.2.10 b), c).

Example 3.2.13. Equations Obtained by Compactification

In the study of polynomial vector fields in the plane $\mathbb{R}^2$, Poincaré used a compactification of $\mathbb{R}^2$ given by a central projection of $\mathbb{R}^2$ into a unit sphere S^2 tangent to the plane $\mathbb{R}^2$ at the origin, when these manifolds are considered as embedded in $\mathbb{R}^3$. This compactification procedure can be extended to construct examples of delay equations on spheres from polynomial delay equations on $\mathbb{R}^n$, $n \geq 1$ (see Oliva [153]).

For the purposes of illustration, let us consider the following differential equation on $\mathbb{R}$:

$$\dot{x}(t) = P\big(x(t-1)\big) \tag{3.3}$$

where P is a polynomial of a certain degree. One can consider the central projection at the line $\{(x, 1) : x \in \mathbb{R}\}$ into the circle $S^1 = \{(y_1, y_2) \in \mathbb{R}^2 : y_1^2 + y_2^2 = 1\}$, given by $(y_1, y_2) = \pm(x, 1)/\Delta(x)$, with $\Delta(x) = (1 + x^2)^{1/2}$.

For $y_2 \neq 0$, we have $x = y_1/y_2$ and therefore

$$(y_2 \dot{y}_1 - y_1 \dot{y}_2)/y_2^2 = \dot{x} = P(*) \cdot [F]$$
$$(y_1 \dot{y}_1 + y_2 \dot{y}_2) = 0,$$

where $P(*)$ denotes the right-hand side of the particular equation in (3.3) which is being considered and $[F]$ denotes an appropriate multiplication factor to be chosen according to the application envisaged. Solving the last system of equations for $\dot{y}_1$ and $\dot{y}_2$, one obtains

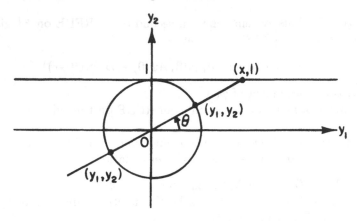

Fig. 3.4.

$$\dot{y}_1 = y_2^2 P(*)y_2[F]$$
$$\dot{y}_2 = -y_1 y_2 P(*)y_2[F].$$

The particular case

$$P(*) = P\big(x(t-1)\big) = -kx(t-1),$$

gives, under the above central projection,

$$P\big(y_1(t-1)/y_2(t-1)\big) = -ky_1(t-1)/y_2(t-1).$$

Choosing for multiplicative factor $[F] = y_2(t-1)/y_2(t)$, one obtains

$$\dot{y}_1(t) = -ky_2^2(t)y_1(t-1)$$
$$\dot{y}_2(t) = ky_1(t)y_2(t)y_1(t-1).$$

This system can be considered as an RFDE on S^1 by a procedure similar to the one used for the examples, 3.2.10 b), c). In terms of the angle coordinate (see Fig. 3.4), the equation can also be written as

$$\dot{\theta}(t) = k\sin\theta(t)\cdot\cos\theta(t-1). \tag{3.4}$$

A different choice for the multiplicative factor $[F]$ could be used to obtain a different equation on S^1. For instance $[F] = \frac{y_2(t-1)}{y_2^2(t)}$, would lead to the RFDE on S^1 given by

$$\dot{y}_1(t) = -ky_2(t)y_1(t-1)$$
$$\dot{y}_2(t) = ky_1(t)y_1(t-1)$$

or, in polar coordinate θ,

$$\dot{\theta}(t) = k \cos \theta(t - 1). \tag{3.5}$$

The multiplicative factor $[F]$ is to be chosen according to the application at hand. If, for instance, the study at infinity in the original coordinate is desired, it is convenient to choose $[F]$ so that the points on S^1 corresponding to $+\infty$ and $-\infty$ in the original coordinate, $\theta = 0$ and $\theta = \pi$, respectively, be invariant under the induced RFDE on S^1. It can be seen from (3.4) and (3.5), that this is the case for the first factor $[F]$ used above, but not for the second.

A similar Poincaré compactification can be used for higher dimensions. In particular, given a delay equation in $\mathbb{R}^2$

$$\dot{x}(t) = Ax(t - 1), \tag{3.6}$$

where A is a 2×2 real matrix, the Poincaré compactification of $\mathbb{R}^2$ into the unit sphere S^2 tangent to $\mathbb{R}^2$ at the origin (with both manifolds considered as embedded in $\mathbb{R}^3$), leads to an RFDE on the unit sphere S^2. This can be accomplished by multiplying (3.6) by the factor

$$\frac{y_3(t-1)}{y_3(t)} = \left(\frac{1 + x_1^2(t) + x_2^2(t)}{1 + x_1^2(t-1) + x_2^2(t-1)} \right)^{1/2}.$$

The delay equation in S^2 obtained in this way is denoted by $\pi(A)$ and it is given by the restriction to S^2 of the following system on $\mathbb{R}^3$

$$\begin{pmatrix} \dot{y}_1(t) \\ \dot{y}_2(t) \\ \dot{y}_3(t) \end{pmatrix} = \begin{pmatrix} 1 - y_1^2(y) & -y_1(t)y_2(t) \\ -y_1(t)y_2(t) & 1 - y_2^2(t) \\ -y_1(t)y_3(t) & -y_2(t)y_3(t) \end{pmatrix} A \begin{pmatrix} y_1(t-1) \\ y_2(t-1) \end{pmatrix} \tag{3.7}$$

The behavior at infinity in $\mathbb{R}^2$ is described by the restriction of (3.7) to the equator S^1, which can be written in polar coordinates for the plane $y_3 = 0$ as

$$\dot{\theta}(t) = (-\sin\theta(t), \cos\theta(t)) A \begin{pmatrix} \cos\theta(t-1) \\ \sin\theta(t-1) \end{pmatrix} \tag{3.8}$$

If $A = (a_{ij})_{i,j=1}^2$, then the initial points of system (3.7) on S^2 are $N = (0,0,1)$, $S = (0,0,-1)$ and the points on the equator S^1 which correspond to solutions of

$$(a_{22} - a_{11}) \sin\theta \cos\theta + a_{21} \cos^2\theta - a_{12} \sin^2\theta = 0.$$

We first give a generic result for $\pi(A)$.

Theorem 3.2.14. *The set $\mathcal{A}$ of all 2×2 real nonsingular matrices A for which $\pi(A)$ on S^2 has all critical points hyperbolic is open and dense in the set $M(2)$ of all real 2×2 matrices. Furthermore, if $A \in \mathcal{A}$ then it is equivalent under a similarity transformation to one of the following types of matrices:*

$$\text{(I)} \quad A = \begin{pmatrix} a_1 & 0 \\ 0 & a_2 \end{pmatrix} \qquad\qquad \text{with } a_1 \neq a_2$$

$$\text{(II)} \quad A = \begin{pmatrix} \alpha & -\beta \\ \beta & \alpha \end{pmatrix}, \qquad\qquad \beta > 0.$$

Proof: One first observes that the critical points in the equator are not hyperbolic if the eigenvalues of A are not distinct. This implies that $\mathcal{A}$ contains either matrices of types I or II. One then shows that the set of all real nonsingular 2×2 matrices A with distinct eigenvalues is open and dense in $M(2)$.

For matrices of type (I), the critical points are N, S and four points in the equator given in terms of the polar angle by $\theta = 0, \pi/2, 3\pi/2$ and π. The hyperbolicity of N and S is equivalent to $-a_1, -a_2 \neq (\pi/2 + 2n\pi)$, $n = 0, \pm 1, \pm 2$. The linear variational equation at the points on the equator given by $\theta = 0, \pi$ can be expressed in spherical coordinates defined by $y_1 = \cos \psi \cos \varphi$, $y_2 = \cos \psi \sin \varphi$, $y_3 = \sin \psi$, as

$$\dot{\psi}(t) = -a_1 \psi(t)$$
$$\dot{\varphi}(t) = -a_1 \varphi(t) + a_2 \varphi(t - 1).$$

The only possibility for characteristic values of these equations to belong to the imaginary axis is to have $|a_1| < |a_2|$ and then the characteristic values $\lambda = iy$ must satisfy $\cos y = a_1/a_2$ and $y^2 = a_2^2 - a_1^2$. Perturbing a_1 and a_2 with $a_1/a_2 = $ constant, we obtain hyperbolicity. The points in the equator given by $\theta = \pi/2, 3\pi/2$ are treated in a similar way.

For matrices of type (II), there are no critical points in the equator and the characteristic values of the linear variational equation at N and S in the imaginary axis, $\lambda = iy$, must satisfy $y^2 = \alpha^2 + \beta^2$ and $\tan y = \pm\alpha/\beta$. Perturbing α and β while maintaining $\alpha/\beta = $ constant we obtain hyperbolicity.

In case (II) of Theorem 3.2.14, by the use of spherical coordinates $y_1 = \cos \psi \cos \varphi$, $y_2 = \cos \psi \sin \varphi$, $y_3 = \sin \psi$, the equation $\pi(A)$ on S^2 can be written

$$\dot{\psi}(t) = -\left(\alpha^2 + \beta^2\right)^{1/2} \sin \psi(t) \cos \psi(t - 1) \cos\left[\varphi_0 - \varphi(t) + \varphi(t - 1)\right]$$
$$\dot{\varphi}(t) = \left(\alpha^2 + \beta^2\right)^{1/2} \frac{\cos \psi(t - 1)}{\cos \psi(t)} \sin\left[\varphi_0 - \varphi(t) + \varphi(t - 1)\right] \tag{3.9}$$

where $0 < \varphi_0 < \pi$ satisfies

$$\cos \varphi_0 = \left(\alpha^2 + \beta^2\right)^{-1/2} \alpha$$
$$\sin \varphi_0 = \left(\alpha^2 + \beta^2\right)^{-1/2} \beta.$$

In the equator of S^2, we have

$$\dot{\varphi}(t) = \left(\alpha^2 + \beta^2\right)^{1/2} \sin\left[\varphi_0 - \varphi(t) + \varphi(t - 1)\right]. \tag{3.10}$$

∎

Theorem 3.2.15. *In case (II) of Theorem 3.2.14, any periodic orbit of equation $\pi(A)$ in the equator of S^2 is given by a periodic solution of constant velocity. If $M = (\alpha^2 + \beta^2)^{1/2} < 1$, then the set of all global solutions in the equator consists of exactly one asymptotically stable hyperbolic periodic solution. There exists a sequence $M_0 < M_1 < M_2 < \cdots$, $M_n \to \infty$, such that for $M_i < M < M_{i+1}$ there exist exactly $2i + 1$ periodic orbits in the equator, their velocities are distinct with the highest velocity increasing to ∞ as M increases, and they are hyperbolic and alternatively asymptotically stable or unstable under the ordering of magnitude of these velocities. If $M = M_i$, $i > 1$, then there exist exactly $2i$ periodic orbits in the equator and all of them, except the one with highest speed are hyperbolic and alternately asymptotically stable or unstable.*

Proof: Let $\varphi(t)$ be a T-periodic solution of (3.10). Then the change of coordinates $u(t) = \varphi(t) - \varphi(t-1)$ satisfies $u(t) = u(t+T)$, and

$$\dot{u}(t) = -M\left[\sin\big(\varphi_0 - u(t)\big) - \sin\big(\varphi_0 - u(t-1)\big)\right] \qquad (3.11)$$

with $M = (\alpha^2 + \beta^2)^{1/2}$. It can be shown that any solution $x(t)$ of $\dot{x}(t) = g\big(x(t)\big) - g\big(x(t-1)\big)$ converges to a limit (finite or infinite) as $t \to \infty$ provided g is continuously differentiable. Besides, we can write

$$x(t) = x(0) - \int_{-1}^{0} g\big(x(\theta)\big)\,d\theta + \int_{t-1}^{t} g\big(x(\tau)\big)\,d\tau.$$

Since (3.11) is an equation of this type with $g(x) = -M\sin(\varphi_0 - x)$, we get that $u(t)$ is bounded and converges to a constant as $t \to \infty$. Since $u(t) = u(t+T)$, $\varphi(t) - \varphi(t-1) = u(t)$ is a constant function, and, from equation (3.10), we have $\dot{\varphi}(t)$ also constant, proving the first statement.

If $\varphi(t)$ is a global solution in the equator, then, from (3.10) with $u(t) = \varphi(t) - \varphi(t-1)$, we have

$$u(t) = M \int_{t-1}^{t} \sin\big[\varphi_0 - u(\tau)\big]\,d\tau.$$

Consider the Banach space $\mathcal{B}$ of all real continuous bounded functions with the sup norm, and let $\mathcal{T} : \mathcal{B} \to \mathcal{B}$ be the map transforming u into the function of t given by the right-hand side of the preceding equation. We have $\|\mathcal{T}(u_1) - \mathcal{T}(u_2)\| \leq M\|u_1 - u_2\|$. Thus, if $M < 1$, $\mathcal{T}$ is a contraction map and therefore there exists a unique fixed point u_0 of $\mathcal{T}$ in $\mathcal{B}$. Any solution ω of

$$\sin(\omega - \varphi_0) = -\omega/M \qquad (3.12)$$

is a fixed point of $\mathcal{T}$ and there exists always at least one solution of this equation. Hence the function $u(t)$ is constant and, therefore, φ is 1-periodic.

To study the hyperbolicity of the periodic orbits in the equator, which we know have constant velocities, we consider the linear variational equation of (3.9) around solutions $\psi(t) = 0$, $\varphi(t) = \omega t$. Clearly, ω must satisfy equation (3.12). It is then easy to prove by analysis of characteristic values that a periodic orbit in the equator with velocity w is hyperbolic if and only if $\cos(\varphi_0 - w) \neq 0$ and $M\cos(\varphi_0 - w) \neq -1$. Analyzing the characteristic equation for the linear variation along the solution, it is easy to prove that all characteristic values have negative real parts if $M\cos(\varphi_0 - w) > 0$, and, therefore, the corresponding periodic orbits of constant velocity w are asymptotically stable.

Since equation (3.12) describes the velocities of periodic orbits in the equator, one has only to study the roots of this equation to conclude the rest of the statement (see Fig. 3.5). ■

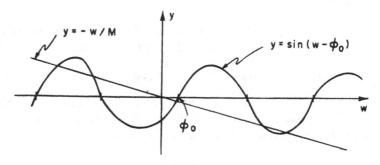

Fig. 3.5.

Remark 3.2.16. It is easy to see that the unstable manifolds of the (unstable) hyperbolic periodic orbits have dimension two. We will return to this type of example in Chapter 6 where its Morse–Smale character is analyzed.

We now consider the scalar equation

$$\dot{x}(t) = -ku(t-1), \quad k \neq 0, \; x(t) \in \mathbb{R}. \tag{3.13}$$

By Poincaré compactification, we can define an RFDE on the circle $S^1 = \{(y_1, y_2) \in \mathbb{R}^2 : y_1^2 + y_2^2 = 1\}$.

In order to obtain an analytic equation on S^1 which leaves the points corresponding to infinity invariant, we multiply (3.13) by the factor

$$\frac{y_2(t-1)}{y_2(t)} = \left(\frac{1 + x^2(t)}{1 + x^2(t-1)}\right)^{1/2}$$

Introducing polar coordinates, we obtain

$$\dot{\theta}(t) = k\sin\theta(t)\cos\theta(t-1). \tag{3.14}$$

There exist four critical points corresponding to $\theta = 0, \pi/2, \pi, 3\pi/2$: $A = (1,0)$, $B = (0,1)$, $C = (-1,0)$, $D = (0,-1)$. The linear variational equation for the points corresponding to infinity, A and C, is $\dot{\theta} = k\theta$ and, therefore, the equation behaves like an ODE close to these points. At the poles B and D, the linear variational equation is precisely the original equation (3.13).

Theorem 3.2.17. *There is a Hopf bifurcation for (3.14) near A and C for k near $(\pi/2 + 2n\pi)$, n integer. If $k > \pi/2$, then (3.14) has periodic solutions of period $T = 4$ satisfying the symmetry conditions $\theta(t) = \pi - \theta(t-2)$, $t \in \mathbb{R}$.*

Proof: The first statement is a standard application of the Hopf bifurcation theorem.

For the second statement, assume $\theta(t)$ is a global solution of (3.14) and let $\varphi(t) = \theta(t) - \pi/2$, $\psi(t) = \varphi(t-1)$. Then $\dot{\psi}(t) = -k\cos\psi(t)\sin\phi(t-2)$. If there exists a solution $\theta(t)$ such that $\theta(t) = -\theta(t-2)$, then $\varphi(t), \psi(t)$ must satisfy

$$\dot{\varphi}(t) = -k\cos\varphi(t)\sin\psi(t)$$
$$\dot{\psi}(t) = k\sin\varphi(t)\cos\psi(t). \tag{3.15}$$

Clearly, this system is Hamiltonian on the torus T^2 with energy function $E = k\cos\varphi\cos\psi$. The phase portrait of this system in the (φ, ψ)-plane has centers at the points $\varphi = m\pi$, $\psi = n\pi$, and saddles at the points $\varphi = \pi/2 + m\pi$, $\psi = \pi/2 + n\pi$, for m, n integers (see Fig. 3.6). The saddle connections are contained in vertical and horizontal lines in the (φ, ψ)-plane. When we go to the torus, we get four saddles and four centers. The limit period of the periodic orbits is $T_\ell = 2\pi/k$ as the orbits approach a center and is $+\infty$ as the orbits approach a saddle. Then there exist always periodic orbits with period greater than $2\pi/k$ and, since $k > \pi/2$, there exist periodic orbits with period four.

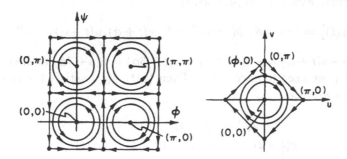

Fig. 3.6.

If we introduce new variables φ and ψ by the relations

$$\begin{aligned} u &= \varphi + \psi \\ v &= \varphi - \psi \end{aligned} \quad \text{or} \quad \begin{aligned} \varphi &= \frac{u+v}{2} \\ \psi &= \frac{u-v}{2} \end{aligned}$$

then system (3.15) becomes the Hamiltonian system

$$\dot{u} = k \sin v = \frac{\partial E}{\partial v}$$

$$\dot{v} = -k \sin u = -\frac{\partial E}{\partial u}$$

(3.16)

where the energy is $E = -k(\cos u + \cos v)$.

We look for periodic solutions of (3.15) satisfying

$$\begin{aligned} \psi(t) &= \varphi(t-1) \\ \varphi(t) &= -\varphi(t+2) \end{aligned} \quad \text{which imply} \quad \begin{aligned} \varphi(t) &= -\psi(t-1) \\ \psi(t) &= -\psi(t+2) \end{aligned}$$

These conditions for a solution of (3.15) are equivalent to finding a periodic solution $\big(u(t), v(t)\big)$ of (3.16) satisfying

$$\begin{aligned} u(t) &= v(t-1) \\ v(t+2) &= -v(t) \end{aligned} \quad \text{which imply} \quad \begin{aligned} u(t-1) &= -v(t) \\ u(t) &= -u(t+2). \end{aligned}$$

Now choose $k > \pi/2$ and, for simplicity, work in the square $|u| + |v| = \pi$ which contains four saddle connections of the (u, v) plane. There exists a $c > 0$ such that the solution defined by $v(1) = c$, $u(1) = 0$, $c < \pi$, has period equal to 4.

Let $u(t), v(t)$ be such a solution. Consider now the functions $\bar{u}(t) = -v(t)$ and $\bar{v}(t) = u(t)$ and verify that $\big(\bar{u}(t), \bar{v}(t)\big)$ satisfy (3.16). But the solutions $\big(u(t), v(t)\big)$ and $\big(\bar{u}(t), \bar{v}(t)\big)$ have the same energy E since $E = -k\big(\cos u(t) + \cos v(t)\big) = -k\big(\cos \bar{u}(t) + \cos \bar{v}(t)\big) = -k(1 + \cos c)$. Thus, for a certain t^*, we have $\bar{u}(t^*) = 0$ and $\bar{v}(t^*) = c$. Both solutions define then the same periodic orbit and there exists $\alpha \in (0, 4)$ such that

$$\big(u(t), v(t)\big) = \big(\bar{u}(t+\alpha), \bar{v}(t+\alpha)\big) = \big(-v(t+\alpha), u(t+\alpha)\big), \quad \forall t \in \mathbb{R}.$$

But $u(t) = -v(t+\alpha) = -u(t+2\alpha) = v(t+3\alpha) = u(t+4\alpha)$ and, since $u(t)$ has period 4, we need to have $\alpha = 1$. Then $\big(u(t), v(t)\big)$ satisfy the conditions required above since

$$u(t) = v(t+3) = v(t-1)$$
$$v(t+2) = u(t+3) = u(t-1) = -v(t).$$

The corresponding 4-periodic functions $\varphi(t), \psi(t)$ are such that $\theta(t)$ is a periodic solution of the equation (3.14) with period $T = 4$ and such that $\theta(t) = \pi - \theta(t-2)$. ∎

The results above, in this example, are due to Oliva [153]. The proof that the solutions of the Equation (3.11) are bounded and approach a limit as $t \to \infty$ follows from more general results of Cooke and Yorke [36]. For the second statement of Theorem 3.2.17 see also [108].

Example 3.2.18. The presence of collision points
   We show now that RFDE may present collision between two C^∞ global
orbits. For this we start with the following C^∞ retarded delay differential
equation introduced by Hale and Lin (see [84]):

$$\dot{x} = \alpha(x)x(t) + \beta(x)x(t-1),$$

where $\alpha : \mathbb{R} \to \mathbb{R}$ and $\beta : \mathbb{R} \to \mathbb{R}$ are suitable C^∞ functions described below.
This means that $f : C^0([-1,0],\mathbb{R}) \to \mathbb{R}$ defines an RFDE given by

$$f(\phi) = \alpha(\phi(0))\phi(0) + \beta(\phi(0))\phi(-1).$$

The C^∞-functions α and β satisfy

$$\alpha(x) = \frac{2e-1}{e-1} \quad and \quad \beta(x) = \frac{-e^2}{e-1} \quad for \quad |x| \leq 1;$$

$$\alpha(x) = 1 \quad and \quad \beta(x) = 0 \quad for \quad |x| \geq 2;$$

$$\alpha(x) + \frac{\beta(x)}{e} = 1, \quad otherwise.$$

It can be checked that the equation has three critical points including zero
which is a hyperbolic point with a two-dimensional local unstable manifold.
Moreover the function $x = ae^t$ (a constant) is a global solution, that is,
satisfies the equation for all $t \in \mathbb{R}$ and $x = be^{2t}$ satisfies the equation for
t such that $0 < be^{2t} \leq 1$. For $t > 0$ sufficiently large the solution curve
$x_t \in C^0([-1,0],\mathbb{R})$, corresponding to the function $x = ae^t$, leaves the ball of
radius 2 and then the RFDE, in this region, reduces to the ODE $\dot{x} = x$, so,
over there, the unstable set corresponding to the local unstable manifold of
zero has dimension one. But for t close to $-\infty$ both ae^t and be^{2t} are inside
the local unstable manifold of zero. Thus, the (global) unstable set of zero
has a piece with dimension two but along the solution ae^t passes to a piece
with dimension one which proves the existence of collision.
   In [55], Fichmann and Oliva also describe a situation of collision for an
RFDE on $\mathbb{R}$; in their example the colliding orbits are bounded and the authors were able in reformulating the equation to obtain RFDE on compact
manifolds (on the circle and on the torus). These last examples are crucial for
the theory developed by the same two authors in [56] where one-to-oneness
and non one-to-oneness (of a semiflow) on a hyperbolic compact invariant
set without critical points were considered in order to study the persistence
of this hyperbolic invariant set under perturbation of the semiflow (see also
Chapter 7 for more details). In order to describe the examples mentioned
above and following [55], we start by recalling the definitions of some auxil-
iary functions. Fix $\varepsilon > 0$ and h, $\varepsilon < h < 1$, such that $(\ln \varepsilon)^{-1} - r > (\ln h)^{-1}$.
Define $j :]0,1[\to \mathbb{R}$ by

$$j(k) = (\ln k)^{-1} - r \text{ for } 0 < k \leq \varepsilon,$$
$$j(k) = (\ln k)^{-1} \text{ for } h \leq k < 1,$$
$$j \text{ decreasing and } C^{\infty}, \text{ otherwise.}$$

So, j is decreasing and C^{∞} on $]0, 1[$. Let $\Psi : \mathbb{R} \to \mathbb{R}$ be the C^{∞} function defined by

$$\Psi(k) = j(k) - (\ln k)^{-1} \text{ for } 0 < k < 1,$$
$$\Psi(k) = 0 \text{ for } k \geq h,$$
$$\Psi(k) = -r \text{ for } k \leq \varepsilon;$$

then Ψ is C^{∞}. Define now a C^{∞} map $z : \mathbb{R} \to \mathbb{R}$ by

$$z(k) = \frac{\exp[j(k)^{-1}] - \exp[(\ln k)^{-1} - r]^{-1}}{k - \exp[(\ln k)^{-1} - r]^{-1}}, \text{ for } 0 < k < 1,$$
$$z(k) = 1 \text{ for } k \geq h,$$
$$z(k) = 0 \text{ for } k \leq \varepsilon.$$

Now, let $\hat{i} : \mathbb{R}^2 \to \mathbb{R}$ be defined as

$$\hat{i}(k, s) = -[(\ln s)^{-1} - \Psi(k)]^{-2} \exp[(\ln s)^{-1} - \Psi(k)]^{-1},$$

for $s > \exp[\Psi(k)]^{-1}$ (or $(\ln s)^{-1} < \Psi(k)$), and

$$\hat{i}(k, s) = 0 \text{ otherwise.}$$

One can see that $\hat{i} : \mathbb{R}^2 \to \mathbb{R}$ is C^{∞} on $\mathbb{R} \setminus (l_1 \cup l_2)$, where $l_1 = \{(k, 1) \in \mathbb{R}^2 \mid k \leq h\}$ and $l_2 = \{(k, 0) \in \mathbb{R}^2 \mid k \geq h\}$.

In fact, $\hat{i}$ is not C^{∞} on $l_1 \cup l_2$. Take two open sets of $\mathbb{R}^2$, O_1 and O_2 containing l_1 and l_2, respectively, and disjoint from the curve

$$l = \{(k, s) \in \mathbb{R}^2 \mid 0 \leq k \leq 1, s = \exp[j(k)]^{-1}\} \cup \{(0, s) \in \mathbb{R}^2 \mid 0 \leq s \leq \exp[-r^{-1}]\}$$

Then $\hat{i}$ is C^{∞} on the closed set $\mathbb{R}^2 \setminus (O_1 \cup O_2)$ (includes the boundary) and can be extended to a C^{∞} function $i : \mathbb{R}^2 \to \mathbb{R}$ (see [111] p.272). For each $\theta \in [-r, 0]$ the evaluation map $\rho_{\theta} : C \to \mathbb{R}$, given by $\rho_{\theta}(\varphi) = \varphi(\theta)$, is C^{∞} since it is linear and continuous. Define

$$S = (z \circ \rho_0)\rho_0 + (1 - z \circ \rho_0)\rho_{-r}$$

that is, $S : C \to \mathbb{R}$ and $S(\varphi) = z(\varphi(0)) \cdot \varphi(0) + [1 - z(\varphi(0))]\varphi(-r)$.

Consider the C^{∞} RFDE of type $\dot{x} = f(x_t)$ given by the C^{∞} function $f : C \to \mathbb{R}$ defined by

$$f(\varphi) = i(\rho_0(\varphi), S(\varphi)). \tag{3.17}$$

The function $v : \mathbb{R} \to \mathbb{R}$, $v(t) \equiv 0$, is a global (constant) C^{∞} solution of the RFDE given by (3.17).

On the other hand the function $u : \mathbb{R} \to \mathbb{R}$ given by $u(t) = \exp[t^{-1}]$ for $t < 0$ and $u(t) = 0$ for $t \geq 0$, is also a global bounded C^∞ solution of the same RFDE. In the sequel we prove that u and v are solutions of (3.17) and that the point $\varphi \in C$, $\varphi \equiv 0$, is a collision point of the global bounded orbits $\{v_t \equiv 0 \mid t \in \mathbb{R}\}$ and $\{u_t \in C \mid t \in \mathbb{R}\}$.

It can be checked that $v \equiv 0$ is a global solution of (3.17). If we take $\varphi \equiv 0 \in C$, $k = \rho_0(\varphi) = 0$, $z(k) = 0$, $\rho_{-r}(\varphi) = 0$, $s = S(\varphi) = 0$, $(0,0) \in l$, then

$$f(\varphi) = i(0,0) = 0 = \varphi'(0).$$

We now show that u is, in fact, a C^∞ global bounded solution of (3.17):

a) For $0 \leq t \leq r$ and $\varphi = u_t \in C$, $k = \rho_0(\varphi) = u(t) = 0$, $z(k) = 0$, $\Psi(k) = -r$, so $s = S(\varphi) = \varphi(-r) = u(t - r) = \exp[t - r]^{-1} \leq \exp[-r]^{-1}$, $(k, s) \in l$, $s \leq \exp[\Psi(k)]^{-1}$, then $f(\varphi) = i(k, s) = 0 = \varphi'(0)$.

b) For $(\ln \varepsilon)^{-1} \leq t < 0$ and $\varphi = u_t \in C$, $k = \rho_0(\varphi) = u(t) = \exp[t]^{-1} \leq \varepsilon$ that implies $z(k) = 0$ and $\Psi(k) = -r$, $s = S(\varphi) = \varphi(-r) = u(t - r) = \exp[t - r]^{-1} = \exp[j(k)]^{-1}$ (in fact $j(k) = (\ln k)^{-1} + \Psi(k) = t - r$). Then $(k, s) \in l$, $(\ln s)^{-1} = t - r$ and so $f(\varphi) = i(k, s) = -[t - r + r]^{-2} \exp[t - r + r]^{-1} = -t^{-2} \exp[t]^{-1} = u'(t) = \varphi'(0)$.

c) For $(\ln h)^{-1} \leq t \leq (\ln \varepsilon)^{-1}$ and $\varphi = u_t \in C$, $\varepsilon \leq k = \rho_0(\varphi) = u(t) = \exp[t]^{-1} \leq h$, $t = (\ln k)^{-1}$, $\varphi(-r) = u(t - r) = \exp[t - r]^{-1}$, $s = S(\varphi) = z(k)k + (1 - z(k)) \exp(t - r)^{-1}$, and so $s = z(k)[k - \exp(t - r)^{-1}] + \exp(t - r)^{-1}$ or

$$s = \frac{\exp j(k)^{-1} - \exp(t - r)^{-1}}{k - \exp(t - r)^{-1}}[k - \exp(t - r)^{-1}] + \exp(t - r)^{-1} = \exp j(k)^{-1}.$$

$j(k) = (\ln s)^{-1} = t + \Psi(k)$, $(k, s) \in l$, then $f(\varphi) = i(k, s) = -[t + \Psi(k) - \Psi(k)]^{-2} \exp[t + \Psi(k) - \Psi(k)]^{-1} = -\frac{1}{t^2} \exp t^{-1} = u'(t) = \varphi'(0)$.

d) For $t \leq (\ln h)^{-1}$ and $\varphi = u_t \in C$, $k = \rho_0(\varphi) = u(t) = \exp t^{-1} > h$ and so $z(k) = 1$, $\Psi(k) = 0$, $s = S(\varphi) = \varphi(0) = k$, $t = (\ln s)^{-1} = (\ln k)^{-1} = j(k)$, so $s = \exp j(k)^{-1}$, $(k, s) \in l$ and then $f(\varphi) = i(k, s) = -t^{-2} \exp t^{-1} = u'(t) = \varphi'(0)$.

In what follows we see how the previous example can be adapted to the case in which the phase space is the set $\tilde{C} = C([-r, 0], S^1)$ of all continuous paths defined on $[-r, 0]$ with values on the unit circle $S^1 \subset \mathbb{R}^2$. Following [55], fix $\delta > 0$ such that $1 + \delta < \pi/2$ and let $g : \mathbb{R} \to \mathbb{R}$ be a C^∞ function such that $g(t) = \arcsin t$ for $t \in]-\delta, \sin(1 + \delta)[$. Replace the map $\rho_\theta : C \to \mathbb{R}$ of the previous example by the C^∞ map $\varphi \in \tilde{C} \to (\varphi(0), \tilde{\rho}_\theta(\varphi)) \in TS^1$ (TS^1 is the tangent bundle of S^1) with $\tilde{\rho}_\theta(\varphi) = g(\sin \varphi(\theta)) \in T_{\varphi(0)} S^1 \equiv \mathbb{R}$, $\theta \in [-r, 0]$ (recall the angle $\varphi(\theta) \in S^1$). Now one obtains a C^∞ RFDE (F) on the compact manifold S^1, $\dot{x}(t) = F(x_t)$, where $F : \tilde{C} \to TS^1$ is such that $F(\varphi) = (\varphi(0), \tilde{f}(\varphi))$, $\tilde{f}(\varphi) \in T_{\varphi(0)} S^1$ is given by $\tilde{f}(\varphi) = i(k, s) = i(\tilde{\rho}_0(\varphi), \tilde{S}(\varphi))$ and $\tilde{S} = (z \circ \tilde{\rho}_0) \cdot \tilde{\rho}_0 + (1 - z \circ \tilde{\rho}_0) \cdot \tilde{\rho}_{-r}$, that is, $s = \tilde{S}(\varphi) = z(k) \cdot k + (1 - z(k)) \cdot g(\sin \varphi(-r))$ and $k = g(\sin \varphi(0))$. The map F is of class C^∞ and since

$g(\sin\varphi(\theta)) = \varphi(\theta)$ when $0 \leq \varphi(\theta) \leq 1$, the computations of the previous example show that the corresponding functions $\tilde{u} : \mathbb{R} \to S^1$, $\tilde{u}(t) = \exp(t^{-1})$ for $t < 0$, $\tilde{u}(t) = 0$ otherwise, and $\tilde{v} : \mathbb{R} \to S^1$ the constant function $\tilde{v}(t) = 0$ $\forall t \in \mathbb{R}$ are global solutions of F such that have $\varphi \equiv 0 \in \tilde{C}$ as a collision point (just observe that we are identifying points in S^1 with angles).

Remark 3.2.19. Also, in [55], the authors present an example of a C^∞ RFDE on the torus $\mathrm{T}^2 \equiv S^1 \times \mathbb{R}/Z$ with collision on a periodic orbit. Let $\hat{C} = C([-r, 0], \mathrm{T}^2)$ be the phase space of all continuous paths defined on $[-r, 0]$ with values on T^2. For $\hat{\varphi} \in \hat{C}$ we have $\hat{\varphi}(\theta) = (\varphi(\theta), \pi(\psi(\theta)))$, $-r \leq \theta \leq 0$, where $\varphi \in \tilde{C}$, $\psi \in C$ and $\pi : \mathbb{R} \to \mathbb{R}/Z$ is the canonical projection. For a fixed $w \in \mathbb{R}$, $w > 0$, define $\hat{F} : \hat{C} \to T\mathrm{T}^2$ by $\hat{F}(\hat{\varphi}) = (\hat{\varphi}(0), \hat{f}(\hat{\varphi}))$ where $\hat{f}(\hat{\varphi}) \in T_{\hat{\varphi}(0)}\mathrm{T}^2 \equiv T_{\varphi(0)}S^1 \times \mathbb{R}$ is given by $\hat{f}(\hat{\varphi}) = (\tilde{f}(\varphi), w)$. In this way, one obtains a C^∞ RFDE $\hat{F}$ on the compact manifold T^2, $\dot{x}(t) = \hat{F}(x_t)$, and from the previous computations, one sees that the functions

$$\hat{u} : \mathbb{R} \to T^2, \quad \hat{u}(t) = (\tilde{u}(t), \pi(wt))$$

and

$$\hat{v} : \mathbb{R} \to T^2, \quad \hat{v}(t) = (0, \pi(wt)), \quad \forall t \in \mathbb{R},$$

are global solutions of $\hat{F}$ with collision on the periodic orbit of $\hat{v}$.

3.3 NFDE on manifolds

An *autonomous functional differential equation (FDE) on a manifold M* is a pair (D, F) of continuous functions

$$F : C^0(I, M) \to TM, \quad D : C^0(I, M) \to M$$

such that $\tau_M F = D$. It is clear that the retarded functional differential equations are special cases of functional differential equations for which the pair (D, F) is equal to (ρ, F). It follows from the definition of FDE that if locally one thinks of TM as a product, then for each $\varphi \in C^0(I, M)$, $F(\varphi) = (D(\varphi), f(\varphi))$ for all φ in a suitable neighborhood of φ.

Definition 3.3.1. *A solution of (D,F) with initial condition φ at t_0 is a continuous function $x(t)$ defined on $t_0 - r \leq t < t_0 + A$, $0 < A \leq \infty$, with values on M, such that if $x_t \in C^0(I, M)$ is (as above) defined by $x_t(\theta) = x(t + \theta)$, $\theta \in I$, $t_0 \leq t < t_0 + A$, one has:*

(i) $x_{t_0} = \varphi$,
(ii) $D(x_t)$ is a C^1-function of $t \in [t_0, t_0 + A)$,
(iii) $(D(x_t), (d/dt)D(x_t)) = F(x_t)$, $t \in [t_0, t_0 + A)$, where $(d/dt)D(x_t)$ denotes the tangent vector to the curve $D(x_t)$ at the point t.

Locally one can write

$$[D(x_t), \frac{d}{dt}D(x_t)] = [D(x_t), f(x_t)] \quad or \quad \frac{d}{dt}D(x_t) = f(x_t),$$

for an appropriate function f. The number $r > 0$ is also called *delay* or *lag* of (D, F). As usual, $x(t)$ is sometimes denoted by $x(t; t_0, \varphi)$ and x_t by $x_t(t_0, \varphi)$ or $x_t(\varphi)$ if $t_0 = 0$.

An *equilibrium (or critical) point* of an FDE (D, F) is a constant function $c \in C^0(I, M)$ such that $f(c) = 0$. A function $g(t)$, $-\infty < t < +\infty$, is said to be a solution of (D, F) on $-\infty < t < +\infty$ if for every $\sigma \in (-\infty, +\infty)$ the solution $x(t; \sigma, g_\sigma)$ of (D, F) exists and satisfies $x_t(\sigma, g_\sigma) = g_t$, $t \geq \sigma$. Such a solution on $-\infty < t < +\infty$ is also called a *global solution*. A *periodic solution* is a global solution such that $g(t + \omega) = g(t)$ for all t and some $\omega > 0$.

Definition 3.3.2. *A C^1 function $D : C^0(I, M) \to M$ is said to be atomic at zero if for any $\varphi \in C^0(I, M)$ the derivative $T_\varphi D$ given by*

$$T_\varphi D(\psi) = \int_{-r}^0 d\eta(\varphi, \theta)\psi(\theta)$$

has nonsingular continuous jump $A(\varphi) = [\eta(\varphi, 0) - \eta(\varphi, 0_-)]$; this means that $\det A(\varphi)) \neq 0$.

Remark 3.3.3. Since the tangent space $T_\varphi C^0(I, M)$ is the set of all $\psi : I \to TM$ such that $\tau_M \psi = \varphi$, it can be identified with $C^0(I, \mathbb{R}^n)$ if we choose, continuously, one frame at $\varphi(\theta)$, for each $\theta \in I$; so the integral considered in the last definition can be computed in a local chart containing φ and the condition $\det A(\varphi)) \neq 0$ does not depend on the choice of that local chart.

An autonomous neutral functional differential equation (NFDE) on M is an FDE (D, F) on M with D atomic at zero. In particular, the case when $D = \rho$, the evaluation map $\rho : \varphi \mapsto \varphi(0)$, the pair (ρ, F), or simply F, is a retarded functional differential equation on M. In fact, we only need to prove that ρ is atomic at zero; since

$$T_\varphi \rho(\psi) = \psi(0) = \int_{-r}^0 d\eta(\varphi, \theta)\psi(\theta),$$

where $\eta(\varphi, \theta) = 0$ for $\theta \in [-r.0)$ and $\eta(\varphi, 0) = I$, then one has $A(\varphi) = I$.

It is also interesting to observe, for later purposes, that the condition $\tau_M F = D$ shows that the smoothness of D is at least equal to the smoothness of F.

3.4 NFDE on $\mathbb{R}^n$

In this section, we define an interesting class of neutral functional differential equations on $\mathbb{R}^n$ and discuss some of the qualitative properties of the solutions, the existence of maximal compact invariant sets and compact global attractors and also the regularity of the solutions on such sets.

We use the same notation as in the previous sections and, furthermore, let $C = C^0([-r, 0], \mathbb{R}^n)$.

3.4.1 General properties

Let us describe, with more details, the properties already mentioned in Chapter 2, Example 2.0.16. Suppose that $D : C \to \mathbb{R}^n$ is a continuous linear operator given as

$$D\varphi = \varphi(0) - \int_{-\delta}^0 d\mu(\theta)\, \varphi(\theta), \qquad (3.18)$$

where $\mu(\theta)$ is an $n \times n$ matrix of bounded variation and there is a continuous nondecreasing function $\gamma(s)$, $s \in [-r, 0]$, $\gamma(0) = 0$ such that, for any $s > 0$,

$$\mathrm{Var}_{[-s,0]}\mu \le \gamma(s). \qquad (3.19)$$

Suppose that $f : C \to \mathbb{R}^n$ is a C^k-function, $k \ge 1$, and is a bounded map; that is, takes bounded sets into bounded sets. A neutral functional differential equation (NFDE) is a relation

$$\frac{d}{dt} Dx_t = f(x_t) \qquad (3.20)$$

where $\frac{d}{dt}$ is the right hand derivative of $x(t)$ at t.

If a solution of (3.20) is continuously differentiable, then it satisfies the relation $D\dot{x}_t = f(x_t)$, an equation in which the derivative occurs with delayed arguments. Our definition of NFDE is a special case of a more general class of FDE where the derivatives occur with delayed arguments in a general form but we will restrict ourselves to the discussion of (3.20) since it is possible in this case to obtain a theory which is as general as the one for RFDE.

For a given function $\varphi \in C$, we say that $x(t, \varphi)$ is a *solution of* (3.20) *on the interval* $[0, \alpha_\varphi)$, $\alpha_\varphi > 0$, *with initial value* φ *at* $t = 0$ if $Dx_t(\cdot, \varphi)$ is defined on $[0, \alpha_\varphi)$, has a continuous right hand derivative on $[0, \alpha_\varphi)$, $x_t(\cdot, \varphi) \in C$ for $t \in [0, \alpha_\varphi)$, $x_0(\cdot, \varphi) = \varphi$ and $x(t, \varphi)$ satisfies (3.20) on the interval $[0, \alpha_\varphi)$.

If a solution of (3.20) exists, then it must satisfy the following equation

$$Dx_t(\cdot, \varphi) = D\varphi + \int_0^t f(x_s(\cdot, \varphi))ds, \quad t \in [0, \alpha_\varphi) \qquad (3.21)$$

and initial condition

$$x_0(\cdot, \varphi) = \varphi. \qquad (3.22)$$

From (3.21), it is clear that each solution $x(t, \varphi)$ of (3.20) has the property that $Dx_t(\cdot, \varphi)$ must be continuously differentiable for $t \in (0, \alpha_\varphi)$ with a right hand derivative at $t = 0$. It is not difficult to show that a solution $x(t, \varphi)$ will be continuously differentiable at $t = 0$ if and only if the initial data φ is continuously differentiable at $\theta = 0$ and $D\frac{d\varphi}{d\theta} = f(\varphi)$.

Notice that we do not require that the function $x(t, \varphi)$ have a right hand derivative, but only that the function $Dx_t(\cdot, \varphi)$ has a right hand derivative. It can be shown that equations of the type (3.20) are very closely related to solutions of undamped wave equations in one space dimension with nonlinear boundary conditions (see, for example, Abolina and Myshkis [1], Brayton [18], Cooke and Krumme [37], Nagumo and Shimura [144]). Our definition of a solution requiring only that $Dx_t(\cdot, \varphi)$ has a right hand derivative corresponds to a weak solution of the corresponding wave equation. For other applications, see Kolmanovski and Myshkis [112].

Theorem 3.4.1. . *If $f \in C^k(C, \mathbb{R}^n)$, $k \geq 1$ (resp., analytic), then, for any $\varphi \in C$, there is a unique solution $x(t, \varphi)$ of (3.20) on a maximal interval of existence $[-r, \alpha_\varphi)$. If $\alpha_\varphi < \infty$, then $|x_t(\cdot, \varphi)| \to \infty$ as $t \to \alpha_\varphi$. Furthermore, $x_t(\cdot, \varphi)$ is continuous in t, φ and is C^k in φ (resp., analytic).*

Proof: The proof of Theorem 3.4.1 follows arguments similar in spirit to the ones used for RFDE (see [72] and [86]) except applied to the integral equation (3.21) written in the form

$$x_0(\cdot, \varphi) = \varphi; \quad x(t) = \int_{-r}^0 [d\mu(\theta)] x(t+\theta) + D\varphi + \int_0^t f(x_s(\cdot, \varphi)) ds, \quad t \in [0, \alpha_\varphi).$$

In this case, it is not possible to use the contraction principle to get a fixed point of a map defined by the right hand side of the equalities in this equation. However, it is possible to show that, on an appropriate class of functions on an interval $[-r, a]$ with a small, one has an α-contraction. This gives existence. The maximal interval of existence is obtained by invoking Zorn's lemma in the usual way. The uniqueness and smoothness properties of the solution in φ require some special arguments.   ∎

Let $T(t)\varphi \equiv T_{D,f}(t)\varphi = x_t(\cdot, \varphi)$ for $t \in [0, \alpha_\varphi)$. If we assume that all solutions of (3.20) are defined for $t \in \mathbb{R}^+$, then the family of maps $\{T(t) \equiv T_{D,f}(t) : C \to C, t \geq 0\}$ is a C^k-semigroup on C (resp., analytic). We will assume in the following that all solutions of (3.20) are defined for $t \geq 0$.

Our next objective is to obtain a representation of the semigroup $T(t), t \geq 0$, as the sum of a conditionally completely continuous operator and a linear semigroup defined on a linear subspace of C. To do this, we make the assumption (only to avoid some technicalities) that the matrix function $\mu(\theta)$ of bounded variation has no singular part; that is,

$$D\varphi = D_0\varphi - \int_{-r}^0 B(\theta)\varphi(\theta)d\theta \qquad D_0\, f = \varphi(0) - \Sigma_{j=1}^\infty B_j\varphi(-r_j), \qquad (3.23)$$

with $r_j > 0$, $1 \leq j$, and where

$$\Sigma_{j=1}^\infty |B_j| < \infty, \qquad \int_{-r}^0 |B(\theta)| d\theta < \infty.$$

To state an important qualitative property of the semigroup $T(t) \equiv T_{D,f}(t)$, $t \geq 0$ of (3.20) with D satisfying (3.23), we need the following nontrivial result which we state without proof.

Lemma 3.4.2. *There are positive constants b, a such that, for any $h \in C([0,\infty), \mathbb{R}^n)$ and any $\varphi \in C$, $D_0\varphi = h(0)$, the solution $x(t, \varphi, h)$ of the equation*

$$D_0 y_t = h(t)$$

satisfies the relation

$$|x(t, \varphi, h)| \leq be^{at}[|\varphi| + \sup_{0 \leq s \leq t} |h(s)|].$$

It is easy to prove that there is a matrix function $\Phi_{D_0} = (\varphi_1^{D_0}, \ldots, \varphi_n^{D_0})$ such that $D\Phi_{D_0} = I$. Finally, we make the observation that the *difference equation*

$$D_0 y_t = 0 \tag{3.24}$$

generates a C^0-semigroup $T_{D_0}(t)$, $t \geq 0$, on the Banach space $C_{D_0} = \{\varphi \in C : D_0\varphi = 0\}$. We say that the operator D_0 is *exponentially stable* if the zero solution of (3.24) is exponentially stable.

Theorem 3.4.3. *Let $T(t) \equiv T_{D,f}(t)$, $t \geq 0$, be the semigroup defined by (3.20) with D satisfying (3.23) and let T_{D_0} be the semigroup on C_{D_0} generated by the difference equation $D_0 y_t = 0$. If $\Phi_{D_0} = (\varphi_1^{D_0}, \ldots, \varphi_n^{D_0})$, $D\Phi_{D_0} = I$ and $\Psi_{D_0} = I - \Phi_{D_0}D_0 : C \to C_{D_0}$, then*

$$T_{D,f}(t) = T_{D_0}(t)\Psi_{D_0} + U_{D,f}(t), \tag{3.25}$$

where $U_{D,f}(t)$ is conditionally completely continuous for $t \geq 0$. If, in addition, the operator D_0 is exponentially stable, then $T_{D,f}(t), t \geq 0$, is a conditional α-contraction.

Proof: We present a sketch of the ideas in the proof. The solutions of (3.20) coincide with the solutions of (3.21). If we let $T(t)\varphi = T_{D,f}(t)\varphi$ and $T(t)\varphi - T_{D_0}(t)\Psi_{D_0}\varphi = U(t)\varphi$, then, for $t \geq 0$,

$$D_0(U(t)\varphi) = \int_{t-r}^{t} B(s - t)(T(s)\varphi)(0)ds + \int_0^t f(T(s)\varphi)ds$$

and $U(0)\varphi = \Phi_{D_0}D_0\varphi$, which varies over a finite dimensional subspace as φ varies over C. If B is a bounded set in C and $T(t)B$ is bounded on $[0, \tau]$, then the right hand side of this equation is uniformly bounded and equicontinuous on $[0, \tau]$. With this information, it is possible to show using Lemma 3.4.2 that $U(t)B$ is a uniformly bounded equicontinuous family. This proves the theorem. For more details, see [86].   ∎

For NFDE, the solutions may not smooth with increasing t. However, if D is exponentially stable, then any solution on a compact invariant set is as smooth as the vector field.

Theorem 3.4.4. *If D in (3.23) is exponentially stable, $f \in C^k(C, \mathbb{R}^n), 0 \le k \le \infty$ (resp. analytic) and J is a compact invariant set of (3.20), then $\varphi \in J$ implies that $\varphi \in C^{k+1}$ (resp. φ is analytic).*

Proof: . We give an indication of the proof which, for the C^k case, is due to Hale [71] and Lopes [126]. The analytic case is due to Hale and Scheurle [92] and is based on an original idea of Nussbaum [146].

To prove the theorem, it is sufficient to prove that every solution $x(t)$, $t \in \mathbb{R}$ on J is C^{k+1} (resp. analytic) in t.

Assuming uniqueness of solutions of (3.20), we first prove that $x(t)$ is a C^1 function if f is C^0. Let Y be the space of functions which are bounded and uniformly continuous in the sup-norm. Since D is assumed to be exponentially stable, for any $h \in X$, there is a unique solution $\mathcal{K}h \in X$ of the equation $Dx_t = h(t)$ which is linear and continuous in h. As a consequence, the operator $\mathcal{K}$ has a representation

$$(\mathcal{K}h)(t) = \int_{-\infty}^{0} d\eta(\theta) \, h(t + \theta)$$

where η is an $n \times n$ matrix of bounded variation.

If $x(t)$ is a solution of (3.20) in J, then we must have

$$x(t) = \int_{-\infty}^{0} d\eta(s) [Dx_0 + \int_{0}^{t+s} f(x_\tau) d\tau].$$

This is enough to show that $x(t)$ is a C^1 function.

An inductive argument will show that x is C^{k+1} if f is C^k.

We now prove the assertion about analyticity. Let X be the complex Banach space of functions $x : (-\infty, \infty) \to \mathbb{C}^n$ which are uniformly continuous, equipped with the sup-norm and let $U = \{x \in X : \sup_t |\operatorname{Im} x(t)| < h\}$ be a neighborhood of the real subspace X. Suppose that f has a holomorphic extension to U which takes closed bounded subsets of U into bounded sets of $\mathbb{C}^n$. Also, extend D to U.

Now, define $\tilde{D} : X \to X$ and $F : U \to X$ by $\tilde{D}x(t) = Dx_t$, $F(x)(t) = f(x_t)$. As noted before, the exponential stability of D implies that the operator $\tilde{D}$ is an isomorphism in X and, for any $x \in X$, is given by

$$\tilde{D}^{-1}x(t) = \int_{-\infty}^{0} d\eta(\theta) \, x(t + \theta).$$

Therefore, $F \circ \tilde{D}^{-1}$ is holomorphic in some neighborhood $U \subset \mathbb{C}$ of X.

Now, consider the ordinary differential equation

$$\frac{d}{ds} x = F(\tilde{D}^{-1}x) \tag{3.26}$$

in U. If y^0 is a solution of (3.20) bounded on $\mathbb{R}$, we have noted above that Dy_t^0 and its derivative are in X. If we define $\xi : \mathbb{R} \to X$ by $\xi(s) = \tilde{D}y_s^0$; that is, $\xi(s)(t) = Dy^0_{s+t}$, then ξ is well defined and differentiable, $\xi(0) = \tilde{D}y^0$. By definition, we have

$$\frac{d\xi}{dt}(s)(t) = \frac{d}{ds}Dy^0_{s+t} = f(y^0_{s+t}) = F(y^0_s)(t) = F(\tilde{D}^{-1}\xi(s))(t)$$

for all $s, t \in \mathbb{R}$. Hence ξ is a solution of (3.26) with initial value $\tilde{D}y^0$. From this fact, it follows that $\tilde{D}y^0$ has a holomorphic extension to a neighborhood of $\mathbb{R}$ and thus y^0 is analytic. This proves the theorem.   ∎

Remark 3.4.5. We could obtain the same result as in Theorem 3.4.4 if we had made the assumption that D_0 is exponentially stable and the matrix B in (3.23) is C^{k+1} (or analytic).

Using the results in Chapter 2 and Theorem 10.1.4, we arrive at the following important result.

Theorem 3.4.6. *Suppose that D is exponentially stable, $T(t), t \geq 0$, is the semigroup defined by (3.20) and $T(t)$ is a bounded map for $t \geq 0$. If $T(t)$ is point dissipative and orbits of bounded sets are bounded, then the following statements hold.*

(i) There is a compact global attractor $\mathcal{A}$ of (3.20).
(ii) There is an equilibrium point (constant solution) of (3.20).
(iii) If $f \in C^k(C, \mathbb{R}^n), 0 \leq k \leq \infty$ (resp. analytic), then $\varphi \in \mathcal{A}$ implies that φ is a C^{k+1} function (resp. an analytic function).
(iv) If f is analytic, then $T(t)$ is one-to-one on $\mathcal{A}$.

3.4.2 Equivalence of point and compact dissipative

In Theorem 2.0.5, we have seen that, for linear C^0-semigroups, point dissipative and compact dissipative are equivalent. It is rather surprising that this also is true for nonlinear NFDE for which the operator D_0 is exponentially stable. The following result is due to Massatt [138].

Theorem 3.4.7. *If D is given by (3.23) with D_0 exponentially stable, then, for equation (3.20), point dissipative and compact dissipative are equivalent. If (3.20) is point dissipative, then there is an equilibrium point of (3.20) and there is a maximal compact invariant set $\mathcal{A}_{D,f}$ which is a local attractor and attracts a neighborhood of each compact set.*

Proof: We only give an outline of the proof. Let $W^{1,\infty} \equiv W^{1,\infty}([-r, 0], \mathbb{R}^n)$ be the space of absolutely continuous functions with derivatives essentially bounded and, for $\varphi \in W^{1,\infty}$, let

$$|\varphi|_{W^{1,\infty}} = \sup_{[-r,0]} |\varphi(\theta)| + \text{ess} \sup_{[-r,0]} |\dot{\varphi}(\theta)|.$$

The space $W^{1,\infty}$ is compactly embedded in C and is dense in C. The idea for the proof is quite simple. We first show that the semigroup $T_{D,f}(t), t \geq 0$, of (3.20) defined on C is also a semigroup on $W^{1,\infty}$. Furthermore, D is exponentially stable on $W^{1,\infty}$. Also, $T_{D,f}(t)$, when restricted to $W^{1,\infty}$, can be represented as in (3.25) (setting $D_0 = D$, $B = 0$) and, in an equivalent norm in C, has the property that $T_D(t)\Psi_D$ is a contraction on $W_D^{1,\infty} = \{\varphi \in W^{1,\infty} : D(\varphi) = 0\}$ and $U_{D,f}(t)$ is conditionally completely continuous on $W^{1,\infty}$. We show also that it has the additional property that B and $U_{D,f}(s)$, $0 \leq s \leq t$, bounded in C implies that $U_{D,f}(t)B$ is bounded in $W^{1,\infty}$. From this fact, we can prove that $T_{D,f}(t), t \geq 0$, is bounded dissipative in $W^{1,\infty}$. This implies that there is a compact global attractor $\mathcal{A}_{D,f}$ for $T_{D,f}(t), t \geq 0$, in $W^{1,\infty}$. The set $\mathcal{A}_{D,f}$ is a compact invariant set in C. Furthermore, if Γ is a compact invariant set in C, then the regularity Theorem 3.4.4 implies that $\Gamma \subset W^{1,\infty}$. It is clearly invariant in $W^{1,\infty}$ and, thus, $\Gamma \subset \mathcal{A}_{D,f}$. This shows that $\mathcal{A}_{D,f}$ is the maximal compact invariant set of $T_{D,f}(t), t \geq 0$, in C. Using the fact that $\mathcal{A}_{D,f}$ is the maximal compact invariant set, the semigroup $T_{D,f}$ is point dissipative and an α-contraction, we deduce that $\mathcal{A}_{D,f}$ attracts compact sets of C (in particular, we have compact dissipative) and $\mathcal{A}_{D,f}$ attracts in C a neighborhood of any compact set in C. For more details see Hale [78].

This proves everything in the theorem except the existence of an equilibrium point. To show this, we need the following general fixed point theorem of Nussbaum [145], Hale and Lopes [85].

Theorem 3.4.8. . *If X is a Banach space and $T : X \to X$ is an α-contraction which is compact dissipative, then T has a fixed point.*

For any $\tau > 0$, there is a fixed point x^τ of $T(\tau)_{D,f}$ and this point must belong to $\mathcal{A}_{D,f}$. For any sequence of $\tau_j \to 0$ as $j \to \infty$, there is a subsequence of the x^{τ_j} which converges to a point x^* which must be an equilibrium point of $T_{D,f}$. This completes the proof of the theorem.   ∎

3.5 An example of NFDE on S^1

Let us start with the following neutral delay differential equation on $\mathbb{R}$

$$\frac{d}{dt}(x(t) - ax(t-1)) = f(x(t)), \quad a > 0. \tag{3.27}$$

Here $f : \mathbb{R} \to \mathbb{R}$ is a C^1 periodic function and a is a constant. This way we have existence and uniqueness of solution in $C^0(I, \mathbb{R})$ (see [72] and [86]).

As we will show, we have two distinct cases: If $a > 0$ is rational, and f is 2π-periodic, equation (3.27) induces an $NFDE$ on S^1. On the other hand when a is not necessarily rational and satisfies $0 < a < 1$ and also f is $\frac{2\pi}{(1-a)}$-periodic, equation (3.27) induces an $RFDE$ on S^1 with infinite delay.

First case: Assume that $a = \frac{p}{q} > 0$, p *and* q positive integers and primes, and $f : \mathbb{R} \to \mathbb{R}$ is C^1 and 2π-periodic. Define $\tilde{f} : S^1 \to \mathbb{R}$ by

$$\tilde{f}(\beta) = f(q\theta), \quad \text{where} \quad \beta = e^{i\theta} \in S^1.$$

One can see that $\tilde{f}$ is well defined (does not depend on the value chosen for θ) since f is 2π-periodic. For $\beta \in S^1$, let u_β be the (positively oriented) unit vector in $T_\beta S^1$ given by $u_\beta = i\beta$. Let us consider on S^1 the neutral functional differential equation:

$$\frac{d}{dt}\left[\frac{y(t)^q}{y(t-1)^p}\right] = \tilde{f}(y(t))[u_{\frac{y(t)^q}{y(t-1)^p}}]. \tag{3.28}$$

Observe that a global solution of (3.28) is a continuous function $y : \mathbb{R} \to S^1$ such that $\frac{y(t)^q}{y(t-1)^p} \in S^1$ is differentiable in t and (3.28) holds for all $t \in \mathbb{R}$. If $x : \mathbb{R} \to \mathbb{R}$ is a global solution of (3.27), $x(0) := \bar{x}$, then $y : \mathbb{R} \to S^1$ given by $y(t) = e^{i\frac{x(t)}{q}}$ is a global solution of (3.28) because

$$e^{i[x(t) - \frac{p}{q}x(t-1)]} = \frac{y(t)^q}{y(t-1)^p} \in S^1$$

and, by derivative, we obtain (3.28). Moreover, $y(0) = e^{i\frac{\bar{x}}{q}} := \bar{y}$.

Conversely, given a global (continuous) solution $y(t)$ of (3.28), $t \in \mathbb{R}$, $y(0) = \bar{y}$, for each $\bar{x}$ such that $\bar{y} = e^{i\frac{\bar{x}}{q}}$, there corresponds a unique continuous lifting $x(t)$, $\bar{x} = x(0)$, that is, $y(t) = e^{i\frac{x(t)}{q}}$ for all $t \in \mathbb{R}$, and since

$$y(t)^q = e^{ix(t)} \quad \text{and} \quad y(t-1)^p = e^{i\frac{p}{q}x(t-1)},$$

we have that

$$\frac{y(t)^q}{y(t-1)^p} = e^{i\left[x(t) - \frac{p}{q}x(t-1)\right]}$$

is differentiable and then

$$\tilde{f}(y(t))[u_{\frac{y(t)^q}{y(t-1)^p}}] = \frac{d}{dt}e^{i\left[x(t) - \frac{p}{q}x(t-1)\right]};$$

thus

$$\frac{d}{dt}e^{i\left[x(t) - \frac{p}{q}x(t-1)\right]} = (i\frac{y(t)^q}{y(t-1)^p})\frac{d}{dt}[x(t) - \frac{p}{q}x(t-1)]$$

so

$$\frac{d}{dt}[x(t) - \frac{p}{q}x(t-1)] = \tilde{f}(y(t)) = f(x(t));$$

the last equality shows that $x(t)$ is a global solution of $\frac{d}{dt}(x(t) - \frac{p}{q}x(t-1)) = f(x(t))$.

Second case: Assume $a \in \mathbb{R}$, $0 < a < 1$, $f \in C^1$ and $\frac{2\pi}{(1-a)}$-periodic and take a global bounded solution $x(t)$ of the $NFDE$ on $\mathbb{R}$

$$\frac{d}{dt}(x(t) - ax(t-1)) = f(x(t)). \qquad (3.29)$$

It defines a (bounded) function $z(t)$ through

$$z(t) := x(t) - ax(t-1). \qquad (3.30)$$

This function $z(t)$ is a global (bounded) solution of the scalar $RFDE$ with infinite delay

$$\dot{z} = f(h(z_t)), \qquad (3.31)$$

evolving on $C_\gamma = \{\varphi \in C((-\infty, 0], \mathbb{R}) : lim_{\theta \to \infty} e^{\gamma\theta}\varphi(\theta) exists\}$, $\gamma \in (0, -ln|a|)$ fixed, where

$$h(z_t) := \sum_{j=0}^{+\infty} a^j z(t-j) \qquad (3.32)$$

and, as usual, $z_t(\theta) := z(t+\theta)$, for all $t, \theta \in \mathbb{R}$. Equation (3.31) defines a semigroup on C_γ which is an α-contraction as was the semigroup for (3.27). We choose the space C_γ rather than $C((-\infty, 0), \mathbb{R})$ in order to have this property. Each globally defined bounded solution of (3.31) is a globally defined bounded solution of (3.27). In fact, given a bounded real function $z(t)$, $t \in \mathbb{R}$, since $0 < a < 1$ one obtains the function $x(t)$ as the sum of the convergent series

$$x(t) := \sum_{j=0}^{+\infty} a^j z(t-j). \qquad (3.33)$$

It can be trivially verified that $x(t)$ is a bounded solution of (3.30). Moreover, that solution of (3.30) is unique for each given bounded function $z(t)$, because, otherwise, one arrives to the condition $v(t) = av(t-1)$, $0 < a < 1$, $t \in \mathbb{R}$, with $v(t)$ bounded, and so, this implies $v(t) = 0$ for all $t \in \mathbb{R}$. Since f is $\frac{2\pi}{(1-a)}$-periodic, one considers $C := C((-\infty, 0], S^1)$ and defines the function $\hat{f}(p) := f(h(w))$ where w is a continuous argument function for $p \in C$. Note that, the $\frac{2\pi}{(1-a)}$-periodicity of f implies that $\hat{f}$ is well defined because if w and $\bar{w}$ are such that $w - \bar{w} = 2k\pi$, k integer, we have

$$h(w) = h(\bar{w} + 2k\pi) = \sum_{j=0}^{\infty} a^j(\bar{w}(-j) + 2k\pi)$$

and so

$$h(w) = \frac{2k\pi}{(1-a)} + \sum_{j=0}^{\infty} a^j \bar{w}(-j) = \frac{2\pi}{(1-a)} + h(\bar{w}).$$

The $RFDE$ on S^1 (with infinite delay and evolving on C) appears when we make $y(t) = e^{iz(t)}$ and by differentiation we obtain

$$\dot{y}(t) = \hat{f}(y_t)u_{y_t} \tag{3.34}$$

where $u_b = ib$ is a unit vector tangent to S^1 at the point $b \in S^1$. So, to each global bounded solution x of (3.29) there corresponds a global solution $y : \mathbb{R} \to S^1$ of (3.34).

The presentation of this section was based on the following question proposed by Hale: *does the NFDE on* $\mathbb{R}$

$$\frac{d}{dt}(x(t) - ax(t-1)) = f(x(t)), \quad a > 0$$

(where $f : \mathbb{R} \to \mathbb{R}$ *is a* C^1 *periodic function) induce an NFDE on* S^1 *?* The first case was obtained after discussions with L. Fichmann and A.L. Pereira; they also showed us the second case.

3.6 A canonical ODE in the Fréchet category

The present example is related with the fact that the global solutions of any autonomous RFDE of class C^∞ are solutions of an ODE defined in a Fréchet space. This observation was made by Fichmann and Oliva in [55].

For each $k \in \mathbb{N}$, consider the set C^k of all $\varphi \in C$, $C = C^0([-r, 0], \mathbb{R})$, such that φ has continuous derivatives up through order k. In particular $C^0 = C$, C^k is a dense subspace of C and, of course, $C^{k+1} \subset C^k$. C^k is a Banach space with norm

$$\|\varphi\|_k = \max_{0 \le j \le k} \sup_{-r \le \theta \le 0} |\varphi^{(j)}(\theta)|$$

where $\varphi^{(j)}$ represents the jth-derivative of $\varphi(\varphi^{(0)} = \varphi$, $\varphi^{(1)} = \varphi'$, etc). At $\theta = -r$ or $\theta = 0$ take the derivatives from the right and from the left, respectively. Consider the dense subspace of C given by $C^\infty = \bigcap_{k=0}^{\infty} C^k$. It is possible to endow C^∞ with the structure of a Fréchet space through the family of seminorms (in fact norms) $\{\| \cdot \|_k\}_{k \in \mathbb{N}}$, that is, with the topology given by the convergence $\varphi_n \xrightarrow{C^\infty} \varphi$ if, and only if, $\|\varphi_n - \varphi\|_k \to 0\ \forall k \in \mathbb{N}$. For a reference see [95]. In that Fréchet space one considers the linear operator $\partial : C^\infty \to C^\infty$ given by $\partial \varphi = \varphi'$. ∂ is a continuous linear operator because $\varphi_n \xrightarrow{C^\infty} \varphi$ implies $\varphi'_n \xrightarrow{C^\infty} \varphi'$. Remark that, if one considers $\partial : C \to C$ with dense domain C^1, it is not continuous; so, if one thinks of C^∞ as a subspace of C, then $\partial : C^\infty \to C^\infty$ is not continuous.

The Fréchet space C^∞ has a classical distance

$$d(\varphi, \psi) = \sum_{k=0}^{\infty} \frac{1}{2^k} \frac{\|\varphi - \psi\|_k}{1 + \|\varphi - \psi\|_k};$$

the topology of C^∞ coincides with that metric topology. We also have $C^\infty \overset{i}{\hookrightarrow} C$ that means the inclusion $i : C^\infty \to C$ is continuous because $\varphi_n \overset{C^\infty}{\longrightarrow} \varphi$ implies $\|\varphi_n - \varphi\|_0 \to 0$, as $n \to \infty$. We are able to define, now, the following autonomous ordinary differential equation.

$$\dot{u}(t) = \partial(u(t)), \quad t \in \mathbb{R}, \quad u(t) \in C^\infty. \tag{3.35}$$

The next result characterizes the global solutions of (3.35):

Proposition 3.6.1. *If $\varphi \in C^\infty$, then $u : \mathbb{R} \to C^\infty$ is a global solution of (3.35) through φ if and only if there exists a C^∞ function $x : \mathbb{R} \to \mathbb{R}$ such that $u(t) = x_t$ for all $t \in \mathbb{R}$. In other words, any global solution of (3.35) passing through φ comes from a C^∞ extension of φ to $\mathbb{R}$ and any C^∞ extension of φ, $x : \mathbb{R} \to \mathbb{R}$, gives rise to a global solution $u : \mathbb{R} \to C^\infty$ of (3.35) such that $u(t) = x_t$ for all $t \in \mathbb{R}$.*

Proof: Let $\varphi \in C^\infty$ and assume that $u : \mathbb{R} \to C^\infty$, satisfying $u(0) = \varphi$, is a (global) solution of (3.35). Then $\dot{u}(t) = \lim\limits_{h \to 0} \frac{u(t+h)-u(t)}{h}$ in C^∞ and $\partial(u(t)) = u(t)^{(1)}$. Therefore $\left\| \frac{u(t+h)-u(t)}{h} - u(t)^{(1)} \right\|_k \overset{h \to 0}{\longrightarrow} 0$ for all $k \in \mathbb{N}$ implies

$$\sup_{-r \leq \theta \leq 0} \left| \frac{u(t+h)^{(k)}(\theta) - u(t)^{(k)}(\theta)}{h} - u(t)^{(k+1)}(\theta) \right| \overset{h \to 0}{\longrightarrow} 0, \text{ for all } k \in \mathbb{N}.$$

In particular, for $k = 0$ and θ fixed, we have

$$\left| \frac{u(t+h)(\theta) - u(t)(\theta)}{h} - u(t)^{(1)}(\theta) \right| \overset{h \to 0}{\longrightarrow} 0.$$

Let $\tilde{u} : \mathbb{R} \times [-r, 0] \to \mathbb{R}$ be such that $\tilde{u}(t, \theta) = u(t)(\theta)$, $(t, \theta) \in \mathbb{R} \times [-r, 0]$. The limit above shows that $\tilde{u}$ is a solution of

$$\frac{\partial \tilde{u}}{\partial t}(t, \theta) = \frac{\partial \tilde{u}}{\partial \theta}(t, \theta), \quad (t, \theta) \in \mathbb{R} \times [-r, 0] \tag{3.36}$$

It is easy to see that the solutions of (3.36) are of the form $\tilde{u}(t, \theta) = x(t+\theta)$ for some $x : \mathbb{R} \to \mathbb{R}$, because the lines $t + \theta =$ constant in $\mathbb{R} \times [-r, 0]$ are level curves of the solutions of (3.36). The map x is of class C^∞ because $\tilde{u}$ is C^∞ in θ. Finally,

$$x(\theta) = \tilde{u}(0, \theta) = \varphi(\theta), \quad \theta \in [-r, 0].$$

Conversely, let $x : \mathbb{R} \to \mathbb{R}$ be a C^∞ extension of φ to $\mathbb{R}$ and define $u(t) \in C^\infty$ by $u(t)(\theta) = x(t + \theta)$. It remains to prove that u is differentiable in $t \in \mathbb{R}$ and that $\dot{u}(t) = \partial(u(t))$, that is, we need to show that

$$\sup_{-r \leq \theta \leq 0} \left| \frac{u(t+h)^{(k)}(\theta) - u(t)^{(k)}(\theta)}{h} - u(t)^{(k+1)}(\theta) \right| \overset{h \to 0}{\longrightarrow} 0, \text{ for all } k \in \mathbb{N}.$$

But

$$\left| \frac{u(t+h)^{(k)}(\theta) - u(t)^{(k)}(\theta)}{h} - u(t)^{(k+1)}(\theta) \right| = |x^{(k+1)}(t+\theta+\xi_\theta) - x^{(k+1)}(t+\theta)|,$$

$|\xi_\theta| < h$, from the mean value theorem. The uniform continuity of $x^{(k+1)}$ on the compact interval $[t-r, t]$ implies what we need. Then u is a global solution of (3.35).   ∎

As a consequence of the last proposition, one concludes that if $f : C \to \mathbb{R}$ is any C^∞ RFDE of type $\dot{x} = f(x_t)$, then each global (so C^∞) solution of that retarded equation is a global solution of (3.35). As usually, the union of all traces (orbits) of all global bounded (in C) solutions of f is denoted by $A(f)$, and then $A(f) \subset C^\infty$ represents part of the flow of (3.35). In a sense, that is a **universal** property of (3.35) relative to all C^∞ RFDE. Another easy consequence of Proposition 3.6.1 is the fact that the ODE (3.35) has existence but does not have uniqueness of solutions and that the phenomenon of collision also happens.

We finish the example by saying that there are global bounded solutions of the universal ODE that cannot correspond to solutions of a continuous RFDE (see [86] and [55]).

4 The Dimension of the Attractor

The purpose of this section is to present results on the "size" of the attractor. This will be given in terms of limit capacity and Hausdorff dimension. The principal results are applicable not only to RFDE but also to some of the abstract dynamical systems considered in Section 2.

Let K be a topological space. We say that K is *finite dimensional* if there exists an integer n such that, for every open covering $\mathfrak{U}$ of K, there exists another open covering $\mathfrak{U}'$ refining $\mathfrak{U}$ such that every point of K belongs to at most $n+1$ sets of $\mathfrak{U}'$. In this case, the *dimension* of K, $\dim K$, is defined as the minimum n satisfying this property. Then $\dim \mathbb{R}^n = n$ and, if K is a compact finite dimensional space, it is homeomorphic to a subset of $\mathbb{R}^n$ with $n = 2 \dim K + 1$. If K is a metric space, its *Hausdorff dimension* is defined as follows: for any $\alpha > 0$, $\varepsilon > 0$, let

$$\mu_\varepsilon^\alpha(K) = \inf \sum_i \varepsilon_i^\alpha$$

where the inf is taken over all coverings $B_{\varepsilon_i}(x_i)$, $i = 1, 2, \ldots$ of K with $\varepsilon_i < \varepsilon$ for all i, where $B_{\varepsilon_i}(x_i) = \{x : d(x, x_i) < \varepsilon_i\}$. Let $\mu^\alpha(K) = \lim_{\varepsilon \to 0} \mu_\varepsilon^\alpha(K)$. The function μ^α is called the *Hausdorff measure of dimension* α. For $\alpha = n$ and K a subset of $\mathbb{R}^n$ with $|x| = \sup |x_j|$, μ^n is the Lebesgue exterior measure. It is not difficult to show that, if $\mu^\alpha(K) < \infty$ for some α, then $\mu^{\alpha_1}(K) = 0$ if $\alpha_1 > \alpha$. Thus,

$$\inf\{\alpha : \mu^\alpha(K) = 0\} = \sup\{\alpha : \mu^\alpha(K) = \infty\}$$

and we define the Hausdorff dimension of K as

$$\dim_H(K) = \inf\{\alpha : \mu^\alpha(K) = 0\}.$$

It is known that $\dim(K) \leq \dim_H(K)$ and these numbers are equal when K is a submanifold of a Banach space. For general K, there is little that can be said relating these numbers. For other properties of Hausdorff dimension, see [107] and [103]. Mallet-Paret [133] showed that negatively invariant sets of compact maps have finite Hausdorff dimension in a separable Hilbert space. Mañé [129] proved more general results in Theorems 4.0.1 and 4.0.2 below.

To define another measure of the size of a metric space K, let $N(\varepsilon, K)$ be the minimum number of open balls of radius ε needed to cover K. Define the *limit capacity* $c(K)$ of K by

$$c(K) = \limsup_{\varepsilon \to 0} \frac{\log N(\varepsilon, K)}{\log(1/\varepsilon)}.$$

In other words, $c(K)$ is the minimum real number such that, for every $\sigma > 0$, there is a $\delta > 0$ such that

$$N(\varepsilon, K) \le \left(\frac{1}{\varepsilon}\right)^{c(K)+\sigma} \quad \text{if } 0 < \varepsilon < \delta.$$

It is not difficult to show that

$$\dim_H(K) \le c(K).$$

For a given Banach space E, let $\mathcal{L}(E)$ be the space of bounded linear maps on E. Another useful property is that, given a Banach space E, a finite dimensional linear subspace S of E with $n = \dim S$, a map $L \in \mathcal{L}(E)$, and using the notation $B_\varepsilon^S(0) = \{v \in S : \|v\| \le \varepsilon\}$, we have

$$N\left(\varepsilon_1, B_{\varepsilon_2}^S(0)\right) \le n2^n \left(1 + \frac{\varepsilon_1}{\varepsilon_2}\right)^n, \quad \text{for all } \varepsilon_1, \varepsilon_2 > 0, \tag{4.1}$$

and

$$N\left((1+\gamma)\lambda\varepsilon, LB_\varepsilon(0)\right) \le n2^n \left(1 + \frac{\|L\| + \lambda}{\lambda\gamma}\right)^n \tag{4.2}$$

for all $\gamma, \varepsilon > 0$, $\lambda > \|L_S\|$, where $B_\varepsilon(0) = B_\varepsilon^E(0)$ and $L_S : E/S \to E/L(S)$ is the linear map induced by S.

Estimates for the limit capacity of the attractor $A(F)$ of an RFDE will be obtained by an application of general results for the capacity of compact subsets of a Banach space E with the property that $f(K) \supset K$ for some C^1 map $f : U \to E$, $U \supset K$, whose derivative can be decomposed as a sum of a compact map and a contraction.

We begin with some notation. For $\lambda > 0$, the subspace of $\mathcal{L}(E)$ consisting of all maps $L = L_1 + L_2$ with L_1 compact and $\|L_2\| < \lambda$ is denoted by $\mathcal{L}_\lambda(E)$. Given a map $L \in \mathcal{L}_\lambda(E)$ we define

$$\nu_\lambda(L) = \min\{\dim S : S \text{ is a linear subspace of } E \text{ and } \|L_S\| < \lambda\}.$$

It is easy to prove that $\nu_\lambda(L)$ is finite for $L \in \mathcal{L}_{\lambda/2}(E)$.

Theorem 4.0.1. *Let E be a Banach space, $U \subset E$ an open set, $f : U \to E$ a C^1 map, and $K \subset U$ a compact set such that $f(K) \supset K$.*

If the Fréchet derivative $D_x f \in \mathcal{L}_{1/4}(E)$ for all $x \in K$, then

$$c(K) \le \frac{\log\left\{\left[2\left(\lambda(1+\sigma) + k^2\right)/\lambda\sigma\right]^\nu\right\}}{\log\left[1/2\lambda(1+\sigma)\right]} \tag{4.3}$$

where $k = \sup_{x \in K} \|D_x f\|$, $0 < \lambda < 1/2$, $0 < \sigma < (1/2\lambda) - 1$, $\nu = \sup_{x \in K} \nu_\lambda(D_x f^2)$. If $D_x f \in \mathcal{L}_1(E)$ for all $x \in K$, then $c(K) < \infty$.

Proof: Assume that $D_x f \in \mathcal{L}_{1/4}(E)$, $x \in K$. Then for some $0 < \lambda < 1/8$, $D_x f^2 \in \mathcal{L}_{\lambda/2}(E)$ for all $x \in K$. By the remark just preceding this theorem, for each $x \in K$, there exists a finite dimensional linear subspace $S(x)$ of E such that $\|(D_x f^2)S(x)\| < \lambda$, and, by continuity, $\|(D_y f^2)S(x)\| < \lambda$ for every y in some neighborhood of x. We construct in this way an open covering of K which can be taken finite, since K is compact. It follows that $\nu = \sup_{x \in K} \nu(D_x f^2) < \infty$. Take $\delta > 1$ and $\sigma > 0$ satisfying $(1 + \sigma)\lambda\delta < 1/2$. By the continuity of f^2, there exists $\varepsilon_0 > 0$ such that $f^2 B_\varepsilon(x) \subset f^2(x) + (D_x f^2)B_{\delta\varepsilon}(0)$ for all $x \in K$, $0 < \varepsilon < \varepsilon_0$. Without loss of generality, we can take $\varepsilon_0 < 1$.

Let $\lambda_0 = (1+\sigma)\delta\lambda$, and $\lambda_1 = \nu 2^\nu \left(1 + \frac{k^2 + \lambda}{\lambda\sigma}\right)^\nu$ where k is as in the statement of the theorem. Then, since $\|D_x f^2\| \le k^2$, the inequality (4.2) gives

$$N(\lambda_0\varepsilon, f^2 B_\varepsilon(x)) \le N(\delta_0\varepsilon, (D_x f^2)B_{\delta\varepsilon}(0)) \le \lambda_1$$

for all $0 < \varepsilon < \varepsilon_0$. Since K is compact, it can be covered by a finite number of balls $B_\varepsilon(x_i)$, $x_i \in K$. It follows that $K \subset f(K) \subset f^2(K) \subset \bigcup_i f^2 B_\varepsilon(x_i)$. Therefore, the last inequality implies

$$N(\lambda_0\varepsilon, K) \le \lambda_1 N(\varepsilon, K) \le \lambda_1 N(\varepsilon/2, K),$$

for all $0 < \varepsilon < \varepsilon_0$. Since $2\lambda_0 < 1$, each ε in the interval $0 < \varepsilon < \lambda_0\varepsilon_0$ can be written as $\varepsilon = (2\lambda_0)^p \bar\varepsilon$ for some $\bar\varepsilon$ in the interval $\lambda_0\varepsilon_0 < \bar\varepsilon < \varepsilon_0/2$ and some integer $p \ge 1$, and, therefore, the last inequality can be applied p times to get

$$N(\varepsilon, K) = N((2\lambda_0)^p\bar\varepsilon, K) \le \lambda_1^p N(\bar\varepsilon, K) \le \lambda_1^p N(\lambda_0\varepsilon_0, K).$$

Writing $\varepsilon_1 = \lambda_0\varepsilon_0$, we have

$$\frac{\log N(\varepsilon, K)}{\log(1/\varepsilon)} \le \frac{p\log\lambda_1 + \log N(\varepsilon_1, K)}{p\log(1/2\lambda_0)} \le \frac{\log\lambda_1}{\log(1/2\lambda_0)} +$$
$$+ \frac{\log N(\varepsilon_1, K)}{\log(1/2\lambda_0)} \frac{\log 2\lambda_0}{\log(\varepsilon/\lambda_0\varepsilon_0)}.$$

Taking the lim sup as $\varepsilon \to 0$, we obtain

$$c(K) \le \frac{\log\lambda_1}{\log(1/2\lambda_0)}.$$

Since this inequality holds for any $\delta > 1$ and $\lambda_0 = (1+\sigma)\delta\lambda$, we get

$$c(K) \le \frac{\log\lambda_1}{\log(1/2(1+\sigma)\lambda)}$$

which is precisely the inequality (4.3) in the first statement in the theorem.

In order to prove the second statement in the theorem, one just notes that, if $D_x f \in \mathcal{L}_1(E)$ for all $x \in K$, then the continuity of $D_x f$ and the compactness of K imply the existence of $0 < \lambda < 1$ such that $D_x f \in \mathcal{L}_\lambda(E)$ for all $x \in K$. Consequently, for every integer $p \geq 1$, $D_x f^p \in \mathcal{L}_{\lambda^p}(E)$ for all $x \in K = \bigcap_{j=0}^p f^{-j}(K)$. Taking p sufficiently large for $\lambda^p < 1/4$, the first statement of the theorem implies $c(K_p) < \infty$. But $K_p \subset K \subset f^p(K_p)$ implies $c(K_p) \leq c(K) \leq c(f^p K_p)$, and, since f^p is a C^1 map, it does not increase the capacity of compact sets. Therefore, $c(K) = c(K_p)$, and the proof of the theorem is complete.   ∎

Theorem 4.0.2. *Let $F \in \mathcal{X}^1$ be an RFDE on a manifold M, and $A_\beta(F) = A(F) \cap \{\varphi \in C^0 : |\varphi| \leq \beta\}$. There is an integer d_β, depending only on M, the delay r and the norm of F, such that*

$$c\big(A_\beta(F)\big) \leq d_\beta \quad \text{for all} \quad \beta \in [0, \infty).$$

Consequently, also $\dim_H A_\beta(F) \leq d_\beta$, $\beta \in [0, \infty)$ *and* $\dim_H A(F) < \infty$ *when M is compact.*

Proof: The case of noneuclidean manifolds M can be reduced to the case of an RFDE defined on $\mathbb{R}^k$, for an appropriate integer k, by the Whitney embedding theorem and considering an RFDE on $\mathbb{R}^k$ defined by an extension of F to $\mathbb{R}^k$ similar to the one constructed in the proof of Theorem 3.1.1. Consequently, we take without loss of generality $M = \mathbb{R}^m$.

The Ascoli Theorem guarantees that $A_\beta(F)$ is compact, and consequently we can take a bounded open neighborhood $U \supset A_\beta(F)$, such that $\Phi_r U$ is precompact. It can be easily shown that $D_x \Phi_r$ is a compact operator for each $x \in U$. On the other hand, from the definition of $A_\beta(F)$ it follows $\Phi_r\big(A_\beta(F)\big) \supset A_\beta(F)$. Consequently, we can apply Theorem 4.0.1 with $K = A_\beta(F)$ and $f = \Phi_r$, $k = \sup_{\varphi \in A_\beta(F)} \|D_\varphi \Phi_r\|$, while taking $\sigma = 1$ and $0 < \lambda < \min(k/4, 1/4)$, to get

$$c\big(A_\beta(F)\big) \leq \frac{\log\left\{\nu\left[(4\lambda + 2k^2)/\lambda\right]^\nu\right\}}{\log[1/4\lambda]} < \infty,$$

The bound of $\dim_H A_\beta(F)$ follows immediately and then it is clear that $\dim_H A(F) < \infty$ when M is compact.   ∎

Another result guarantees that the set $A(F)$ can be "flattened" by any projection of a residual set of projections π from C^0 into a finite dimensional linear subspace of C^0 with sufficiently high dimension, in the sense that the restriction of π to $A(F)$ is one-to-one. This result is included here because it is of possible importance for the study of A-stability and bifurcation. It uses the following:

Theorem 4.0.3. *If E is a Banach space and $A \subset E$ is a countable union of compact subsets K_i of E and there exists a constant D such that $\dim_H(K_i \times K_i) < D$ for all i, then, for every subspace $S \subset E$ with $D + 1 < \dim S < \infty$,* *there is a residual set $\mathcal{R}$ of the space $\mathcal{P}$ of all continuous projections of E onto S (taken with the uniform operator topology) such that the restriction π/A is one-to-one for every $\pi \in \mathcal{R}$.*

Proof: We transcribe the proof given by Mañé (see [129]). Suppose $A = \bigcup_{i=1}^{\infty} K_i$ where each K_i is compact, and take S and $\mathcal{P}$ to be as in the statement of this theorem. Denote

$$P_{i,\varepsilon} = \left\{ \pi \in \mathcal{P} : \operatorname{diam} \left(\pi^{-1}(p) \cap K_i \right) < \varepsilon \text{ for all } p \in S \right\},$$

where diam denotes the usual diameter of a set in a metric space. Clearly, $P_{i,\varepsilon}$ is open and $\mathcal{R} = \bigcap_{i,j=1}^{\infty} P_{i,1/j}$ is the set of projections onto S which are one-to-one when restricted to A. It is, therefore, sufficient to prove that every $P_{i,\varepsilon}$ is dense in $\mathcal{P}$.

Let $Q_{i\varepsilon} = \{ v - w : v, w \in K_i \text{ and } \|v - wi\| \geq \varepsilon \}$. Then $\pi \in p_{i,\varepsilon}$ if and only if $\pi^{-1}(0) \cap Q_{i,\varepsilon} = \emptyset$. Denote by h the canonical homomorphism of E onto the quotient space E/S. The set $h(Q_{i,\varepsilon}) - \{0\}$ is a countable union of the compact sets $\mathcal{J}_j = \{ h(v) : v \in Q_{i,\varepsilon}, \|h(v)\| \geq 1/j \}$, $j = 1, 2 \ldots$. Therefore there exists a sequence $L_k : E/S \to \mathbb{R}$ of continuous linear maps such that $L_k(x) = 0$ for all $k \in \mathbb{N}$, implies $x \notin h(Q_{i,\varepsilon}) - \{0\}$. Let

$$Q_{i,\varepsilon,k,j} = \left\{ v \in Q_{i,\varepsilon} : \left| L_k \big(h(v) \big) \right| \geq 1/j \right\}$$

and

$$P_{i,\varepsilon,k,j} = \left\{ \pi \in \mathcal{P} : \pi^{-1}(0) \cap Q_{i,\varepsilon,k,j} = \emptyset \right\}.$$

Then each $P_{i,\varepsilon,k,j}$ is open and $P_{i,\varepsilon} = \bigcap_{k,j} P_{i,\varepsilon,k,j}$. Consequently, it is sufficient to show that every $P_{i,\varepsilon,k,j}$ is dense in $\mathcal{P}$.

Let $\pi_0 \in \mathcal{P}$, $C = \{ v \in S : \|v\| = 1 \}$ and define $\zeta : S - \{0\} \to C$ by $\zeta(v) = v/\|v\|$. Then

$$\dim_H \zeta(\pi_0 Q_{i,\varepsilon}) \leq \sup_{\delta > 0} \dim_H \zeta \left[(\pi_0 Q_{i\varepsilon}) \cap (S - B_\delta(0)) \right].$$

But the restriction of ζ to $[(\pi_0 Q_{i\varepsilon}) \cap (S - B_\delta(0))]$ is Lipschitz and, therefore,

$$\dim_H \zeta \left[(\pi_0 Q_{i\varepsilon}) \cap (S - B_\delta(0)) \right] \leq \dim_H \pi_0 Q_{i,\varepsilon}.$$

It follows that $\dim_H \zeta(\pi_0 Q_{i,\varepsilon}) \leq \dim_H Q_{i,\varepsilon} \leq \dim_H (K_i \times K_i)$. Since $\dim_H C = \dim S - 1 > \dim_H (K_i \times K_i)$, there exists $u \in C$ such that $u \notin \zeta(\pi_0 Q_{i,\varepsilon})$. Given $\delta > 0$ and an integer k, let us consider $\pi_{\delta,k} \in \mathcal{P}$ given by $\pi_{\delta,k} = \pi_0 + \delta u L_k \circ h$. Assume $\pi_{\delta,k}(x) = 0$ and $x \in Q_{i,\varepsilon,k,j}$ then $\pi_0(x) = -\delta L_k \big(h(x) \big) u$ and $L_k \big(h(x) \big) \neq 0$ and $\pi_0(x) \neq 0$. Consequently,

$$u = - \left[\delta L_k \big(h(x) \big) \right]^{-1} \pi_0(x)$$

and, then, $u = \zeta(u) = \zeta \big(\pi_0(x) \big) \in \zeta(\pi_0 Q_{i,\varepsilon})$ contradicting the choice of u. This proves that $\pi_{\delta,k}^{-1}(0) \cap Q_{i,\varepsilon,k,j} = \emptyset$ and, therefore, $\pi_{\delta,k} \in P_{i,\varepsilon,k,j}$. Since $\pi_{\delta,k} \to \pi_0$ as $\delta \to 0$, this proves $P_{i,\varepsilon,k,j}$ is dense in $\mathcal{P}$. ∎

An important remark on the last proof, made by Mañé (private communication), is the following: If $c(A)$ denotes the capacity of A and A is compact, the inequalities $dim_H(A) \leq c(A)$ and $c(A \times A) \leq 2c(A)$ are obvious. Then

$$dim_H(A \times A) \leq c(A \times A) \leq 2c(A).$$

This is important because the aim was to prove that compact invariant sets of compact (non linear) maps have finite capacity (whose proof is completely independent of the last Theorem). Therefore, if A is such an invariant set, by the previous remark the conclusion of Theorem 4.0.3 holds if $2c(A) + 1 < dim S < \infty$.

Theorem 4.0.4. *Let $F \in \mathcal{X}^1$ be an RFDE on $\mathbb{R}^m$. There is an integer d, depending only on m, the delay r and the norm of F, such that, if S is a linear subspace of C^0 with $d \leq \dim S < \infty$, then there is a residual set $\mathcal{R}$ of the space of all continuous projections of C^0 onto S, such that the restriction $\pi/A(F)$ is one-to-one for every $\pi \in \mathcal{R}$.*

Proof: Apply Theorem 4.0.3 with $E = C^0$, $A = A(F) = \bigcup_{\beta=1}^{\infty} A_\beta(F)$, taking into account Theorem 4.0.2 and the fact that $A_\beta(F)$ is compact for every $\beta > 0$ (see [129]). ∎

If M is a compact manifold, it is possible to obtain more information on the dimension of the attractor set using algebraic topology (see Mallet-Paret [134] and also Mañé [129]).

Lemma 4.0.5. *Suppose M is a compact manifold. Then the map $\rho|A(F)$: $\varphi \in A(F) \mapsto \varphi(0)$, induces an injection $(\rho|A(F))^* : H^*(M) \to H^*(A(F))$ on Čech cohomology.*

Proof: Define $\psi : M \to C^0$ by $\psi(p)(t) = p$ for all $-r \leq t \leq 0$. Then $\rho\psi$ is the identity and $\psi\rho$ is homotopic to the identity. Therefore, $\rho^* : H^*(M) \to H^*(C^0)$, the induced map on Čech cohomology, is the identity. But, if i : $A(F) \to C^0([-r, 0], M)$ denotes the inclusion map, we have $(\rho/A(F))^* = (\rho i)^* = i^*\rho^*$. Thus, we have reduced the problem to the injectivity of i^* which, by the continuity property of Čech cohomology, is reduced to showing that if $K_n = C\ell\Phi_r^n(C^0)$ and $i_n : K_n \to C^0$ is the inclusion map then i_n^* is injective for all n (recall that $\bigcap_{n \geq 0} K_n = A(F)$). But we can write $\Phi_r^n = i_n g_n$, with $g_n : C^0 \to K_n$, and then $\Phi_r^{*n} = g_n^* i_n^*$. Now observe that if $\Phi_t^{(\lambda)}$, $t > 0$, is the solution map on C^0 defined by the RFDE $\dot{x}(t) = \lambda F(x_t)$, then $\Phi_r^{(1)} = \Phi_r$, $\Phi_r^{(0)} = \psi\rho$ and the maps $\psi\rho$ and Φ_r are homotopic. Hence $g_n^* i_n^* = \Phi^{*n}_T = (\psi\rho)^{*n} = I$ and i_n^* is injective. ∎

A consequence of Lemma 4.0.5 is the following (see Kurzweil [116] for a delay differential equation and Mallet-Paret [134] for the general case):

Theorem 4.0.6. *Let $F \in \mathcal{X}^1$ be an RFDE on a compact manifold M. Then $\dim A(F) \geq \dim M$, and the map $\rho : \varphi \to \varphi(0)$ maps $A(F)$ onto M, that is, through each point of M passes a global solution.*

Proof: Let $n = \dim M$. Since $H^{*k}(M) = 0$ for $k > m$, $H^{*m}(M)$ is nontrivial and $A(F)$ is compact, the first and last statements of the theorem follow from the preceding lemma. Suppose ρ does not take $A(F)$ onto M. Then there is a p in M such that $\rho\big(A(F)\big) \subseteq M \backslash \{p\}$. But $H^m(M) = \mathbb{Z}_2$ and $H^m\big(M \backslash \{p\}\big) = 0$, which is a contradiction. ∎

The following example due to Oliva (see [149]) shows how easily one can construct examples where the evaluation map ρ is not one-to-one on $A(F)$.

Example 4.0.7. Let S^1 be the circle. Any point p in S^1 is determined by an angle x and given p, x is only determined up to a multiple of 2π. The unit tangent vector u_p at the point p is equal to

$$u_p = -(\sin x) \cdot \mathbf{i} + (\cos x) \cdot \mathbf{j}$$

The function $g : S^1 \to R$ given by

$$g(p) = \frac{\pi}{2}(1 - \cos x)$$

defines an RFDE on S^1 in the following way:

$$\varphi \in C^0(I, S^1) \mapsto f(\varphi) = \big[g(\varphi(0)) + g(\varphi(-1))\big] \cdot u_{\varphi(0)}.$$

A solution $p(t)$ satisfies $\dot{p}(t) = f(p_t)$ where $p(t) = 0 + \cos x(t)\mathbf{i} + \sin x(t)\mathbf{j}$ and then

$$\dot{p}(t) = \big(-\sin x(t)\mathbf{i} + \cos x(t)\mathbf{j}\big) \cdot x(t) = \big[g(p(t)) + g(p(t-1))\big] u_{p(t)}$$

or

$$\dot{x}(t) = \frac{\pi}{2}\big(1 - \cos x(t)\big) + \frac{\pi}{2}\big(1 - \cos x(t-1)\big). \tag{4.4}$$

The constant solutions of (4.4) must satisfy $2 = \cos x(t) + \cos x(t-1)$. Thus $x(t) = 2k\pi$, $k = 0, \pm 1, \pm 2, \ldots$ The only critical point is $P = 0 + \mathbf{i}$. On the other hand, $x(t) = \pi t$ is a solution of (4.4) and on S^1 the corresponding periodic solution is given by $p(t) = 0 + (\cos \pi t)\mathbf{i} + (\sin \pi t)\mathbf{j}$. Thus, ρ is not one-to-one on $A(F)$.

Theorem 4.0.6 does not hold when M is not compact, as one can see in the next example introduced by Popov [168] and also discussed in Hale [72]. Consider in $M = \mathbb{R}^3$ the system

$$\begin{aligned}
\dot{x}(t) &= 2y(t) \\
\dot{y}(t) &= -z(t) + x(t-1) \\
\dot{z}(t) &= 2y(t-1)
\end{aligned} \tag{4.5}$$

A simple computation shows that for $t \geq 1$ one obtains $\ddot{y}(t) = \dddot{x}(t) = 0$ and any solution $(x(t), y(t), z(t))$ must lie in the plane

$$x(t) - 2y(t) - z(t) = 0.$$

The finite-dimensionality of the sets $A_\beta(F)$ implies the finite dimensionality of the period module of any almost periodic solution of F, generalizing what happens for ordinary differential equations. Let us recall the definition of period module. Any almost periodic function $x(t)$ has a Fourier expansion

$$x(t) \sim \sum a_n e^{-i\lambda_n t}$$

where $\sum |a_n|^2 < \infty$; the *period module* of $x(t)$ is the vector space $\mathcal{M}$ spanned by $\{\lambda_n\}$ over the rationals. The fact that the period module is finite dimensional implies that the almost periodic solution is quasiperiodic. The next Corollary is due to Mallet-Paret [133]:

Corollary 4.0.8. *Let $F \in \mathcal{X}^1$ be an RFDE on a manifold M. Then there is an integer N depending only on the delay r, the norm of F and on M such that, for any almost periodic solution $x(t)$ of F, the period module $\mathcal{M}$ of x has finite-dimension $\leq N$; that is, there are only finitely many rationally independent frequencies in the Fourier expansion for x.*

Proof: An easy modification of a result given by Cartwright [24] for ordinary differential equations shows that $\dim \mathcal{M}$ equals the topological dimension of the hull $\mathcal{H}$ of x. Clearly, $\mathcal{H}$ is homeomorphic to the set of initial data at $t = 0$ for its elements, which is a subset of $A_\beta(F)$, where β is a bound on the solution x. Since $A_\beta(F)$ is finite-dimensional by Theorem 4.0.1, so are $\mathcal{H}$ and $\mathcal{M}$. ∎

The set $A(f)$ may not have finite dimension if f is only in $\mathcal{X}^0$. In fact, let Q_L be the set of functions $\gamma : \mathbb{R}^n \to \mathbb{R}^n$ with global Lipschitz constant L. For each $\gamma \in Q_L$, each solution of $\dot{x}(t) = \gamma(x(t))$ is defined for all $t \in \mathbb{R}$. One can prove the following result (see Yorke [207]):

Theorem 4.0.9. *For each $L > 0$ there is a continuous RFDE(f) on $\mathbb{R}^n$, depending only on L, such that, for every $\gamma \in Q_L$, every solution of $\dot{x}(t) = \gamma(x(t))$ is also a solution of the RFDE(f). In particular, $A(f)$ has infinite dimension.*

5 Stability and Bifurcation

As for ordinary differential equations, the primary objective in the qualitative theory of RFDE is to study the dependence of the flow $\Phi_t = \Phi_t^F$ on F. This implicitly requires the existence of a criterion for deciding when two RFDE and, more generally, two semigroups, are equivalent. A study of the dependence of the flow on changes of a certain parameter (in particular the semigroup itself) through the use of a notion of equivalence based on a comparison of all orbits is very difficult and is likely to give too small equivalence classes. The difficulty is associated with the infinite dimensionality of the phase space and the associated smoothing properties of the solution operator as well as with the appearance of special phenomena that do not make sense in the setting of flows on finite dimensional manifolds. For example, the case of two global trajectories, i.e. both defined for all $t \in \mathbb{R}$, with a collision point and, of course, after the collision they keep evolving together; that can happen even in a C^∞ system and with two C^∞ global solutions, as we saw in Chapter 3.

In order to compare all orbits of two semigroups one needs to take into account the changes in the range of the solution map Φ_t, for each fixed t, a not so easy task due to the difficulties associated with backward continuation of solutions. Therefore, it is reasonable to begin the study by considering a notion of equivalence which ignores some of the orbits of the semigroups to be compared.

As in ODE, the equilibrium points and periodic orbits play a very important role in the qualitative theory. In showing that two ODE are equivalent, a fundamental role is played by linearization of the flow near equilibrium points and linearization of the Poincaré map near a periodic orbit—the famous Hartman-Grobman theorem. What is the generalization of this result for RFDE and also for more general situations? To see some of the difficulties, we consider equilibrium points in some detail.

Suppose p_F is an equilibrium point of an RFDE(F), $F \in \mathcal{X}^1$. If p_F is hyperbolic as a solution of F, then an application of the Implicit Function Theorem guarantees the existence of neighborhoods U of F in $\mathcal{X}^1$ and V of p_F in $C^0(I, M)$ such that, for each $G \in U$, there is a unique equilibrium point p_G in V and it is hyperbolic. Furthermore, the local stable manifold $W_{\text{loc}}^s(p_F)$ and local unstable manifold $W_{\text{loc}}^u(p_F)$ of p_F are diffeomorphic to the corresponding ones for G. The fact that these sets are diffeomorphic does not necessarily imply that the flows are equivalent in the sense that all orbits of F near p_F can be mapped by a homeomorphism onto orbits of G near p_G. The smoothing property of the flow generally prevents such a homeomorphism from being constructed. This implies the Hartman-Grobman theorem will not be valid; that is, the flow cannot be linearized near p_F. On the other hand, the local unstable manifolds are finite dimensional and, consequently, the restriction of the flows to them can also be described by ordinary differential equations. It follows that the flows $\Phi_t^F/W_{\text{loc}}^u(p_F)$ and $\Phi_t^G/W_{\text{loc}}^u(p_G)$ are diffeomorphisms, and therefore one can find a homeomorphism $h : W_{\text{loc}}^u(p_F) \to W_{\text{loc}}^u(p_G)$ which preserves orbits.

The proof that such an h exists follows along the same lines as the proof of the classical Hartman-Grobman theorem making use of the analytic representation of the unstable manifolds $W_{\text{loc}}^u(p_F)$ and $W_{\text{loc}}^u(p_G)$ in terms of a coordinate system on the linearized unstable manifolds. It is also possible to define the global unstable set $W^u(p_F)$ of p_F by taking the union of the orbits through $W_{\text{loc}}^u(p_F)$. However, if $\Phi_t(F)$ is not one-to-one, then the manifold structure may be destroyed. On the other hand, if $\Phi_t(F)$ and $D\Phi_t(F)$ are one-to-one when restricted to the set $A(f)$ of all global bounded solutions, then $W^u(p_F)$ is a finite dimensional immersed submanifold of $C^0(I, M)$.

The set $A(F)$ contains all ω-limit and α-limit points of bounded orbits of F, as well as the equilibrium points, the periodic orbits and the bounded unstable manifolds of both. As a matter of fact, $A(F)$ consists of all of the points of orbits of solutions that have a backward continuation and, thus, it is reasonable to begin the qualitative theory by agreeing to make the definitions of equivalence relative to $A(F)$.

If $A(F)$ is compact and Φ_t is one-to-one on $A(F)$, then Φ_t is a group of $A(F)$. This implies that the solution operator does not smooth on $A(F)$. Therefore, one can attempt to modify several of the important ideas and concepts from ordinary differential equations so they are meaningful for RFDE and other dynamical systems. These remarks suggest the following definition.

Definition 5.0.1. *Two RFDE F and G defined on manifolds are said to be equivalent, $F \sim G$, if there is a homeomorphism $h : A(F) \to A(G)$ which preserves orbits and sense of direction in time. An RFDE(F) defined on a manifold is said to be A-stable if there is a neighborhood V of F such that $G \sim F$ if $G \in V$.*

As mentioned in Chapter 3, every ordinary differential equation on a manifold M can be considered as an RFDE on M with phase space $C^0(I, M)$. In particular, if X is a vector field on M and $\rho : C^0(I, M) \to M$ is the evaluation map $\rho(\varphi) = \varphi(0)$, the function $F = X \circ \rho$ is an RFDE on M. For each point $p \in M$ there is a solution of the ordinary differential equation defined by the vector field X which passes through p at $t = 0$. The map $\Sigma_X : M \to C^0(I, M)$ such that $\Sigma_X(p)$ is the restriction to $I = [-r, 0]$ of the solution of X through p at $t = 0$, is a cross-section with respect to ρ and the attractor of F is a manifold diffeomorphic to M and given by $A(F) = \Sigma_X(M)$. Clearly, the qualitative behavior of the flow of F on $A(F)$ is in direct correspondence with the qualitative behavior of the flow of the ordinary differential equation defined by X on M. It follows that all the bifurcations that occur for ordinary differential equations also occur for RFDE. In this sense, the definition of A-stability given above is a generalization of the usual definition for ordinary differential equations.

In ordinary differential equations, we have the classical Hartman-Grobman Theorem which asserts that the flow near a hyperbolic equilibrium is topologically equivalent to its linearization. For RFDE which possess a compact global attractor Sternberg [195], [194], has given an appropriate extension of this theorem with the topological equivalence being relative to the solutions on the attractor.

We end this section with some examples from FDE and elementary PDE illustrating how the set $A(F)$ may vary with F and, in particular, how elementary bifurcations (non A-stable F) influence the behavior of $A(F)$. These special examples are chosen because they are nontrivial and yet it is still possible to discuss $A(F)$. Also, they illustrate the importance that the form of the equations play in the generic theory.

Let us come back to the Levin-Nohel equation on $\mathbb{R}$ ([123], see Chapter 3). Suppose that a and g are functions which satisfy the following conditions:

$$a \in C^2([0,\delta],\mathbb{R}), a(0) = 1, a(\delta) = 0, a(s) > 0, \quad for \ \ s \in [0,\delta), \qquad (5.1)$$

where $g \in C^1(\mathbb{R},\mathbb{R})$, and define $G(x) := \int_0^x g(s)ds$. Consider the equation

$$\dot{x}(t) = -\int_{-\delta}^0 a(-\theta)g(x(t+\theta))d\theta. \qquad (5.2)$$

The equation (5.2) is a special case of a $RFDE$ on $\mathbb{R}$ defined on the space $C = C([-\delta,0],\mathbb{R})$ by

$$f(\varphi) = -\int_{-\delta}^0 a(-\theta)g(\varphi(\theta))d\theta.$$

We remark that there is no loss in generality in taking $a(0) = 1$. In fact, since a is positive on $[0,\delta]$, we must have $a(0) > 0$ and this value can be incorporated into g. For every $\varphi \in C$, suppose that the solution $x(t,\varphi)$ exists for all $t \geq 0$ and define the semigroup $T_{a,g} : \mathbb{R}^+ \times C \to C$, $(t,\varphi) \mapsto T_{a,g}(t)\varphi := x_t(\cdot,\varphi)$. Also, suppose that there is a metric space $U \subset C^2([0,\delta],\mathbb{R}) \times C^1(\mathbb{R},\mathbb{R})$ of functions (a,g) satisfying (5.1) such that $T_{a,g}$ has a compact global attractor $\mathcal{A}_{a,g}$ and the sets $\mathcal{A}_{a,g}$ are upper semicontinuous in (a,g).

A basic problem is to characterize the set U of (a,g) for which the above properties hold true and then to determine those $(a,g) \in U$ for which the flow on $\mathcal{A}(a,g)$ is stable as well as discussing properties of the bifurcation points. This seems to be a very difficult problem for general (a, g). However, we are able to characterize the flow for a rather large class of (a, g). We will describe this class and then attempt to study general perturbations of the flow for perturbations of these special (a, g).

We know that the globally defined solutions $x(t)$ of (5.2) on the attractor $\mathcal{A}(a, g)$ are C^2 functions. A few elementary calculations shows that a globally defined solution $x(t)$ on $\mathcal{A}(a, g)$ is a solution of the equation

$$\ddot{x}(t) + g(x(t)) = -a'(\delta)H_{\delta,g}(x_t) + \int_{-\delta}^{0} a''(-\theta)H_{-\theta,g}(x_t)d\theta, \qquad (5.3)$$

where we are using the notation

$$H_{s,g}(\psi) = \int_{-s}^{0} g(\psi(\theta))d\theta, \quad s \in [0, \delta]. \qquad (5.4)$$

This relationship leads to a class of (a, g) for which we can obtain an explicit characterization, in the analytic case, of the nonwandering set $\Omega(a, g)$ of (5.2) for a class of (a, g). To accomplish this, define $V : C \to \mathbb{R}$ as

$$V(\varphi) = G(\varphi(0)) - \frac{1}{2}\int_{-\delta}^{0} a'(-\theta)[H_{-\theta,g}(\varphi)]^2 d\theta,$$

where $G(x) = \int_0^x g(u)du$. The derivative of V along solutions of (5.2) is given by

$$\dot{V}(\varphi) = \frac{1}{2}a'(\delta)[H_{\delta,g}(\varphi)]^2 - \frac{1}{2}\int_{-\delta}^{0} a''(-\theta)[H_{-\theta,g}(\varphi)]^2 d\theta. \qquad (5.5)$$

Let us now impose the following additional conditions on (a, g):

$$a'(s) \leq 0, \quad a''(s) \geq 0, \quad s \in [0, \delta], \quad G(x) = \int_0^x g(u)du \to \infty \text{ as } |x| \to \infty. \qquad (5.6)$$

From (5.6), we see that $\dot{V}(\varphi) \leq 0$ for all $\varphi \in C$. Therefore, for any solution $x(t)$ of (5.2), we have $V(x_t) \leq V(x_0)$ for all $t \geq 0$ for which x_t is defined and bounded. Furthermore, the hypotheses on a in (5.1),(5.6) imply that $V(x_t) \geq G(x(t))$ for all $t \geq 0$ for which x_t is defined. As a consequence of the hypothesis on g in (5.6), we see that each solution of (5.2) exists for all $t \geq 0$ and is bounded for $t \geq 0$. This shows that the semigroup $T_{a,g}$ on C is well defined.

Since each positive orbit $\gamma^+(\varphi)$ of (5.2) is bounded, we know that $\omega(\varphi)$ is compact, connected and invariant. If a negative orbit $\gamma^-(\varphi)$ exists and is precompact, then $\alpha(\varphi)$ is a compact, connected invariant set. The following result characterizes the nonwandering set $\Omega(a, g)$ of (5.2) with (a, g) satisfying (5.1) and (5.6).

Theorem 5.0.2. *Suppose that (a, g) satisfy (5.1) and (5.6). Then the non-wandering set $\Omega(a, g) := \Omega(g)$ is invariant. If a is not linear, then $\Omega(g)$ is equal to the set E_g of equilibrium points, that is, the zeros of g. If a is linear, then $\Omega(g) = E_g \cup \mathrm{Per}_g(\delta)$, where $\mathrm{Per}_g(\delta)$ is the set of periodic orbits of period δ. These periodic orbits coincide with the δ-periodic solutions of the second order ordinary differential equation*

$$\ddot{x}(t) + g(x(t)) = 0. \tag{5.7}$$

Proof: Our first observation is that, if φ belongs to either a ω-limit set or a compact α-limit set, then (5.5) and the invariance of these sets imply that the function $V(T(t)\varphi) = V(\varphi)$ for all $t \geq 0$. Therefore, $\dot{V} = 0$ and such sets are contained in the set

$$(4.8) \quad S = \{\psi \in C : H_{\delta,g}(\psi) = 0 \text{ if } a'(\delta) \neq 0, H_{s,g}(\psi) = 0 \text{ if } a''(s) \neq 0\},$$

where $H_{s,g}(\psi)$ is defined in (5.4). As a consequence of this remark and (5.3), it follows that each (compact) ω-limit set or compact α-limit set is generated by the bounded solutions of the ordinary differential equation (5.7) which have the property that $x_t \in S$ for all $t \in \mathbb{R}$. If a is not a linear function, then there are an s_0, $a''(s_0) \neq 0$ and an interval I_{s_0} containing s_0 such that $a''(s) \neq 0$ for $s \in I_{s_0}$. If $x(t)$ is a bounded solution of (5.7) with $H_{s,g}(x_t) = 0$ for $t \in (-\infty, \infty)$, $s \in I_{s_0}$, then $\dot{x}(t) = \dot{x}(t - s)$ for $s \in I_{s_0}$. Therefore, $\dot{x}(t)$ is a constant. Since $x(t)$ is bounded, it follows that $x(t)$ is an equilibrium point of (4.2). If a is a linear function, then $a(\delta) = 0$, $a(0) = 1$, imply that $a'(\delta) \neq 0$. If $x(t)$ is a bounded solution of (5.7) with $H_{\delta,g}(x_t) = 0$ for $t \in (-\infty, \infty)$, then $\dot{x}(t) = \dot{x}(t - \delta)$ for all t; that is, $x(t) = kt+$ (a periodic function of period δ) for some constant k. Since $x(t)$ is bounded, this implies that $x(t)$ is a δ-periodic function. Since V is constant on any compact invariant set and $\dot{V}$ is negative off these sets, it follows, provided that (a, g) are real analytic functions, that $\Omega(a, g)$ is independent of g and given as stated above. This completes the proof of the theorem. ∎

Let us discuss the implications of Theorem 5.0.2 for linear systems. For $\delta = 1$ and $g_\gamma(x) = \gamma^2 x$, where γ is a positive constant, equation (5.2) is

$$\dot{x}(t) = -\gamma^2 \int_{-\delta}^{0} a(-\theta)x(t + \theta)d\theta. \tag{5.8}$$

The asymptotic behavior of the solutions of (5.8) is determined by the eigenvalues, which are the solutions of

$$\lambda = -\gamma^2 \int_{-\delta}^{0} a(-\theta)e^{\lambda\theta}d\theta. \tag{5.9}$$

If $\mathrm{Re}\,\lambda < 0$ for each solution of (5.9), then the origin is the compact global attractor. If there are a pair of imaginary eigenvalues $\pm i\omega$ with the remaining ones having negative real parts, then there is a two dimensional subspace P_γ of C such that each orbit on P_λ is $(2\pi/w)$-periodic. The orbits can be indexed as Γ_α, $\alpha \in \mathbb{R}^+$. Each periodic orbit Γ_α has a strongly stable manifold $W^s(\Gamma_\alpha)$ of codimension two and these manifolds form a foliation of C.

For a satisfying (5.1), (5.6) with a not linear, Theorem 5.0.2 implies that all eigenvalues of (5.8) have negative real parts. If a is a linear function, then there are purely imaginary eigenvalues if and only if there is an integer m such that $\gamma = m$, and, if $\gamma = m$, then there is the 2-dimensional subspace of periodic solutions of period 1. If γ is not an integer, then the origin is the compact global attractor.

Let us now return to the general equation (5.2) with a, g satisfying (5.1) and (5.6). We want to show that each ω-limit set (as well as each compact α-limit set) is either a singleton in E_g or a single periodic orbit. To do this, we need some information about the eigenvalues of the linear variational equation about an equilibrium point. If $g(c) = 0$, then the linear variational equation about c is

$$\dot{y}(t) = -\int_{-\delta}^0 a(-\theta)g'(c)y(t+\theta)d\theta.$$

The eigenvalues satisfy the equation

$$\lambda + g'(c)\int_{-\delta}^0 a(-\theta)e^{\lambda\theta}d\theta = 0.$$

Since $a > 0$, it follows that $\lambda = 0$ is an eigenvalue if and only if $g'(c) = 0$ and, in this case, it is simple. This remark together with the results in Section 2 implies that, if the ω-limit set of a solution of (5.2) belongs to the set E_g of equilibria, then it is singleton.

Theorem 5.0.3. *If (a, g) satisfies (5.1) and (5.6), then the ω-limit set (resp. α-limit set) of a positive orbit (precompact negative orbit) of (5.2) is either a single equilibrium point or a single periodic orbit in C generated by a δ-periodic solution of (5.7). If a is not linear (that is, there is an s_0 such that $a''(s_0) \neq 0$), then the ω-limit set is always an equilibrium point and the system is gradient.*

Proof: The conclusions about equilibrium points has been proved in the remarks before the statement of the theorem. To complete the proof, we need only consider the case where $a(s)$ is the linear function, $a(s) = (\delta - s)/\delta$ and to consider the case where the ω-limit set does not contain an equilibrium point. It is easy to verify that the δ-periodic solutions of (5.7) are solutions of (5.2). Since each solution of (5.2) must lie on one of the curves

$$\frac{\dot{x}^2}{2} + G(x) = \text{constant},$$

any periodic orbit must be symmetrical with respect to the x-axis. Let $u(t, \alpha)$, $u(0, \alpha) = \alpha$, $\dot{u}(0, \alpha) = 0$, be a nonconstant periodic solution of (5.7) of least period p and $mp = \delta$ for some integer m. If there is an interval of periodic orbits in C of $\omega(\varphi)$, then p is independent of α since the minimal period is continuous in α. Furthermore,

$$V(u_t(\cdot, \alpha)) - G(\alpha) = V(u_0(\cdot, \alpha)) - G(\alpha)$$

$$= \frac{1}{2\delta} \int_{-\delta}^{0} [\int_{\theta}^{0} g(u(s, \alpha)) ds]^2 d\theta = \frac{1}{2\delta} \int_{-\delta}^{0} \dot{u}^2(\theta, \alpha) d\theta$$

$$= \frac{1}{2mp} \int_{-mp}^{0} \dot{u}^2(\theta, \alpha) d\theta = \frac{1}{2p} \int_{-p}^{0} \dot{u}^2(\theta, \alpha) d\theta$$

$$= \frac{1}{p} \int_{0}^{p/2} \dot{u}^2(\theta, \alpha) d\theta = \frac{1}{p} \int_{0}^{p/2} [G(\alpha) - G(u(\theta, \alpha))] d\theta$$

$$= \frac{\sqrt{2}}{p} \int_{\alpha}^{\gamma(\alpha)} [G(\alpha) - G(\tau)]^{1/2} d\tau,$$

where $\gamma(\alpha) = u(\alpha, p/2)$. On the other hand, the derivative of

$$G(\alpha) + \frac{\sqrt{2}}{p} \int_{\alpha}^{\gamma(\alpha)} [G(\alpha) - G(\tau)]^{1/2} d\tau$$

with respect to α is not zero. Therefore, $V(u_t(\cdot, \alpha))$ is not constant for α in an interval. This implies that $\omega(\varphi)$ is a single periodic orbit.

If a is not linear, then the system is gradient with the function V above being the Lyapunov function. This completes the proof of the theorem. ∎

If (a, g) satisfy (5.1) and (5.6) and $\Omega(a, g) \equiv \Omega(g) = E_g \cup \text{Per}_g(\delta)$ is compact, then (5.2) is point dissipative and there is the compact global attractor $\mathcal{A}_{a,g}$. The following result gives a representation of $\mathcal{A}_{a,g}$.

Theorem 5.0.4. *Suppose that (a, g) satisfy (5.1) and (5.6), $\Omega(g)$ is compact and let $\mathcal{A}_{a,g}$ be the compact global attractor of (5.2). If $W^u(\text{Per}_g(\delta))$ (resp. $W^u(E_g)$) denotes the unstable set for $\text{Per}_g(\delta)$ (resp. E_g), then*

$$\mathcal{A}_{a,g} = W^u(E_g) \cup W^u(\text{Per}_g(\delta)). \tag{5.10}$$

If, in addition, each equilibrium point of (5.2) is hyperbolic, then

$$\mathcal{A}_{a,g} = (\cup_{c \in E_g} W^u_{a,g}(c)) \cup W^u(\text{Per}_g(\delta)). \tag{5.11}$$

If the function a is nonlinear, then hyperbolicity of an equilibrium point c is equivalent to $g'(c) \neq 0$ with c being hyperbolically stable if $g'(c) > 0$ and a hyperbolic saddle-point with a one dimensional unstable manifold if $g'(c) < 0$. If the function a is nonlinear, E_g is compact and each element of E_g is hyperbolic, then the system (5.2) is gradient and

$$\mathcal{A}_{a,g} = \cup_{c \in E_g} W^u_{a,g}(c), \quad \dim \mathcal{A}_{a,g} = 1. \tag{5.12}$$

Proof: Since every element in $\mathcal{A}_{a,g}$ has a globally defined orbit through it and $\mathcal{A}_{a,g}$ is compact, relations (5.8) and (5.9) follow from Theorem 5.0.3. A careful analysis of the characteristic equation for an equilibrium points leads to the other conclusions in the theorem.   ∎

Remark 5.0.5. If a, g satisfy (5.1) and (5.6), then no periodic orbit can be hyperbolic. In fact, if a is a linear function and there is a periodic orbit Γ of period δ which is hyperbolic, then there would be a hyperbolic periodic orbit $\Gamma_{\bar{a}}$ for any $\bar{a}$ close to a in the C^2-topology. If we perturb the linear function a to a function $\bar{a}$ which satisfies (5.1) and (5.6) and there is an s_0 for which $a''(s_0) > 0$, then the system corresponding to $\bar{a}$ is gradient and there are no periodic orbits. Therefore, there can be no hyperbolic periodic orbits of (5.2) when a is linear.

Remark 5.0.6. From the previous remark, in the class of functions (a, g) satisfying (5.1), (5.6), there is no generic Hopf bifurcation. The alternative is to enlarge the choice of allowable perturbations in the function a, perhaps considering those a for which $a''(s)$ changes sign. In this case, we show that a generic Hopf bifurcation can occur.

Suppose that $\alpha, \epsilon, \beta, \gamma$ are constants with $\gamma > 0$, $\epsilon \in (0, 3)$, $\beta \neq 0$ and let

$$a_{\alpha,\epsilon}(s) = (1 - s)(\alpha s^2 + \alpha(1 - \epsilon)s + 1),$$

$$g_{\beta,\gamma}(s) = 4\pi^2 + \beta s^2 + \gamma s^3.$$

The parameters ϵ, β, γ will be fixed and α will be the bifurcation parameter varying in a small neighborhood of zero.

We assume that β, γ are such that 0 is the only zero of $g_{\beta,\gamma}$ and the equation (5.7) with g replaced by $g_{\beta,\gamma}$ has no periodic solution of minimal period 1. For $\alpha = 0$, Theorem 5.0.3 implies that the origin is the compact global attractor for $(a_{0,\epsilon}, g_{\beta,\gamma})$.

The linear variational equation about the origin has two eigenvalues $\pm 2\pi i$ and the remaining ones have negative real parts. For α small, there are two eigenvalues $\lambda_{\alpha,\epsilon}, \bar{\lambda}_{\alpha,\epsilon}$ with $\lambda_{0,\epsilon} = 2\pi i$. Furthermore, a few computations will show that $\mathrm{Re}\lambda_{\alpha,\epsilon} = -\alpha K_\epsilon + O(\alpha^2)$ as $\alpha \to 0$ with $K_\epsilon = 1 - \frac{2\epsilon}{3}$. As a consequence, if $\epsilon \neq \frac{3}{2}$, these eigenvalues cross the imaginary axis transversaly as α increases through zero. The crossing is from left to right if $\epsilon \in (\frac{3}{2}, 3)$ and from right to left otherwise.

There is a center manifold M_α, smooth in α, and existing for $|\alpha| < \alpha_0$ with α_0 sufficiently small. Using normal form theory (see Chapter 8, Section 8.8) it is possible to show that the flow on M_0 in polar coordinates (ρ, ξ) is given by

$$\dot{\rho} = -L_\beta \rho^3 + O(\rho^4)$$
$$\dot{\xi} = 2\pi + O(|\rho|)$$

as $\rho \to 0$, where $L_\beta = \frac{\beta^2}{18\pi^2}$.

This fact together with the above observations about the eigenvalues of the linear variational equation imply that the flow on the center manifold M_α in polar coordinates (ρ, ξ) is given by

$$\dot{\rho} = -\alpha K_\epsilon \rho - L_\beta \rho^3 + O(\alpha \rho^2 + \alpha \rho^3 + \rho^4)$$
$$\dot{\xi} = 2\pi + O(|\alpha|^2 + |\rho|)$$

as $\rho \to 0$, $\alpha \to 0$.

This implies that there is a generic subcritical Hopf bifurcation if $\epsilon \in (0, \frac{3}{2})$ and a generic supercritical Hopf bifurcation otherwise.

We can always choose the center manifold M_α for $|\alpha| < \alpha_0$ such that there is neighborhood U of 0 with smooth boundary such that the vector field on M_α points inside $U \cap M_\alpha$ for all $|\alpha| < \alpha_0$. As a consequence, in the supercritical case, the compact global attractor is diffeomorphic to a disk D_α with the boundary being an hyperbolic stable periodic orbit. In the subcritical case, there is a hyperbolic periodic orbit but it is unstable. Therefore, in the subcritical case, there must be another periodic orbit. To show that there is only one, we must compute the fifth degree terms of the vector field on the center manifold. This involves many computations and we have not done them.

The assertion in Theorem 5.0.3 about the ω-limit sets is essentially contained in the paper of Levin and Nohel [123]. The proof in the text is due to Hale [70]; in this paper, Hale introduced the right geometric approach to the theory of functional differential equations and gave a presentation to a national SIAM meeting in Denver in 1963 on just this topic. Onuchic, in 1965, developed a series of seminars at the University of São Paulo on the contents of [70] that correspond to the beginning of the Brazilian school on the analytic theory of FDE. The presentation above of Theorem 5.0.3 follows some unpublished notes of Hale (see also, Hale, Magalhães and Oliva [87]). The computations for the Hopf bifurcation are due to T. Faria.

The next question of interest is to discuss the flow on the attractor $\mathcal{A}_{a,g}$ under the assumption that $E \cup \text{Per}_g(\delta)$ is compact with each element in E_g being hyperbolic. The representation (5.10) states that we need to discuss the limiting behavior of the unstable sets. We also would like to understand the sensitivity of the flow to the kernel function a and the nonlinear function g. The simplest question is the following:

Q : If (a, g) satisfy (5.1), (5.6) and a is not a linear function, is there a residual set $\mathcal{S}$ in $C^2(\mathbb{R}, \mathbb{R})$ such that, for each $g \in \mathcal{S}$, the stable and unstable manifolds of elements of E_g are transversal?

Although this is not an easy question to answer, it is interesting to study because it sheds light upon some of the complexities of the flow defined by RFDE.

Let us now fix a as a nonlinear function satisfying (5.1), (5.6) and investigate the dependence of the attractor $\mathcal{A}_{a,g}$ upon g in the class for which all equilibria remain hyperbolic; that is, the zeros of g are simple. From Theorem 5.0.4, we know that the dimension of $\mathcal{A}_{a,g}$ is one. To understand the flow on $\mathcal{A}_{a,g}$, it is only necessary to know the limit sets of the unstable manifolds of the saddle points.

If there is only one equilibrium point of (5.2), then it is the compact global attractor.

If there are only three equilibria, $c_1 < c_2 < c_3$, then c_1, c_3 are hyperbolically stable and c_2 is a saddle with a one dimensional unstable manifold. From Theorem 5.0.4 and the connectedness of $\mathcal{A}_{a,g}$, the set $\mathcal{A}_{a,g}$ is a one-dimensional manifold with boundary given by the closure of $W^u_{a,g}(c_2)$, that is, there are orbits $\gamma_1, \gamma_3 \subset W^u_{a,g}(c_2)$ such that $\alpha(\gamma_1) = x_2 = \alpha(\gamma_3)$, $\omega(\gamma_1) = c_1$, $\omega(\gamma_3) = c_3$. The point c_2 'lies between' c_1 and c_2 on the attractor $\mathcal{A}_{a,g}$. The order of the equilibrium points on the real line are preserved on the attractor.

If there are exactly five equilibrium points $c_1 < c_2 < c_3 < c_4 < c_5$, then c_2, c_4 are saddles with one dimensional unstable manifolds and the other points are stable. From Theorem 5.0.4 and the connectedness of $\mathcal{A}_{a,g}$, the set $\mathcal{A}_{a,g}$ is the closure of the union of the unstable manifolds of c_2 and c_4. If the flow on the attractor preserves the order on the real line of the equilibria, then $\omega(W^u_{a,g}(c_2)) = \{c_1, c_3\}$ and $\omega(W^u_{a,g}(c_4)) = \{c_3, c_5\}$. Intuitively, this should be the case, but it is not. To state a precise result, it is convenient to introduce the symbol $j[k, \ell]$, where j is one of the saddle points c_2, c_4, to designate that there exist orbits $\gamma_{jk}, \gamma_{j\ell}$ in $W^u_{a,g}(c_j)$ such that $\omega(\gamma_{jk}) = c_k$, $\omega(\gamma_{j\ell}) = c_\ell$; that is, there is an orbit of (5.2) which connects the equilibrium point c_j to c_k and an orbit which connects c_j to c_ℓ. The following result is due to Hale and Rybakowski (see [91]):

Theorem 5.0.7. *Suppose that a is a given nonlinear function satisfying (5.1), (5.6), let $c_1 < c_2 < c_3 < c_4 < c_5$ be given real numbers and consider the class $\mathcal{G}$ of functions g satisfying (5.1), (5.6) with the zeros of g being the set $\{c_j, 1 \leq j \leq 5\}$ with each zero being simple (and therefore corresponding to a hyperbolic equilibrium point of (5.2)). In the class $\mathcal{G}$, each of the following situations can be realized in $\mathcal{A}_{a,g}$:*

- *(i) 2[1,3],   4[3,5],*
- *(ii) 2[1,4],   4[3,5],*
- *(iii) 2[1,5],   4[3,5],*
- *(iv) 2[1,3],   4[2,5]*
- *(v) 2[1,3],   4[1,5].*

The only flow which preserves the natural order on the reals of the equilibria is case (i). Case (ii) has a nontransverse intersection of the unstable manifold $W^u_{a,g}(c_2)$ of c_2 and the stable manifold $W^s_{a,g}(c_4)$ of c_4 and case (iv) of $W^s_{a,g}(c_2)$ and $W^u_{a,g}(c_4)$. For cases (i), (iii) and (v), we have the transversality property of the stable and unstable manifolds, but we do not preserve the natural order of the equilibria in the flow on $\mathcal{A}_{a,g}$ in case (iii) and (v). See [91] for the proof of Theorem 5.0.7. That proof shows the existence of the nontransverse intersection of two saddle points with the functions g being analytic. However, the proof does not show that the transversality of the stable and unstable manifolds is generic in this class; that is, in cases (ii) and (iv), it has not been shown that a small perturbation of g will break the connection between the saddles. One would expect this to be the case and it would be interesting to prove or disprove it.

If it can be shown that it is not possible to break the connections between saddles by perturbing g, then we should try to enlarge the class of functions a and consider perturbations in a and perhaps also g. An obvious more general choice is the one mentioned in Remark 5.0.6; namely, those functions a for which $a''(s)$ changes sign.

Theorem 5.0.7 shows some of the unexpected things that can happen even in the gradient case. In case a is linear, there exist g for which there are periodic orbits of period δ. Very little is known about the compact global attractor in this case.

Remark 5.0.8. We should like to mention that equation (5.2) gives an interesting example on the unit circle $S^1 \subset \mathbb{R}^2$ if we assume that the function g is periodic in x of period 2π and we assume only that a is C^2 and g is C^1. Since S^1 is compact without boundary, we know that the semigroup T is defined and has the compact global attractor $\mathcal{A}_{a,g}$ and $\mathcal{A}_{a,g}$ is upper semicontinuous in (a, g). Of course, the discussion of the flow on $\mathcal{A}_{a,g}$ is as difficult as before.

The separate situations for $A_{b,g}$ when one considers the case mentioned in the last remark can then be depicted if we identify two of the zeros a_1 and a_5 of g (see Figure 5.1).

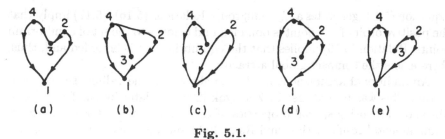

(a)   (b)   (c)   (d)   (e)

Fig. 5.1.

As another important example, let us consider a simple parabolic equation. Take the scalar equation corresponding to the Chafee–Infante problem (see Chapter 10 below for more details)

$$u_t = u_{xx} + \lambda f(u), \qquad 0 < x < \pi,$$
$$u = 0 \qquad \text{at } x = 0, \pi \tag{5.13}$$

with $\lambda > 0$ being a real parameter and $f(u)$ being a given non-linear function of u. If

$$V(\varphi) = \int_0^\pi \left[\varphi_x^2 - \lambda F(\varphi)\right] dx, \quad F(u) = \int_0^u f, \tag{5.14}$$

and $u(t, x)$ is a solution of (5.13), then

$$\frac{d}{dt} V\big(u(t, x)\big) = - \int_0^\pi u_t^2 dx \leq 0. \tag{5.15}$$

We note that Kening Lu in [127] proved the following result:

Theorem 5.0.9. *The Hartman-Grobman theorem holds in $H_0^1(0, \pi)$ for the origin of system* (5.13).

Also, with a suitable hypothesis, Equation (5.13) has a compact global attractor:

Theorem 5.0.10. *If one assumes*

$$F(u) \to -\infty \quad as \quad u \to \pm\infty \tag{5.16}$$

then Equation (5.13) generates a C_0-semigroup $T_\lambda(t)$, $t \geq 0$, on $X = H_0^1(0, \pi)$, in which every orbit is bounded and has ω-limit set as an equilibrium point. Moreover, there is a compact global attractor A_λ for $T_\lambda(t)$ (see Chapter 2) and, if $\varphi \in A_\lambda$, the α-limit set of φ is an equilibrium point.

Proof: The equilibrium points of (5.13) are the solutions of the equation

$$\begin{aligned} u_{xx} + \lambda f(u) = 0, \quad & 0 < x < \pi, \\ u = 0 \quad & \text{at } x = 0, \pi \end{aligned} \tag{5.17}$$

Equation (5.13) generates a C_0-semigroup. Relations (5.16), (5.17) imply that the positive orbit of any point is bounded and belongs to the set of equilibrium points. Relation (5.16) implies that the equilibrium set is bounded and, thus, there exists the compact global attractor.

An additional argument is needed to show that the ω-limit set is a singleton. This was proved first by Zelenyak [209] and later by Chafee [25] and Matano [139], using special properties of the PDE. It also follows from more a fundamental result of Hale and Raugel [90] on the convergence of positive orbits to a singleton in more general evolutionary equations. For the above example, the essential thing about the equation is that the eigenvalues of a linear second order ODE with separated boundary conditions are simple.

∎

An equilibrium point, u_0 is *hyperbolic* if no eigenvalue of the operator $\partial^2/\partial x^2 + \lambda f'(u_0)$ on X is zero and it is called *stable (hyperbolic)* if all eigenvalues are negative. The *unstable manifold* $W^u(u_0)$ is the set of $\varphi \in X$ such that $T_\lambda(t)\varphi$ is defined for $t \leq 0$ and $\to u_0$ as $t \to -\infty$. The *stable manifold* $W^s(u_0)$ is the set of $\varphi \in X$ such that $T_\lambda(t)\varphi \to u_0$ as $t \to \infty$. The set $W^u(u_0)$ is an embedded submanifold of X of finite dimension m (m being the number of positive eigenvalues of the above operator). The set $W^s(u_0)$ is an embedded submanifold of codimension m. These manifolds are tangent at u_0 to the stable and unstable manifolds of the linear operator $\partial^2/\partial x^2 + f'(u_0)$ on X (see Henry [99]).

The following remark is a simple but important consequence of Theorem 5.0.10.

Corollary 5.0.11. *If (5.16) is satisfied and there are only a finite number of hyperbolic equilibrium points $\varphi_1, \varphi_2, \ldots, \varphi_k$ of (5.13) with each being hyperbolic, then*

$$A_\lambda = \bigcup_{j=1}^{k} W^u(\varphi_j).$$

Corollary 5.0.11 states that A_λ is the union of a finite number of finite dimensional manifolds. The complete dynamics on A_λ will only be known when we know the specific way in which the equilibrium points are connected to each other by orbits.

To state a specific result, let us consider the special case of equation (5.13) where

$$f(0) = 0, \quad f'(0) = 1$$
$$\limsup f(u)/u \le 0, \quad uf''(u) < 0 \text{ if } u \neq 0.$$
$$(5.18)$$

Theorem 5.0.12. *If f satisfies (5.18) and $\lambda \in (n^2, (n+1)^2)$, n an integer, then there are exactly $2n + 1$ equilibrium points $\alpha_\infty = 0$, α_j^+, α_j^-, $j = 0, 1, \ldots, n-1$, where α_j^+, α_j^- have j zeros in $(0, \pi)$, $\dim W^u(\alpha_j^\pm) = j$, $0 \le j \le n-1$, $\dim W^u(\alpha_\infty) = n$ and*

$$A_\lambda = \left(\bigcup_j W^u(\alpha_j^\pm) \right) \cup W^u(\alpha_\infty).$$

For $n^2 < \lambda < (n+1)^2$, $n = 0, 1, 2, 3$, the attractor A_λ has the form shown in Figure 5.2. For $n > 3$, the picture is similar but in n-dimensions.

Historical references for Equation (5.13) can be found in Chafee and Infante [26], Henry [99] and Hale [78]. There is a general method for determining the connecting orbits due to Fiedler and Rocha [57], [58] which we return to in Chapter 10.

Another interesting example comes from a model for the transverse motion of an elastic beam with ends fixed in space which is given by the nonlinear equation

$$u_{tt} + \alpha u_{xxxx} - \left[\lambda + k \int_0^\ell u_s^2(s,t)ds \right] u_{xx} + \delta u_t = 0 \qquad (5.19)$$

where α, β, δ and λ are positive constants and the boundary conditions are stated for hinged or clamped ends. In each case, the equation defines a flow in

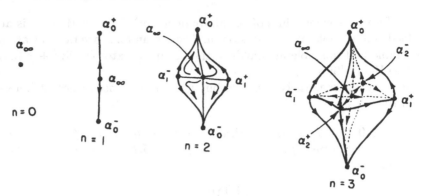

Fig. 5.2.

a suitable Banach space and the semigroup is asymptotically smooth. Also, the system is gradient and there is a compact global attractor $\mathcal{A}_\lambda$ (see, for example, Hale [78] for some details and references). New ideas seem to be required to give the detailed structure of the flow on the attractor.

The examples just mentioned are a good illustration of what we may call Morse–Smale systems in infinite dimensions. We will make this concept precise in the next Chapter where we will introduce a quite general class of Morse–Smale maps and semiflows together with the proofs of openness and A-stability for these dynamical systems. Let, for example, $\{\Phi_F(t), t \geq 0\}$ be the family of mappings considered in Theorem 3.1.2 corresponding to a $RFDE$ on a manifold M. For any hyperbolic equilibrium point x of this system, one can define in the usual way the local stable manifold $W^s_{\mathrm{loc}}(x)$ and the local unstable manifold $W^u_{\mathrm{loc}}(x)$. Assume that the manifold $W^u_{\mathrm{loc}}(x)$ has finite dimension; also, for any hyperbolic periodic orbit γ, one can analogously define local stable manifold $W^s_{\mathrm{loc}}(\gamma)$ and local unstable manifold $W^u_{\mathrm{loc}}(\gamma)$ of γ, with the latter being finite dimensional. One can consider the global unstable sets $W^u(x)$ and $W^u(\gamma)$ by taking the union of the orbits through points in $W^u_{\mathrm{loc}}(x)$, $W^u_{\mathrm{loc}}(\gamma)$ respectively. In general, these sets will not be manifolds. To be certain that $W^u(x)$, $W^u(\gamma)$ are immersed submanifolds, we suppose that $D\Phi_F(t)$ is one-to-one on the tangent space of $C^0(I, M)$ at points of the set $A(f)$ and that $\Phi_F(t)$ is also injective on $A(f)$, for all t. We can now give the definition of Morse–Smale $RFDE$:

Definition 5.0.13. *The dynamical system* $\{\Phi_F(t), \ t \geq 0\}$, *is said to be* Morse–Smale *if*

(i) *the nonwandering set* $\Omega(F)$ *is the union of a finite number of equilibrium points and periodic orbits, all hyperbolic;*

(ii) *the local stable and global unstable manifolds of all equilibrium points and periodic orbits intersect transversaly.*

The situations depicted in Figure 5.1 a), c), e) for Equation (5.2) are Morse–Smale systems as well as the ones shown in Figure 5.2 for Equation (5.13) and $\lambda \in \left(n^2, (n+1)^2\right)$, $n = 0, 1, 2, 3$. Other examples are given in next Chapter 6.

There are important physical situations with high complexity on the corresponding flows as, for example, in hydrodynamical problems. Besides this, Ladyzenskaya [119] proved that the set $A(F)$ corresponding to the semigroup $T_F(t)$ generated by the Navier-Stokes equation F in a two dimensional domain is a compact set and $f = T_F(t)$, $t > 0$, is a compact map. The dimension of $A(F)$ may increase as the Reynolds number Re increases and it seems that the dynamical system $F = F(\text{Re})$ is in fact the object of investigation in turbulence theory dealing with flows at large values of Re. How does $A(F)$ change as Re $\rightarrow +\infty$?

We finish this Section by remarking that L. Fichmann (see [54]) constructed a counter example for the Hartman-Grobman theorem; he introduced a compact dissipative dynamical system defined by a difference equation with diffusion for which the Hartman-Grobman result is not true. Let us recall now his construction. A discrete-time version of the Chaffee-Infante problem:

$$u_t = u_{xx} + f(u), \qquad (t,x) \in (0, \infty) \times [0, 1],$$
$$u(t,0) = u(t,1) = 0 \qquad \text{for } t > 0 \tag{5.20}$$

was treated in [155] (see also subsection 10.4.2 below) by taking a sequence $\{u_n\}_{n \in \mathbb{N}}$ of functions $u_n \in H_0^1$ satisfying (for $\Delta u = u_{xx}$ and a given $\epsilon > 0$):

$$u_{n+1} - u_n = \epsilon \Delta u_{n+1} + \epsilon f \circ u_n, \quad n \in \mathbb{N}. \tag{5.21}$$

In [155] the authors studied the iteration of the map $\Phi_\epsilon : H_0^1 \rightarrow H_0^1$ given by $\Phi_\epsilon = (I - \epsilon \Delta)^{-1} \circ (I + \epsilon \hat{f})$ (where $\hat{f}(p) = f \circ p$ for $p \in H_0^1$ and I is the identity of H_0^1), since $\{u_n\}_{n \in \mathbb{N}}$ satisfies (5.21) if and only if $u_{n+1} = \Phi_\epsilon(u_n)$, $n \in \mathbb{N}$.

Fichmann considered a version of (5.21) with continuous argument, given by

$$u(t,x) = u(t-r,x) + \epsilon u_{xx}(t,x) + \epsilon f(u(t-r,x)), \qquad (t,x) \in (0, \infty) \times [0, 1],$$
$$u(t,0) = u(t,1) = 0 \qquad \qquad \text{for } t > 0, \tag{5.22}$$

with r, ϵ positive constants, f of class C^1 with $f(0) = 0$ and $\sup_{y \in \mathbb{R}} |f'(y)| = K < \infty$.

It seems reasonable that (5.22) is an approximation of (5.20) when $r = \epsilon \rightarrow 0$. Consider the dynamical system defined by (5.22) in the Hilbert space $L_2^H := L_2([-r, 0], H_0^1)$. It was shown in [54] that (5.22) defines a nonlinear semigroup of operators from L_2^H to itself, $\{S(t), t \geq 0\}$, strongly continuous in time. Taking f of class C^2 with bounded second derivative, for each $t > 0$, $S(t) : L_2^H \rightarrow L_2^H$ is Lipschitzian and has a strongly continuous Hadamard derivative, though it is nowhere Fréchet differentiable. For dissipativity, it was assumed also that $\epsilon K < 1$ and $\lim_{|y| \rightarrow \infty} \sup f(y)/y < \pi^2$. Then it can be found a closed bounded attractor $\mathcal{A} \subset L_2^H$, which attracts every compact set. In case $K \geq \pi^2$ one has $\mathcal{A} = \{0\}$ and it attracts all bounded sets of L_2^H, but if there is a non-zero equilibrium (for example, if $f'(0) > \pi^2$), $\mathcal{A}$ is not compact. There are finitely many equilibria and in general (for most choices of f) they are all hyperbolic: the linearization of the nonlinear semigroup is a C^0-linear semigroup with a hyperbolic splitting (and, typically, infinite dimensional unstable space). But when there are some non-zero equilibrium, the equilibria are *not* isolated invariant sets: every neighborhood of each equilibrium contains non-constant periodic orbits (with period r). Indeed, every neighborhood contains uncountably many hyperbolic r-periodic orbits: fixed points of $S(r)$ such that the derivative has spectrum disjoint from the unit circle.

As was mentioned by Fichmann, this radical departure from the expected behavior near a hyperbolic element is presumably due to the use of a Hadamard derivative, rather than a Fréchet derivative. Perhaps the infinite dimensional unstable space plays a role. In any case, there is no hope for a Hartman-Grobman result. We already noted that K. Lu, in [127], proved the Hartman-Grobman theorem for system (5.20) in H_0^1.

For the sake of completeness we recall the definition of Hadamard derivative for an operator $S : E \to F$, with E, F Banach spaces. One says that S is Hadamard-differentiable at $q \in E$ if there exists a linear map $S'(q) : E \to F$ such that for each $h \in E$ one can write $S(q + tk + th) = S(q) + tS'(q)h + tR(t, k, h)$ where $R(t, k, h) \to 0$ as $(t, k) \to (0, 0)$ in $\mathbb{R} \times E$. In this case the (unique) map $S'(q)$ is then called the Hadamard derivative of S at q (see [41] for more details).

6 Stability of Morse–Smale Maps and Semiflows

6.1 Morse–Smale maps

We start this Chapter by dealing with maps and Section 6.1 corresponds, precisely, to Section 10 of [87]. Later, in Section 6.2, we will present the case of semiflows. Let us consider smooth maps $f : B \to E$, B being a Banach manifold embedded in a Banach space E. The maps f belong to $C^r(B, E)$, the Banach space of all E-valued C^r-maps defined on B which are bounded together with their derivatives up to the order $r \geq 1$. Let $C^r(B, B)$ be the subspace of $C^r(B, E)$ of all maps leaving B invariant, that is, $f(B) \subset B$. Denote by $A(f)$ the set

$$A(f) = \{x \in B : \text{ there exists a sequence } (x = x_1, x_2, \ldots) \in B,$$
$$\sup_j \|x_j\| < \infty \text{ and } f(x_j) = x_{j-1}, \ j = 2, 3, \ldots\}.$$

Special subspaces $KC^r(B, B)$ of $C^r(B, B)$ will be introduced satisfying the following compactness and reversibility properties: "any $f \in KC^r(B, B)$ is reversible, has $A(f)$ compact and given a neighborhood U of $A(f)$ in B, there exists a neighborhood $\mathcal{W}(f)$ of f in $KC^r(B, B)$ such that $A(g) \subset U$ for all $g \in \mathcal{W}(f)$" (reversibility for a C^1 map f means $f/A(f)$ and $df/A(f)$ are injective maps). The choice of the classes $KC^r(B, B)$ depends on the problems in view. In each case we need to assume appropriate hypothesis on the data in order to obtain the required compactness and reversibility properties for the elected $KC^r(B, B)$.

Global unstable manifolds of hyperbolic periodic orbits of a map $f \in KC^r(B, B)$ are introduced using the reversibility of f. The nonwandering set $\Omega(f)$ is the set of all $z \in A(f)$ such that given a neighborhood V of z in $A(f)$ and $n_0 \in N$, there exists $n > n_0$ such that $f^n(V) \cap V \neq \emptyset$. If $f \in KC^r(B, B)$, $\Omega(f)$ is compact and invariant and contains all $\omega(x)$ and $\alpha(x)$ of all $x \in A(f)$. Morse–Smale maps will be introduced (see Definition 6.1.16) and we denote by MS the set of Morse–Smale maps of $KC^r(B, B)$.

From the dynamics point of view, we will see that a Morse–Smale map f exhibits the simplest orbit structure, specially the "gradient like" ones, that is, the $f \in MS$ for which there exists a continuous Lyapunov function

$V : B \to R$ such that if $x \in B$ and $f(x) \neq x$, then $V(f(x)) < V(x)$. In this case $\Omega(f)$ is equal to Fix(f), the set of all fixed points of f.

Important stability theorems for (and existence of) Morse–Smale diffeomorphisms defined on a compact manifold M are well known. They say that any Morse–Smale diffeomorphism f is stable. That is, there exists a neighborhood $\mathcal{V}(f)$ of f in Diff$^r(M)$, the set of all C^r-diffeomorphisms of M, $r \geq 1$, such that for each $g \in \mathcal{V}(f)$ corresponds a homeomorphism $h = h(g) : M \to M$ and $h \cdot f = g \cdot h$ holds on M.

We say that $f \in KC^r(B, B)$ is A-stable if there exists a neighborhood $\mathcal{V}(f)$ of f in $KC^r(B, B)$ such that to each $g \in \mathcal{V}(f)$ corresponds a homeomorphism h, $h = h(g) : A(f) \to A(g)$ and $h \cdot f = g \cdot h$ holds on $A(f)$. The main results of this section can be summarized as follows:

"The set MS is open in $KC^r(B, B)$ and any $f \in MS$ is A-stable".

Let $x = f(x)$ be a fixed point of a C^r-map, $f : B \to B$, $r \geq 1$. The fixed point is said to be *hyperbolic* if the spectrum $\sigma(df(x))$ of the derivative $df(x)$ is disjoint from the unit circle of the complex plane. Under the above hypothesis one can define local unstable and local stable C^r manifolds denoted by $W^u_{loc}(x)$ and $W^s_{loc}(x)$, respectively and they are always transversal at the point x. One defines also the invariant sets (see Chapter 2) $W^u(x) = \bigcup_{i \geq 0} f^i(W^u_{loc}(x))$, $W^s(x) = \bigcup_{i \geq 0} f^{-i}(W^s_{loc}(x))$ and $W^s(x) \cap A(f)$.

Proposition 6.1.1. *Let $x = f(x)$ be a hyperbolic fixed point of a reversible C^r map $f : B \to B$, $r \geq 1$. The set $W^u(x) = \bigcup_{i \geq 0} f^i(W^u_{loc}(x))$ is an injectively immersed C^r-submanifold of B.*

The manifold $W^u(x)$ is the (global) unstable manifold of the hyperbolic fixed point x. We already mentioned that $W^u(x)$ is invariant under $f \in C^r(B, B)$ so $W^u(x) \subset A(f)$.

If $g = f^n$, $n \geq 1$, is a power of a bounded map $f : B \to B$, it is easy to see that $A(f) = A(g)$. If $f/A(f)$ is injective then $g/A(g)$ is also injective. If f is compact, g is compact and if f is reversible, g is reversible.

An element $x \in B$ is a periodic point of f if it is a fixed point of some iterate of f; the smallest integer $m > 0$ with $f^m(x) = x$ is the period of x. It is clear that the orbit $\mathcal{O}(x) = \{x, f(x), f^2(x), \ldots, f^{m-1}(x)\}$ of a periodic point x is a finite set with m points. Fix(f) and Per(f) will denote, respectively, the set of all fixed points and of all periodic points of f. We have, obviously, Fix$(f) \subset$ Per$(f) \subset \Omega(f)$.

A periodic point x with period m is said to be a hyperbolic periodic point if $\mathcal{O}(x)$ is hyperbolic, that is, if all points $y \in \mathcal{O}(x)$ are hyperbolic fixed points of f^m. We can talk about $W^u_{loc}(y)$, $W^s_{loc}(y)$ for all $y \in \mathcal{O}(x)$. The unstable manifold of y is $W^u(y) = \bigcup_{i \geq 0} f^{mi}(W^u_{loc}(y))$.

Definition 6.1.2. *A hyperbolic periodic point x of f is a source if $W^s_{loc}(x) \cap A(f) = \{x\}$; is a sink if $W^u_{loc}(x) = \{x\}$; otherwise x is a saddle.*

Proposition 6.1.3. *Let f be a smooth C^0-reversible map, that is, $f/A(f)$ is injective, and x be a hyperbolic periodic source (sink; saddle). Then $y \in \mathcal{O}(x)$ is also a source (sink; saddle)*

Let x be a hyperbolic fixed point of a smooth map $f : B \to B$. From now on we will assume $\dim W^u_{\mathrm{loc}}(x) < \infty$, by technical reasons; the *local λ-lemma* (see Proposition 6.1.6) which is a fundamental tool in the present section, requires finite dimensionality for the unstable manifold, at least when f is only C^1.

Assume, for instance, that the given C^r map $f : B \to B$ is compact. Then the derivative $df(x)$ at the hyperbolic fixed point x is a linear compact operator and, as we saw in Chapter 4, $A(f)$ has finite Hausdorff dimension so $W^u(x)$ is also finite dimensional.

Suppose that x is not a sink; then there exists an open disc B^u in $W^u_{\mathrm{loc}}(x)$ such that $\mathrm{Cl}\, B^u \subset W^u_{\mathrm{loc}}(x)$ and f^{-1}/B^u is a contraction. It follows that $f^{-1}(B^u) \subset B^u$. As usually, a *fundamental domain* for $W^u_{\mathrm{loc}}(x)$ is the compact set $G^u(x) = \mathrm{Cl}\, B^u - f^{-1}(B^u)$. If $y \in W^u_{\mathrm{loc}}(x) - \{x\}$, there exists an integer k such that $f^k(y) \in G^u(x)$. Any neighborhood $N^u(x)$ of $G^u(x)$ such that $N^u(x) \cap W^s_{\mathrm{loc}}(x) = \emptyset$ is called a *fundamental neighborhood for* $W^u_{\mathrm{loc}}(x)$.

If the hyperbolic fixed point is not a source we will consider a neighborhood $V = B^s \times B^u$ of x, B^s being an open disc in $W^s_{\mathrm{loc}}(x)$ such that f/B^s is a contraction and $\mathrm{Cl}\, B^s \subset W^s_{\mathrm{loc}}(x)$.

We define the *fundamental domain* for $W^s_{\mathrm{loc}}(x)$ as

$$G^s(x) = \mathrm{Cl}[B^s \cap A(f)] - f(B^s \cap A(f)).$$

If $A(f)$ is compact and f is C^0-reversible, then $f/A(f)$ is a homeomorphism and $G^s(x)$ is compact. It is clear that $x \notin G^s(x)$ so $W^u_{\mathrm{loc}}(x) \cap G^s(x) = \emptyset$ and there exists a neighborhood $N^s(x)$ of $G^s(x)$ which does not intersect $W^u_{\mathrm{loc}}(x)$; $N^s(x)$ is called a *fundamental neighborhood for* $W^s_{\mathrm{loc}}(x)$.

Remark 6.1.4. Any point of $W^s \cap A(f) = [\bigcup_{i>0} f^{-i}(W^s_{\mathrm{loc}}(x))] \cap A(f)$ reaches $B^s \cap A(f)$ after finitely many iterations of $f/A(f)$ or its inverse.

Remark 6.1.5. Given $y \in B^s \cap A(f) - \{x\}$, there exists an integer k such that $\tilde{f}^k(y) \in G^s(x)$, $\tilde{f}$ being the restriction of f to $A(f)$.

In fact, if $y \notin f(B^s \cap A(f))$ there is nothing to prove. If $y \in f(B^s \cap A(f))$ one considers the sequence $y = y^0, y^1, y^2, \ldots$, $f(y^i) = y^{i-1}$, $i \geq 1$, and there exists a first integer i_0 such that $y^{i_0} \notin B^s \cap A(f)$ (if $y^i \in B^s \cap A(f)$ for all $i \geq 1$ then $y \in W^u_{\mathrm{loc}}(x) \cap B^s = \{x\}$). If $y^{i_0} \in \mathrm{Cl}[B^s \cap A(f)]$, $y^{i_0} \in G^s(x)$; if $y^{i_0} \notin \mathrm{Cl}[B^s \cap A(f)]$ then $y^{i_0-1} \in B^s \cap A(f) - f(B^s \cap A(f)) \subset G^s(x)$.

Given two submanifolds $i_1 : W_1 \to B$ and $i_2 : W_2 \to B$ one says that W_1 and W_2 are $\varepsilon - C^1$ close manifolds if there exists a diffeomorphism $\gamma : W_1 \to W_2$ such that $i_1 : W_1 \to B$ and $i_2 \circ \gamma : W_1 \to B$ are $\varepsilon - C^1$ close maps.

Proposition 6.1.6 (local λ-lemma). *Let x be a hyperbolic fixed point of a smooth map $f : B \to B$, $\dim W^u_{loc}(x) < \infty$, and B^u be an embedded open disc in $W^u_{loc}(x)$, containing x. Let q be a point of $W^s_{loc}(x)$, $q \neq x$, and D^u be a disc centered at q, transversal to $W^s_{loc}(x)$, such that $\dim D^u = \dim W^u_{loc}(x)$. Then there is an open set V of B containing B^u such that given $\varepsilon > 0$ there exists $n_0 \in N$ such that if $n > n_0$ the connected component of $f^n(D^u) \cap V$ through $f^n(q)$ and the open disc B^u are $\varepsilon - C^1$ close manifolds.*

It is interesting to remark that we do not need to assume compactness or reversibility for the smooth map f but, when f is only of class C^1, the available proofs use, strongly, the finite dimensionality of $W^u_{loc}(x)$ (see [176], p.50).

In the same hypothesis of the local λ-lemma, let x be a hyperbolic fixed point of a smooth map $f : B \to B$ and $W^u_{loc}(x)$ be the local finite dimensional unstable manifold of x. The unstable set is the union $W^u(x) = \bigcup_{n \geq 0} f^n(W^u_{loc}(x))$. The *topological boundary* $\partial W^u(x)$ of the invariant set $W^u(x)$ is defined as $\partial W^u(x) = \omega(W^u(x) \setminus \{x\})$ (here $\omega(C)$ for a set C is the usual ω-limit set of C, $\omega(C) = \bigcap_{k \geq 0} \mathrm{Cl}(\bigcup_{n \geq k} f^n(C))$). It is easy to prove that this is equivalent to the set of all $y \in B$ such that $y = \lim f^{n_i}(y_i)$, $n \to \infty$ as $i \to \infty$, the y_i belonging to a fundamental domain $G^u(x)$ for $W^u_{loc}(x)$. It is clear that if $A(f)$ is compact, then $\partial W^u(x)$ is an invariant set.

Proposition 6.1.7. *Let x be a hyperbolic fixed point of a smooth map f and assume $\dim W^u_{loc}(x) < \infty$. Then $W^u(x)$ is invariant and $\mathrm{Cl}\, W^u(x) = \partial W^u(x) \cup W^u(x)$. If, in addition, the set $A(f)$ is compact, then $\partial W^u(x)$ and $\mathrm{Cl}\, W^u(x)$ are invariant sets.*

We say that $x = f^n(x)$ is a non-degenerate n-periodic point if n is the period of x and $1 \notin$ spectrum of $df^n(f)$.

As an application of the Implicit Function Theorem and classical results on stable and unstable manifolds, one obtains (see [176], p.27)

Proposition 6.1.8. *Let $x = x(f)$ be a non-degenerate n-periodic point of a C^r map $f : B \to B$, $r \geq 1$. There exist neighborhoods U of x in B and $\mathcal{V}(f)$ of f in $C^r(B,B)$ such that any $g \in \mathcal{V}(f)$ has in U only one n-periodic point $x(g)$ and no other m-periodic point with $m \leq n$. Moreover, if x is hyperbolic, the local stable and unstable manifolds depend continuously on $g \in \mathcal{V}(f)$; in particular if $W^u_{loc}(x(f))$ is finite dimensional, one has $\dim W^u_{loc}(x(f)) = \dim W^u_{loc}(x(g))$ for all $g \in \mathcal{V}(f)$.*

Proposition 6.1.9. *Let P be a hyperbolic periodic point of a smooth map f, $\dim W^u_{loc}(P) < \infty$, and $N^u(P)$ a fundamental neighborhood for $W^u_{loc}(P)$. Then, there exists a neighborhood W of P such that*

$$\bigcup_{n \geq 0} f^{-n}(N^u(P)) \cup W^s_{loc}(P) \supset W.$$

Proof: Let p be the period of P and $h = f^p$. If the proposition is not true, there exists a sequence $x_\nu \to P$ as $\nu \to \infty$, such that $x_\nu \notin W^s_{loc}(P)$ and $x_\nu \notin \bigcup_{n \geq 0} f^{-n}(N^u(P))$. Let $V = B^s \times B^u$ be a neighborhood of P considered in the construction of $N^u(P)$. Let k_ν be the first integer such that $Z_{k_\nu} = h^{k_\nu}(x_\nu) \notin V$; such a first integer does exist, otherwise $x_\nu \in W^s_{loc}(P)$. The sequence $k_\nu \to \infty$ as $\nu \to \infty$; in fact, if $k_\nu \leq M$ for all $\nu \geq 1$, since $h^{k_\nu}(P) = P$ and h^{k_ν} is continuous there exists a neighborhood $\tilde{V}$ of P, $\tilde{V} \subset V$ such that $h^{k_\nu}(\tilde{V}) \subset V$ for all $k_\nu \leq M$ which is absurd because the $x_\nu \in \tilde{V}$ for all $\nu \geq \nu_0$ imply $h^{k_\nu}(x_\nu) \in V$ giving a contradiction.

We may assume the neighborhood V is chosen such that $h(x_s, x_u) = \big(L_s x_s + \Phi_s(x_s, x_u), L_u x_u + \Phi_u(x_s, x_u)\big)$ verifies $\|L_s\|, \|L_u^{-1}\| < a < 1$, $\left|\frac{\partial \Phi_i}{\partial x_j}\right| \leq k$, $a + k < 1$, $i, j = u, s$. Since $x_\nu \to P$, there exists ν_1 such that for all $\nu \geq \nu_1$ one has $\|x_\nu\| \leq \frac{1}{\nu}$ and if $h(x_s, x_u) = (\bar{x}_s, \bar{x}_u)$, $\bar{x}_s = B^s$, $\bar{x}_u \in B^u$, one obtains, also, $\|\bar{x}_s\| = \|L_s x_s + \Phi_s(x_s, x_u)\| \leq a\big(\frac{1}{\nu}\big) + k\big(\frac{1}{\nu}\big) < \frac{1}{\nu}$. The canonical projections of $h^{k_\nu - 1}(x_\nu)$ on B^u and B^s are $\pi_u\big(h^{k_\nu - 1}(x_\nu)\big)$ and $\pi_s\big(h^{k_\nu - 1}(x_\nu)\big)$, respectively; since Cl B is compact, there is a limit point Z for $\pi_u\big(h^{k_\nu - 1}(x_\nu)\big)$, that is, at least for a subsequence one has $\pi_u\big(h^{k_\nu - 1}(x_\nu)\big) \to Z$ as $\nu \to \infty$ and $\|\pi_s\big(h^{k_\nu - 1}(x_\nu)\big) < \frac{1}{\nu}$ for all $\nu \leq \nu_1$. The above argument shows us that there exists a subsequence of $h^{k_\nu - 1}(x_\nu)$ which has $Z \in$ Cl B^u as a limit and $h^{k_\nu}(x_\nu) = h\big(h^{k_\nu - 1}(x_\nu)\big) \to h(Z)$. That limit point Z is not P because $h(Z) \notin V$, then Z reaches $N^u(P)$ after a finite number of iterations of h and, by continuity, each $h^{k_\nu - 1}(x_\nu)$, for large ν, reaches $N^u(P)$ after a finite number of iterations of h, that is, $x_\nu \in \bigcup_{n \geq 0} f^{-n}(N^u(P))$ which is a contradiction. The proof is then, complete. ∎

Proposition 6.1.10. *Let P be a hyperbolic p-periodic point of a map $f \in C^r(B, B)$, $r \geq 1$, $\dim W^u_{loc}(P) < \infty$, and $N^u(P)$ a fundamental neighborhood for $W^u_{loc}(P)$. Then, there exist neighborhoods U of P in B and $\mathcal{V}(f)$ of f in $C^r(B, B)$ such that $N^u(P)$ is a fundamental neighborhood for $W^u_{loc}(P^*)$, $P^* = P^*(g)$ being the unique hyperbolic p-periodic point in U corresponding to $g \in \mathcal{V}(f)$. Moreover, there exists a neighborhood W_0 of P such that for all g in $\mathcal{V}(f)$ one has*

$$\bigcup_{n \geq 0} g^{-n}(N^u(P)) \cup W^s_{loc}(P^*(g)) \supset W_0.$$

Proof: The first statement is a consequence of Proposition 6.1.8. Assume that the remaining statement is not true; then there exist sequences $x_\nu \to P$ and $g_\nu \to f$ such that

$$x_\nu \notin W^s_{loc}(P^*(g_\nu))$$

and

$$x_\nu \notin \bigcup_{n \geq 0} g_\nu^{-n}(N^u(P)).$$

As before, let k_ν be the first integer such that $g_\nu^{pk_\nu}(x_\nu) \notin V = B^s \times B^u$; such a first integer does exist because $x_\nu \notin W_{loc}^s(P^*(g_\nu))$. Call $h = f^p$ and $h_\nu = g_\nu^p$; if $k_\nu \leq M$ for $\nu \geq 1$, $h^{k_\nu}(P) = P$ implies the existence of $\tilde{V}$, neighborhood of P, such that $h^{k_\nu}(\tilde{V}) \subset V$; then $f^{pk_\nu}(x_\nu) \in V$ for large ν and $g_\nu \to f$ implies $g_\nu^{pk_\nu}(x_\nu) \in V$ which contradicts the definition of k_ν. We know that $g_\nu^{pk_\nu}(x_\nu) = h_\nu^{k_\nu}(x_\nu) \notin V$ but $g_\nu^{p(k_\nu-1)}(x_\nu) = h_\nu^{k_\nu-1}(x_\nu) \in V$ for all $\nu \geq 1$. The convergence $h_\nu \to h$ is in the C^1-norm then we can choose V such that $h_\nu(x_s, x_u) = \left(L_s^\nu x_s + \Phi_s^\nu(x_s, x_u), L_u^\nu x_s + \Phi_u^\nu(x_s, x_u) \right)$, $\|L_s\| < a < 1$, $\left| \frac{\partial \Phi_s}{\partial x_j} \right| \leq k$, $0 < a + k < 1$ and given $\varepsilon > 0$, $\exists \nu_0$ such that $\nu \geq \nu_0$ implies $\|L_s - L_s^\nu\| < \varepsilon$ and $\left| \frac{\partial \Phi_s^\nu}{\partial x_j} - \frac{\partial \Phi_s}{\partial x_j} \right| < \varepsilon$, $j = u, s$. It follows that for a suitable $\varepsilon > 0$, $\|L_s\| \leq a + \frac{\varepsilon}{2} < 1$ and $\left| \frac{\partial \Phi_s^\nu}{\partial x_j} \right| < k + \frac{\varepsilon}{2}$, $(a + \frac{\varepsilon}{2}) + (k + \frac{\varepsilon}{2}) = (a + k + \varepsilon) < 1$. The same argument used in the last Proposition 6.1.9 shows that $h_\nu^{k_\nu-1}(x_\nu) \to \bar{Z}$. If $\bar{Z} = P$, $h(\bar{Z}) = h(P) = P = h\left(h_\nu^{k_\nu-1}(x_\nu)\right)$ and since $h_\nu \to h$, $\left| h_\nu\left(h_\nu^{k_\nu-1}(x_\nu)\right) - h\left(h_\nu^{k_\nu-1}(x_\nu)\right) \right| < \varepsilon$ that is $\left| h_\nu^{k_\nu}(x_\nu) - P \right| < \varepsilon$ which is absurd since $h_\nu^{k_\nu}(x_\nu) \notin V$. Since $\bar{Z} \neq P$ and $\bar{Z} \in W_{loc}^u(P)$, with a finite number of iterations of $\bar{Z}$ by $h/W_{loc}^u(P)$ or its inverse one reaches $N^u(P)$ and with the same number, for large ν, x_ν reaches $N^u(P)$ by using h_ν and $x_\nu \in \bigcup_{n \geq 0} g_\nu^{-n}\left(N^u(P)\right)$ which is a contradiction. The proof is complete. ∎

To state dual results corresponding to Propositions 6.1.9 and 6.1.10 we need to assume reversibility for f and some compactness hypothesis in the set of maps to be considered.

Proposition 6.1.11. *Let P be a hyperbolic periodic point of a C^0-reversible and smooth map f such that $A(f)$ is compact; let $N^s(P)$ be a fundamental neighborhood for $W_{loc}^s(P)$. Then, there exists a neighborhood W of P such that*

$$\bigcup_{n \geq 0} f^n\left(N^s(P)\right) \cup W_{loc}^u(P) \supset W \cap A(f).$$

Proof: Let p be the period of P ($f^p(P) = P$) and V the neighborhood used in the construction of $N^s(P)$. If Proposition 6.1.11 is not true, there exists a sequence $x_\nu \to P$, $x \in A(f)$, $x_\nu \notin W_{loc}^u(P)$ and $x_\nu \notin \bigcup_{n \geq 0} f^n\left(N^s(P)\right)$. Each x_ν defines a unique sequence $(x_\nu = x_0, x_1, x_2, \ldots) \in A(f)$, $f^p(x_\nu^i) = x_\nu^{i-1}$, $i \geq 1$. Let k_ν be the first integer such that $x_\nu^{k_\nu} \notin V$ (if $x_\nu^i \in V$ for all i, $x_\nu \in W_{loc}^u(P)$). See that $k_\nu \to \infty$ as $\nu \to \infty$; if $k_\nu \leq M$ let $\tilde{f} = f/A(f)$ be the homeomorphism obtained restricting f to $A(f)$, $\tilde{f}^{-pk_\nu}(P) = P$, and given $V \cap A(f)$, there exists $\tilde{V} = \tilde{V}(P)$ such that $\tilde{f}^{pk_\nu}\left(\tilde{V} \cap A(f)\right) = V \cap A(f)$ and for large $\nu > \nu_0$, $x_\nu \in \tilde{V} \cap A(f)$ then $\tilde{f}^{-pk_\nu}(x_\nu) \in V \cap A(f)$ which is a contradiction. We have limit points $x_\nu^{k_\nu-1} \to x$, $x_\nu^{k_\nu} \to y$, $\tilde{f}^p(y) = x$ and $x \neq P$ (if $x = P \Rightarrow y = P$ (contradiction since $y \notin V$)). It is easy to see that $x \in A(f) \cap W_{loc}^s(P)$ since there exists a sequence $x, f^p(x), f^{2p}(x), \ldots$, constructed using $x_\nu^{k_\nu-1}, x_\nu^{k_\nu-2}$, and so on. With a finite number of iterations by $\tilde{f}^p$,

x reaches $N^s(P)$. Since $x_\nu^{k_\nu-1} \to x$ the same happens with $x_\nu^{k_\nu-1}$ and $x_\nu \in \bigcup_{n\geq 0} f^n(N^s(P))$ and we obtain a contradiction. The proof is complete. ∎

Consider now a (not unique) topological subspace $KC^r(B,B)$ as defined in the beginning of the section.

Proposition 6.1.12. *Let P be a hyperbolic p-periodic point of a map $f \in KC^r(B,B)$, $r \geq 1$, and $N^s(P)$ a fundamental neighborhood for $W_{loc}^s(P)$. Then, there exist neighborhoods U of P in B and $\mathcal{V}(f)$ of f in $KC^r(B,B)$ such that $N^s(P)$ is a fundamental neighborhood for $W_{loc}^s(P^*)$, $P^* = P^*(g)$ being the unique hyperbolic p-periodic point in U corresponding to $g \in \mathcal{V}(f)$. Moreover, there exists a neighborhood W_0 of P such that for all $g \in \mathcal{V}(f)$ one has*

$$\bigcup_{n\geq 0} g^n(N^s(P)) \cup W_{loc}^u(P^*(g)) \supset W_0 \cap A(g).$$

Proof: The first statement follows from Proposition 6.1.8. If the second statement is not true there exist sequences $x_\nu \to P$ and $g_\nu \to f$ such that $x_\nu \in A(g_\nu)$,

$$x_\nu \notin W_{loc}^u(P^*(g_\nu)) \quad \text{and} \quad x_\nu \notin \bigcup_{n\geq 0} g_\nu^n(N^s(P)).$$

Following the same arguments as in the proof of Proposition 6.1.11, each x_ν defines a unique sequence

$$(x_\nu = x_\nu^0, x_\nu^1, x_\nu^2 \ldots) \in A(g_\nu), \quad g_\nu^p(x_\nu^i) = x_\nu^{i-1}, \quad i > 1.$$

Let k_ν be the first integer such that $x_\nu^{k_\nu} \notin V$, V being a neighborhood of P used in the construction of $N^s(P)$ (if $x_\nu^i \in V$ for all $i \geq 0$, $x \in W_{loc}^u(P^*(g_\nu))$). The sequence g_ν may be chosen in order to obtain $A(g_\nu)$ in a $(1/\nu)$-neighborhood of $A(f)$, then $x_\nu^{k_\nu}$ approaches $A(f)$ as $\nu \to \infty$. Since $A(f)$ is compact there exists a sequence $y_\nu \in A(f)$, each y_ν giving the minimum for the distances between the $x_\nu^{k_{n\nu}}$ and $A(f)$. The sequence y_ν has a limit point $y \in A(f)$ and it is clear that $x_\nu^{k_\nu} \to y$ as $\nu \to \infty$. then $y \notin V$. See that $k_\nu \to \infty$ as $\nu \to \infty$, (if $1 \leq k_\nu \leq M$ and since $x_\nu^{k_\nu-1} \to f^p(y)$, $x_\nu^{k_\nu-2} \to f^{2p}(y)$ etc., one obtains $f^{Mp}(y) = P$ which implies $y = P \in V$—contradiction) then $P \in \omega(y)$ with respect to f^p and $y \in A(f) \cap W_{loc}^s(P)$, $y \neq P$. With a finite number of iterations of y by $f^p/A(f)$ one reaches $N^s(P)$ and with the same number, for large ν, $x_\nu^{k_\nu}$ reaches $N^s(P)$ using g_ν^p, then $x_\nu \in \bigcup_{n\geq 0} g_\nu^n(N^s(P))$, which is a contradiction. The proof is complete. ∎

Proposition 6.1.13. *Assume it is given a topological subspace S of $C^r(B,B)$ such that any $f \in S$ is reversible, has $A(f)$ compact and admits a neighborhood $\mathcal{V}(f)$ in S such that $\bigcup_{g\in\mathcal{V}(f)} A(g)$ is relatively compact. Then S has the properties of a $KC^r(B,B)$.*

Proof: If the proposition were not true, there would exist a neighborhood U_0 of $A(f)$, a sequence $f_\nu \to f$, $f_\nu \in S$ and points $x_\nu \in A(f_\nu)$ such that $x_\nu \notin U_0$. But the elements of the sequence (x_ν) eventually belong to $\bigcup_{g \in \mathcal{V}(f)} A(g)$ for a suitable neighborhood $\mathcal{V}(f)$ of f in S; so (x_ν) has a limit point x^0. Since $x_\nu \in A(f_\nu)$ there exists a sequence

$$(x_\nu = x_\nu^0, x_\nu^1, x_\nu^2, \ldots, x_\nu^i, \ldots) \in A(f_\nu)$$

such that $f_\nu(x_\nu^i) = x_\nu^{i-1}$ for $i \geq 1$. Choosing an appropriate subsequence of the (x_ν^0), hence the (x_ν^i) for each i, one obtains a sequence of limit points

$$(x^0, x^1, x^2, \ldots), \quad x_\nu^i \to x^i, \ i \geq 0.$$

But $\|f(x^1) - x^0\| \leq \|f(x^1) - f(x_\nu^1)\| + \|f(x_\nu^1) - f_\nu(x_\nu^1)\| + \|x_\nu^0 - x^0\|$, that is, $f(x^1) = x^0$; analogously $f(x^i) = x^{i-1}$, $i \geq 2$, and $x^0 \in A(f) - U_0 = \Phi$ which is a contradiction. ∎

Remark 6.1.14. Let M be a compact manifold and $B = C^0(I, M)$. Let $S \subset C^r(B, B)$ be the set of all analytic flow maps f_F of analytic (RFDE) $F \in \mathcal{X}^r$ defined on the manifold M. Then S is a particular $KC^r(B, B)$, $r \geq 1$. The reversibility of f_F follows from the analyticity of F, and each f_F being compact implies $A(f_F)$ is a compact set. The "continuity" of $A(f_F)$ with respect to f_F follows from Proposition 6.1.13.

In fact, the map $f : F \in \mathcal{X}^r \to f_F \in C^r(B, B)$ is well defined, continuous and injective. Moreover, $\mathcal{X}^r$ is homeomorphic to $f(\mathcal{X}^r)$ with the relative topology since $f_{F_\nu} \to f_F$ (in the topology of $C^r(B, B)$) implies $F_\nu \to F$ in $\mathcal{X}^r$. By Arzela's theorem and the above homeomorphism we see that the topological subspaces $\tilde{S} = f(\mathcal{X}^r)$ and S satisfy the hypotheses of Proposition 6.1.13.

The choice of the class $KC^r(B, B)$ depends on the case we are studying; for instance, maps arising from retarded functional differential equations, neutral functional differential equations, semi-linear parabolic equations, hyperbolic equations, can be considered. In each case we need to assume the appropriate hypotheses on the equations in order to obtain the compactness properties of $KC^r(B, B)$.

Proposition 6.1.15. *Let $P, Q \in \text{Per}(f)$ be distinct hyperbolic periodic points of a reversible map f such that $\dim W_{loc}^u(P)$, $\dim W_{loc}^u(Q) < \infty$. If $A(f)$ is compact and $\text{Cl}\, W^u(Q) \cap W^u(P) \neq \Phi$, then there exists $x \in \text{Cl}\, W^u(Q) \cap W_{loc}^s(P)$ such that $x \notin \mathcal{O}(P)$.*

Proof: From the hypothesis it follows that $P \in \text{Cl}\, W^u(Q)$ so there exists a sequence $Z_i = f^{n_i}(y_i) \to P$ with $n_i \to \infty$ as $i \to \infty$, $y_i \in G^u(Q)$. The points P and Q are fixed points of the power $g = f^{[p,q]}$, p and q being the periods of P and Q and $[p, q]$ its least common multiple. Since Q is a limit point of a sequence $(Z_i = Z_i^0, Z_i^1, Z_i^2, \ldots) \in W^u(Q)$, $g(Z_i^\nu) = Z_i^{\nu-1}$, $\nu \geq 1$, there exists a fixed integer k_i such that $Z_i^{k_i} \notin \bar{U}_0$, $\bar{U}_0$ being a suitable bounded and closed neighborhood of P, chosen together with closed neighborhoods $\bar{U}_n$ of $f^n(P)$, $1 \leq n \leq p-1$, satisfying the condition $g(\bar{U}_i) \cap \bar{U}_j = \Phi$, $0 \leq i \neq j \leq p-1$. Since $Z_i^{k_i} \in W^u(Q) \subset A(f)$ then the sequence $Z_i^{k_i}$ has a limit point x; but $Z_i^{k_i} \notin Int\bar{U}_0$, that is, $Z_i^{k_i} \in A(f) - Int\bar{U}_0$ which is closed, then $x \in A(f) - Int\bar{U}_0$ and $x \neq P$. We remark now that $k_i \to \infty$ as $i \to \infty$ because if $k_i \leq M$ for all $i \geq 1$, there exists a neighborhood $\tilde{V}$ of P such that $g^{-k_i}(\tilde{V}) \subset \bar{U}_0$ for all $k_i \leq M$. But $Z_i \to P$ implies $Z_i \in \bar{U}_0$ for large i and since $g^{k_i}(Z_i^{k_i}) = Z_i$ one obtains $Z_i^{k_i} \in \tilde{V}$ which is a contradiction. Given $\ell \geq 1$, $g^\ell(x) \in \bar{U}_0$; in fact, for large i, $k_i > \ell$ and $g^\ell(x) = g^\ell(\lim Z_i^{k_i}) = \lim g^\ell(Z_i^{k_i}) = \lim(Z_i^{k_i-\ell}) \in \bar{U}_0$, then $x \in \mathrm{Cl}\, W^u(Q) \cap W_{loc}^s(P)$. Finally, since $g(x) \in \bar{U}_0$, $x \notin \mathcal{O}(P) - \{p\}$; otherwise $x \in \bar{U}_j$ for some $0 < j < p-1$ which implies $g(x) = x \in \bar{U}_j \cap \bar{U}_0 \neq \Phi$. ∎

Definition 6.1.16. *Let f be an element of the topological space $KC^r(B,B)$, $r \geq 1$. We say that f is a* Morse–Smale map *($f \in MS$) if:*

1) $\Omega(f)$ is finite (then $\Omega(f) = \mathrm{Per}(f)$).
2) If $P \in \mathrm{Per}(f)$, P is hyperbolic and $\dim W^u(P) < \infty$.
3) If P and Q belong to $\mathrm{Per}(f)$ then $W^u(Q) \cap W_{loc}^s(P)$ ($\cap$ means transversal).

Remark 6.1.17. It is clear that if f is C^0-reversible one has $\Omega(f)$ finite implies $\Omega(f) = \mathrm{Per}(f)$. In fact, $\Omega(f)$ is invariant, then the negative orbit of $x_0 \in \Omega(f) - \mathrm{Per}(f)$ has an infinite number of points otherwise $x_0 \in \mathrm{Per}(f)$. This is a contradiction. But, even without assuming C^0-reversibility one has: $\Omega(f)$ finite implies $\Omega(f) = \mathrm{Per}(f)$. For otherwise there exists $x_0 \in \Omega(f) - \mathrm{Per}(f)$ and $x_i = f^i(x_0)$ ($i \geq 0$) are in $\Omega(f)$ ($f(\Omega(f)) \subset \Omega(f)$, always, by continuity of f). Since $\Omega(f)$ is finite, the x_i ($0 \leq i \leq m$) are distinct but $x_m \in \{x_0, x_1, \ldots, x_{m-1}\}$. Since $x_0 \notin \mathrm{Per}(f)$, $x_m \neq x_0$ and $x_m = x_{m-p}$ for some $p \in \{1, \ldots, m-1\}$ and then $\{f^i(x_0) : i \geq m-p\} = \{x_{m-p}, \ldots, x_{m-1}\}$ is bounded away from x_0 and $x_0 \notin \Omega(f)$ which is a contradiction.

As a corollary of the local λ-lemma (see Proposition 6.1.6) one can easily prove the following:

Proposition 6.1.18 (global λ-lemma). *Let $f : B \to B$ be a smooth reversible map and assume $A(f)$ is compact. Let $W^u(P)$ be the global unstable manifold of a hyperbolic fixed point P, $\dim W^u(P) = r$, and $N \subset A(f)$ be an injectively immersed invariant submanifold of B with a point q of transversal intersection with $W_{loc}^s(P)$. Then, for any given cell neighborhood B^r embedded in $W^u(P)$, centered in P, and any $\varepsilon > 0$, there exists one r-cell in N, $\varepsilon - C^1$ close to B^r.*

Proof: The reversibility of f enables us to define the global unstable manifold $W^u(P) \subset A(f)$ and $f/A(f)$ is a homeomorphism. The proof follows from the local λ-lemma. ∎

Remark 6.1.19. We do not need to assume N embedded since $N = \bigcup_{k \geq 0} N_k$, $N_0 \subset N_1 \subset \ldots$, with each N_k embedded.

Corollary 6.1.20. *Let $P_i \in \mathrm{Per}(f)$, $i = 1, 2, 3$, be hyperbolic points. If $W^u(P_1)$ and $W^s_{loc}(P_2)$, $W^u(P_2)$ and $W^u_{loc}(P_3)$ have $Q_1, Q_2 \notin \mathrm{Per}(f)$ of transversal intersections then $W^u(P_1)$ and $W^s_{loc}(P_3)$ also have a point $Q_3 \notin \mathrm{Per}(f)$ of transversal intersection.*

Corollary 6.1.21. *Let $P \in \mathrm{Per}(f)$ be hyperbolic. If $W^u(P)$ meets $W^s_{loc}(P)$ in a point $Q \notin \mathcal{O}(P)$ of transversal intersection, then $\Omega(f)$ is not finite.*

Let us introduce now the set MR of all elements of the topological space $KC^r(B, B)$, $r \geq 1$, such that

1) $\Omega(f)$ is finite (then $\Omega(f) = \mathrm{Per}(f)$).
2) $P \in \mathrm{Per}(f) \Longrightarrow P$ is hyperbolic and $\dim W^u(P) < \infty$.
3) If $P, Q \in \mathrm{Per}(f)$ and $W^u(P) \cap W^s_{loc}(Q) \neq \Phi$ then there exists a point of transversal intersection.

It is clear that $MS \subset MR$ and if $f \in MR$, $A(f)$ is the union of all unstable manifolds of $P \in \mathrm{Per}(f)$.

Proposition 6.1.22. *If $f \in MR$, there exist in $\mathrm{Per}(f)$ at least one sink and at least one source. Moreover, $A(f) = \bigcup_{P \in \mathrm{Per}(f)} W^u(P)$.*

Proof: It is possible the source and the sink be identical, case in which $A(f)$ is a single point. If there are no sources in $\mathrm{Per}(f) = \Omega(f)$ then there exists a cycle with transversal intersections and unstable manifolds with the same dimension. Using the global λ-lemma and their corollaries one concludes that $\Omega(f)$ is not finite. The same argument shows the existence of a sink.   ∎

Proposition 6.1.23. *Let $f \in MR$ and $P, Q \in \mathrm{Per}(f)$ such that $P \neq Q$ and $\mathrm{Cl}\, W^u(Q) \cap W^u(P) \neq \Phi$. Then there exists a sequence $P_1, P_2, \ldots, P_n \in \mathrm{Per}(f)$, $P_1 = P$, $P_n = Q$, such that*

$$W^u(P_{i+1}) \cap W^s_{loc}(P_i) \neq \Phi, \quad 1 \leq i \leq n-1.$$

Proof: We start with some remarks:

a) If $x \in \mathrm{Cl}\, W^u(Q) \cap W^u(P)$, x is assumed to be in $W^u_{loc}(P)$.
b) $\mathrm{Cl}\, W^u(Q) \cap W^u(P) \neq \Phi$ if and only if $\partial W^u(Q) \cap W^u(P) \neq \Phi$.
c) P cannot be a source $(W^s_{loc}(P) \cap A(f) = \{P\})$. In fact, Proposition 6.1.15 implies that $\mathrm{Cl}\, W^u(Q) \cap W^u(P) \neq \Phi \Longrightarrow W^s_{loc}(P) \cap A(f) \neq \{P\}$.
d) If P is a sink $(W^u_{loc}(x) = \{x\})$ it is enough to define $P = P_1$ and $Q = P_2$.

Finally, P is a saddle, then by Proposition 6.1.15 there exists $x \in$ $\mathrm{Cl}\, W^u(Q) \cap W^s_{\mathrm{loc}}(P)$ and $x \notin \mathcal{O}(P)$; we may assume

$$x \in \partial W^u(Q) \cap W^s_{\mathrm{loc}}(P)$$

otherwise we are done. But $\partial W^u(Q) \subset A(f)$ then $x \in A(f)$, that is, $x \in$ $W^u(P_2)$ for some $P_2 \in \mathrm{Per}(f)$ which implies

$$W^u(P_2) \cap W^s_{\mathrm{loc}}(P) \neq \Phi \quad \text{and} \quad W^u(P_2) \cap \partial W^u(Q) \neq \Phi.$$

If $P_2 = Q$ the proposition is proved. If $P_2 \neq Q$ we repeat the argument and get the sequence $(P_1 = P, P_2, P_3, \ldots)$. Remark that in this sequence $P_i \neq P_j$ otherwise $\Phi \neq W^u(P_i) \cap W^s_{\mathrm{loc}}(P_i) \neq \mathcal{O}(P_i)$ and $\Omega(f)$ is not finite by Corollary 6.1.21. Since $\Omega(f)$ is finite we reach the given point Q. ∎

Proposition 6.1.24. *Let $f \in MR$. Then for each $P \in \mathrm{Per}(f)$, $W^u(P)$ is embedded in B. In particular, f as a map from $W^u(P)$ into itself is differentiable.*

Proof: If $W^u(P)$ is not embedded we have $\partial W^u(P) \cap W^u(P) \neq \Phi$ and then there exists $x \in W^u(P) \cap W^s_{\mathrm{loc}}(P)$, $x \notin \mathcal{O}(P)$, with transversality ($f \in MR$), then $\Omega(f)$ is not finite. ∎

We introduce in the set of orbits of periodic points a partial order using the following definition:

Definition 6.1.25. *Let $f \in MR$ and $P, Q \in \mathrm{Per}(f)$. Then $\mathcal{O}(P) \leq \mathcal{O}(Q)$ if $\mathrm{Cl}\, W^u(\mathcal{O}(Q)) \cap W^u(\mathcal{O}(P)) \neq \emptyset$. Here, $W^u(\mathcal{O}(Q)) = \bigcup_{x \in \mathcal{O}(Q)} W^u(x)$.*

The above definition does not depend on the choice of the particular representatives of $\mathcal{O}(P)$ and $\mathcal{O}(Q)$. If $P_1 \in \mathcal{O}(P)$ and $Q_1 \in \mathcal{O}(Q)$ we see that $\mathrm{Cl}\, W^u(Q) \cap W^u(P) \neq \emptyset$ if and only if $W^u(Q) \cap W^s_{\mathrm{loc}}(P) \neq \emptyset$, if and only if $W^u(Q_1) \cap W^s_{\mathrm{loc}}(P_1) \neq \emptyset$. The relation $\mathcal{O}(P) \leq \mathcal{O}(Q)$ is obviously reflexive and transitive by using the global λ-lemma and their corollaries. Finally if $W^u(Q) \cap W^s_{\mathrm{loc}}(P) \neq \emptyset$ and $W^u(P) \cap W^s_{\mathrm{loc}}(Q) \neq \emptyset$ for $Q \notin \mathcal{O}(P)$ we obtain a kind of cycle and the global λ-lemma shows that $\Omega(f)$ is infinite which is a contradiction. Then, $\mathcal{O}(P) = \mathcal{O}(Q)$ and $\leq$ is a partial order.

The set of orbits of all periodic points of a map $f \in MR$ together with the above defined partial order is called the *phase diagram* $D(f)$ of f. For $P, Q \in \mathrm{Per}(f)$, a chain connecting Q to P in the phase diagram of f is a sequence $P_0, \ldots, P_n$ with $P_i \in \mathrm{Per}(f)$, $P_i \notin \mathcal{O}(P_{i+1})$, $P_0 = P$ and $P_n = Q$, such that $W^u(P_{i+1}) \cap W^s_{\mathrm{loc}}(P_i) \neq \emptyset$. The integer n is the length of the chain. Q is said to have k-behavior relative to P (write $\mathrm{beh}(Q|P) = k$) if the maximum length of chains connecting Q to P is $k \in N$; complete the definition by setting $\mathrm{beh}(Q|P) = 0$ if and only if $W^u(Q) \cap W^s_{\mathrm{loc}}(P) = \Phi$. If $Q \in \mathcal{O}(P)$ then $\mathrm{beh}(Q|P) = 0$ but not conversely because if P, Q are fixed points and sinks we have $\mathrm{beh}(Q|P) = 0$ and $Q \notin \mathcal{O}(P)$. It is also clear that for distinct orbits $\mathcal{O}(P) \leq \mathcal{O}(Q)$ implies $\mathrm{beh}(Q|P) > 0$.

We will show that given $f \in MR$, there is a neighborhood $\mathcal{V}(f)$ of f in $KC^r(B,B)$ such that $g \in \mathcal{V}(f)$ implies $g \in MR$ and there is an *isomorphism between phase diagrams*, that is, a bijection $\rho(g) : D(f) \to D(g)$ between the phase diagrams of f and g which is ordering and indices preserving, that means: $P, Q \in \mathrm{Per}(f)$, $\mathcal{O}(P) \leq \mathcal{O}(Q)$, implies $\mathcal{O}(\rho(g)P) \leq \mathcal{O}(\rho(g)Q)$ and $\dim W^u(P) = \dim W^u(\rho(g)P)$.

Since $f \in MR$, by Proposition 6.1.8 each $g \in \mathcal{V}_1(f)$ defines a map

$$\rho(g) : \mathrm{Per}(f) \to \mathrm{Per}(g) \subset \Omega(g)$$
$$P \longmapsto P^* = \rho(g)P.$$

We will construct neighborhoods V of $A(f)$ and $\mathcal{V}(f)$ of f such that

$$\Omega(g) \cap V = \rho(g)[\mathrm{Per}(f) \cap V]$$

for all $g \in \mathcal{V}(f)$. We will proceed by induction on the phase diagram of f.

For each sink S_i of f, choose a neighborhood $V_0(S_i) \subset W^s_{\mathrm{loc}}(S_i)$ and $\varepsilon_0(S_i) > 0$ such that if $|g - f|_r < \varepsilon_0(S_i)$ then $V_0(S_i) \subset W^s_{\mathrm{loc}}(S_i^*)$, where $S_i^* = \rho(g)S_i$. Let $V_0 = \bigcup V_0(S_i)$ and $\varepsilon_0 = \min\{\varepsilon_0(S_i) \mid S_i \text{ is a sink of f}\}$. In V_0 we trivially have $\Omega(g) \cap V_0 = \rho(g)[\mathrm{Per}(f) \cap V_0]$ for all $|g - f|_r < \varepsilon_0$. If, now, S is a saddle near sinks ($\mathrm{beh}(S|S_i) \leq 1$ for all sinks S_i), by the compactness of the fundamental domain $G^u(S)$, there exist n_0 and a fundamental neighborhood $N^u(S)$ such that given $x \in N^u(S)$, $f^n(x) \in V_0$ for some $n \leq n_0$. The same happens with g near f; by Proposition 6.1.10, $\bigcup_{n \geq 0} g^{-n}(N^u(S)) \cup W^s_{\mathrm{loc}}(S^*)$ contains a neighborhood $U_1(S)$ of S in B, for all g belonging to a suitable $\varepsilon_1(S)$-neighborhood of f in $KC^r(B,B)$. Consider $V_1(S) = V_0 \cup [\bigcup_{n=1}^{n_0} f^{-n}(V_0)] \cup U_1(S)$ and $\varepsilon_1(S)$ for each saddle S near sinks and finally $V_1 = \bigcup V_1(S)$ and $\varepsilon_1 = \min\{\varepsilon_1(S)\}$ for all saddles near sinks. In V_1 we have

$$\Omega(g) \cap V_1 = \rho(g)[\mathrm{Per}(f) \cap V_1].$$

By induction, assume now that we have constructed V_k, ε_k corresponding to the points in $\mathrm{Per}(f)$ whose behavior with respect to sinks of f is $\leq k$, so that $\Omega(g) \cap V_k = \rho(g)[\mathrm{Per}(f) \cap V_k]$ for $|g - f|_r < \varepsilon_k$. Let P_{k+1} be a point next to these in the phase-diagram of f. Again, by the compactness of $G^u(P_{k+1})$ there exists $n_1(P_{k+1})$ such that $f^n(x) \in V_k$ for all $x \in G^u(P_{k+1})$ and some $1 \leq n \leq n_1(P_{k+1})$. Using inverse images of V_k by f one defines $N^u(P_{k+1})$ and $\varepsilon_{k+1}(P_{k+1})$; for $|g - f|_r < \varepsilon_{k+1}(P_{k+1})$ the same happens with g. Use again Proposition 6.1.10 to obtain $U_{k+1}(P_{k+1}) = $ neighborhood of $P_{k+1} \subset W^s_{\mathrm{loc}}(P^*_{k+1}) \cup \bigcup_{n \geq 0} g^{-n}(N^u(P_{k+1}))$. Define $U_{k+1} = \bigcup U_{k+1}(P_{k+1})$ and $\varepsilon_{k+1} = \min\{\varepsilon_{k+1}(P_{k+1})\}$, $n_1 = \max\{n_1(P_{k+1})\}$; finally

$$V_{k+1} = V_k \cup \left[\bigcup_{n=1}^{n_1} f^{-n}(V_k)\right] \cup U_{k+1}$$

and in V_{k+1} we have

$$\Omega(g) \cap V_{k+1} = \rho(g)[\text{Per}(f) \cap V_{k+1}]$$

for all $|g - f|_r < \varepsilon_{k+1}$. The induction is complete. Remark that in V_{k+1} there are no other non-wandering points besides $P_i \in \text{Per}(f)$ and the corresponding P_i^* of g. The procedure reaches the sources and we define the above mentioned neighborhoods V of $A(f)$ and $\mathcal{V}(f)$ of f such that

$$\Omega(g) \cap V = \rho(g)[\text{Per}(f) \cap V]$$

for all $g \in \mathcal{V}(f)$. But $f \in KC^r(B,B)$ and we reduce $\mathcal{V}(f)$, if necessary, and obtain $A(g) \subset V$ for all $g \in \mathcal{V}(f)$. Then, since $\Omega(g) \subset A(g) \subset V$, it follows that $\Omega(g) = \text{Per}(g)$ for all $g \in \mathcal{V}(f)$ and we have finished the proof of the following:

Theorem 6.1.26. *The set MR is open in $KC^r(B,B)$, $r \geq 1$. Moreover, if $f \in MR$ there is a neighborhood $\mathcal{V}(f)$ of f in $KC^r(B,B)$ such that for each $g \in \mathcal{V}(f)$ the map $\rho(g) : \text{Per}(f) \to \text{Per}(g)$ considered above is a diagram isomorphism. In particular, f is Ω-stable.*

Consider again a smooth map $f \in MR$. If $P_k, P_{k+1} \in \text{Per}(f)$ satisfy $\text{beh}(P_k|P_{k+1}) = 1$ and if $G^s(P_{k+1})$ is a fundamental domain (then compact) for $W_{\text{loc}}^s(P_{k+1})$ we have that $W^u(P_k) \cap G^s(P_{k+1})$ is also compact. In fact, if $x_\nu \to x$, $x_\nu \in W^u(P_k) \cap G^s(P_{k+1})$, it is clear that $x \in G^s(P_{k+1})$ and if $x \notin W^u(P_k)$ (then $x \in \partial W^u(P_k)$), there exists $\tilde{P} \in \text{Per}(f)$ such that $x \in W^u(\tilde{P})$, $\tilde{P} \neq P_{k+1}$ and $\tilde{P} \neq P_k$; but by Proposition 6.1.23 $\text{Cl}\, W^u(P_k) \cap W^u(\tilde{P}) \neq \Phi$ implies $W^u(P_k) \cap W_{\text{loc}}^s(\tilde{P}) \neq \emptyset$, then $\text{beh}(P_k|P_{k+l}) > 1$ giving us a contradiction, that is, $x \in W^u(P_k)$.

Proposition 6.1.12 combined with Theorem 6.1.26, Proposition 6.1.6 and the arguments of transversality of manifolds prove the following:

Proposition 6.1.27. *Let $f \in MS$, $P \in \text{Per}(f)$ and $\dim W^u(P) = m$. Fix a cell neighborhood B^m of P in $W_{\text{loc}}^u(P)$. Given $\varepsilon > 0$, there exist neighborhoods V of P, and $\mathcal{V}(f)$ of f in $KC^r(B,B)$, $r \geq 1$, such that if for some $Q \in \text{Per}(f)$, $W^u(Q^*(g)) \cap V \neq \emptyset$ then $W^u(Q^*(g)) \cap V$ is fibered by m-cells $\varepsilon - C^1$ close to B^m, $g \in \mathcal{V}(f)$ and $Q^*(g) = \rho(g)Q$.*

From Theorem 6.1.26 and Proposition 6.1.27 we obtain the following result.

Theorem 6.1.28. *The set MS of all r-differentiable Morse–Smale maps is open in MR (then in $KC^r(B,B)$), $r \geq 1$. Moreover, if $f \in MS$, then its phase-diagram is stable (up to a diagram isomorphism) under small C^r perturbations of f in $KC^r(B,B)$.*

Remark 6.1.29. In proving Proposition 6.1.27, we really have an *Unstable Foliation* of $U = V \cap A(f)$ for $f \in MS$ at $P \in \text{Per}(f)$, that is, a continuous foliation $\mathcal{F}^u(P) : x \in U \to \mathcal{F}_x^u(P)$ such that:

a) the leaves are C^1 discs, varying continuously in the C^1 topology and $\mathcal{F}_P^u(P) = W^u(P) \cap U$,

b) each leaf $\mathcal{F}_x^u(P)$ containing $x \in U$, is contained in U,

c) $\mathcal{F}^u(P)$ is f-invariant; that is, $f(\mathcal{F}_x^u(P)) \supset \mathcal{F}_{f(x)}^u(P)$, x and $f(x)$ in U.

Moreover, using the reversibility property of the MS maps, this unstable foliation can be easily globalized through saturation by f. The same happens for g in a suitable neighborhood $\mathcal{V}(f)$ of f in MS (then in $KC^r(B, B)$).

By induction on the phase diagram of $f \in MS$ and using the global λ-lemma we easily obtain a so-called *compatible system* of global unstable foliations $\mathcal{F}^u(P_1), \mathcal{F}^u(P_2), \ldots, \mathcal{F}^u(P - n)$, for any maximal chain $(P_1, P_2, \ldots, P_n) \in \text{Per}(f)$, $\mathcal{O}(P_i) \geq \mathcal{O}(P_{i+1})$, $i = 1, 2, \ldots, n-1$, P_1 being a source and P_n being a sink. The compatibility means that "if a leaf F of $\mathcal{F}^u(P_k)$ intersects a leaf $\tilde{F}$ of $\mathcal{F}^u(P_\ell)$, $k < \ell \leq n$, then $F \supset \tilde{F}$; moreover, the restriction of $F^u(P_\ell)$ to a leaf of $\mathcal{F}^u(P_k)$ is a C^1 foliation."

In the sequel we will prove a stability theorem for Morse–Smale maps.

Definition 6.1.30. *A map f in $KC^r(B, B)$ is A-stable if there exists a neighborhood $\mathcal{V}(f)$ of f in $KC^r(B, B)$ such that to each $g \in \mathcal{V}(f)$ one can find a homeomorphism $h = h(g) : A(f) \to A(g)$ satisfying the conjugacy condition $h \cdot f = g \cdot h$ on $A(f)$.*

The properties of $f \in MS$, specially the reversibility of f and the compactness of $A(f)$, the finite dimensionality of the unstable manifolds $W^u(P)$, $P \in \text{Per}(f)$, the existence of compatible systems of global unstable foliations and the parametrized version of the Isotopy Extension Theorem are the main tools to be used in the proof of the next Theorem 6.1.32.

In order to recall the Isotopy Extension Theorem (IET) one needs some more notation. Let N be a C^r compact manifold, $r \geq 1$ and A an open set of R^s. Let N be a C^∞ manifold with $\dim M > \dim N$. We indicate by $C_A^k(N \times A, M \times A)$ the set of C^k mappings $f : N \times A \to M \times A$ such that $\pi = \pi' \cdot f$, endowed with the C^k topology, $1 \leq k \leq r$. Here, π and π' denote the natural projections $\pi : N \times A \to A$, $\pi' : M \times A \to A$. Let $\text{Diff}_A^k(M \times A)$ be the set of C^k diffeomorphisms φ of $M \times A$ such that $\pi' = \pi \cdot \varphi$, again with the C^k topology.

Lemma 6.1.31 (Isotopy Extension Theorem). *Let $i \in C_A^k(N \times A, M \times A)$ be an embedding and A' a compact subset of A. Given neighborhoods U of $i(N \times A)$ in $M \times A$ and V of the identity in $\text{Diff}_A^k(M \times A)$, there exists a neighborhood W of i in $C_A^k(N \times A, M \times A)$ such that for each $j \in W$ there exists $\varphi \in V$ satisfying $\varphi \cdot i = j$ restricted to $N \times A'$ and $\varphi(x) = x$ for all $x \notin U$.*

Theorem 6.1.32. *Any Morse–Smale map f in $KC^r(B, B)$ is A-stable.*

Proof: By Theorem 6.1.28 (openness) there exists a neighborhood of f in $KC^r(B, B)$ containing only Morse–Smale maps. We saw, also, that if $P_k, P_{k+1} \in \mathrm{Per}(f)$ satisfy $\mathrm{beh}(P_k|P_{k+1}) = 1$ then $W^u(P_k) \cap G^s(P_{k+1})$ is compact. If P_1 is a source and $\mathrm{beh}(P_1, P_{k+1}) = k$, there exists a maximal chain $(P_1, P_2, \ldots, P_{k+1})$ such that $\mathrm{beh}(P_i, P_{i+1}) = 1$, $i = 1, 2, \ldots, k$. Recall that $G^s(P_{k+1}) = \mathrm{Cl}[B^s \cap A(f)] - f(B^s \cap A(f))$. Since the compact set $A(f)$ is equal to the union of all global unstable manifolds of periodic points of f (Prop. 6.1.22) and $\Omega(f) = \mathrm{Per}(f)$ is finite, we may assume that $B^s = B^s(P_{k+1})$ have been chosen in such a way that $A(f)$ is transversal to ∂B^s (besides being transversal to B^s). This requires explanation since ∂B_s is not generally differentiable. We may however choose B_s so ∂B_s— near $A(f)$—is a differentiable manifold and transverse to $A(f)$. The crucial property we need is that, given $x \in \partial B_s \cap W^u(Q)$, $Q \in \mathrm{Per}(f)$, there exist $x', x'' \in W^u(Q) \cap W^s_{\mathrm{loc}}(P_{k+1})$ arbitrarily close to x, with $x' \in B_s$, $x'' \notin \bar{B}_s$. Call $S_E = S_E(P) = \partial B^s \cap G^s(P)$; we have also $S_E = \partial B^s \cap A(f)$. In fact, $S_E \subset \partial B^s \cap \mathrm{Cl}[B^s \cap A(f)] \subset \partial B^s \cap A(f)$ trivially. For the reverse inclusion, let $x \in \partial B^s \cap A(f)$; since $f(B^s) \subset B^s$ and $x \notin B^s$, $x \notin f(B^s \cap A(f))$ while $x \subset \mathrm{Cl}\, B^s \cap A(f)$, so we only need to prove $x \notin \mathrm{Cl}[B^s \cap A(f)]$. For some Q, $x \subset W^u(Q) \cap \partial B^s$, and these meet transversaly so there exist $x' \subset W^u(Q) \cap B^s$ arbitrarily close to x, i.e., $x \subset \mathrm{Cl}[W^u(Q) \cap B^s] \subset \mathrm{Cl}[A(f) \cap B^s]$. We have incidentally proved $\mathrm{Cl}[A(f) \cap B^s] = A(f) \cap \mathrm{Cl}\, B^s$, which will be needed later. Remark finally that, using the relative topology of $A(f) \cap W^s_{\mathrm{loc}}(P)$, we have $\partial G^s(P) = S_E \cup S_I$, $S_I = S_I(P) = F(S_E)$, "∂" relative to $A(f) \cap W^s_{\mathrm{loc}}(P)$. In fact, $G^s(P) = (B^s \cup \partial B^s) \cap A(f) - f(B^s \cap A(f)) = [B^s \cap A(f) - f(B^s \cap A(f))] \cup S_E = [(\mathrm{Int}\, G^s(P)) \cup f(S_E)] \cup S_E = \mathrm{Int}\, G^s(P) \cup (S_E \cup S_I)$.

The stable set $W^s(P)$ is the set of all points $x \in B$ such that $\omega(x) = \{P\}$. Any point $z \in W^s(P) \cap A(f)$ reaches $G^s(P) - S_I(P)$ after a finite number of iterations of $\bar{f}$ or $(\bar{f})^{-1}$, $\bar{f} = f/A(f)$.

Given any bounded embedded disc $D \subset W^u(P) = W^u(P; f)$, for g C^1-close to f there is a disc $D^* \subset W^u(P^*(g)) = W^u(P^*; g)$ C^1-close to D, $P^* \in D^*$, where $P^* = \rho(g)P$; we say $W^u(P^*; g)$ is C^1-close to $W^u(P; f)$ "on compact sets."

Let P_2 be a periodic point of f with behavior ≤ 1 with respect to sources and consider a pair (P_1, P_2) such that P_1 is a source and $\mathrm{beh}(P_1/F_2) = 1$. The manifolds $W^u(P_1; f)$ and $W^u(P_1^*; g)$ are C^1-close on compact sets and let h_1' be the corresponding diffeomorphism; also $W^s_{\mathrm{loc}}(P_2; f)$ and $W^s_{\mathrm{loc}}(P_2^*; g)$ are C^1-close for g in a suitable neighborhood of f, $P_2^* = \rho(g)P_2$. By the implicit function theorem and the existence of transversality for the intersections $W^u(P_1; f) \cap W^s_{\mathrm{loc}}(P_2; f)$ and $W^u(P_1^*; g) \cap W^s_{\mathrm{loc}}(P_2^*; g)$, there is a well defined diffeomorphism h_2 from $G^s(P_2; f) \cap W^u(P_1; f)$ into $W^s_{\mathrm{loc}}(P_2^*; g) \cap W^u(P_1^*; g)$. Define a differentiable map h_2 from $G^s(P_2; f) \cap W^u(P_1; f)$ into $W^u(P_1^*, g)$ equal to $\bar{h}_2$ on $W^u(P_1; f) \cap S_E(P_2)$ and equal to $\tilde{h}_2 = g \cdot \bar{h}_2 \cdot f^{-1}$ on $W^u(P_1; f) \cap S_I(P_2)$. To construct h_2 we use the IET (Lemma 6.1.31) just observing that $(h_1')^{-1} \cdot \bar{h}_2$ maps $W^u(P_1; f) \cap S_E(P_2)$ into $W^u(P_1; f)$ and $(h_1')^{-1} \cdot \tilde{h}_2 = (h_1')^{-1} \cdot$

$(g \cdot \bar{h}_2 \cdot f^{-1})$ maps $W^u(P_1; f) \cap S_I(P_2)$ into $W^u(P_1; f)$, both are near the corresponding inclusion maps and so can be extended to an embedding of $G^s(P_2; f) \cap W^u(P_1; f)$ into $W^u(P_1; f)$. The property we obtain for h_2 is that $h_2 f(x) = g h_2(x)$ for all $x \in W^u(P_1; f) \cap S_E(P_2)$; in fact, $g h_2(x) = g \bar{h}_2(x)$ and $h_2 f(x) = h_2(f(x)) = \bar{h}_2(f(x)) = g\bar{h}_2 f^{-1}(f(x)) = g\bar{h}_2(x)$. This map h_2 can be extended to $z \in W^s(P_2; f) \cap W^u(P_1; f)$ since there exists a unique $n \in \mathbb{Z}$ such that

$$(\bar{f})^n(z) \in (G^s(P_2; f) - S_I(P_2)) \cap W^u(P_1; f):$$

define $h_2(z) = g^{-n}(h_2(f^n(z)))$ and $h_2(P_2) = P_2^*$.

We do the same with all sources $F_i \in \mathrm{Per}(f)$ such that $\mathrm{beh}(F_i|P_2) = 1$ and h_2 is defined on $W^s(P_2; f) \cap A(f)$. For the remaining points $\tilde{P}_2 \in \mathrm{Per}(f)$ with behavior ≤ 1 with respect to sources proceed analogously and obtain h_2 defined on $W^s(\tilde{P}_2; f) \cap A(f)$ satisfying $h_2 f = g h_2$ and $h_2(\tilde{P}_2) = \tilde{P}_2^*$.

The next step is the consideration of $P_3 \in \mathrm{Per}(f)$ with behavior ≤ 2 with respect to sources and we will construct a homeomorphism h_3 on $W^s(P_3; f) \cap A(f)$ starting with $G^s(P_3) - S_I(P_3)$. For the sources with behavior 1 relative to P_3 the procedure is equal to that above. Let now P_1 be a source in $\mathrm{Per}(f)$ such that $\mathrm{beh}(P_1|P_3) = 2$. We have at least one sequence $(P_1 P_2 P_3)$ such that $\mathrm{beh}(P_1|P_2) = \mathrm{beh}(P_2|P_3) = 1$. Since $\mathrm{beh}(P_2|P_3) = 1$ we define a diffeomorphism $\bar{\bar{h}}_3$ on $G^s(P_3; f) \cap W^u(P_2; f)$ exactly as we did above with h_2. But $W^u(P_1; f)$ approaches $W^u(P_2; f)$ and it is well defined a foliation on $W^u(P_1; f)$ induced by $W^u(P_2; f)$; the same happens with $W^u(P_1^*; g)$ relatively to $W^u(P_2^*; g)$ for g near f in MS. The existence of a compatible system of global unstable foliations guarantees that $W^u(P_1; f)$ intersects $W^s_{\mathrm{loc}}(P_3; f)$ with its leaves accumulating in the (compact) set $W^u(P_2; f) \cap G^s(P_3; f)$. To each leaf $\mathcal{F}_x$ of $W^u(P_1; f) \cap G^s(P_3; f)$ near $W^u(P_2; f) \cap G^s(P_3; f)$ corresponds a unique point $x \in W^s_{\mathrm{loc}}(P_2; f) \cap W^u(P_1; f)$ near P_2. Using h_2 (defined in the P_2 level), to $\mathcal{F}_x$ corresponds a unique leaf $\mathcal{F}^*_{h_2(x)}$ of $W^u(P_1^*; g) \cap G^s(P_3^*; g)$. Consider the map $\bar{h}_3$ defined on $G^s(P_3; f) \cap W^u(P_2; f)$ and use the C^1-closeness on compact sets of $W^u(P_2; f)$ with the leaves of $W^u(P_1; f)$ [respectively of $W^u(P_2^*; g)$ with the leaves $W^u(P_1^*; g)$] to obtain a diffeomorphism $i_x : \mathcal{F}_x \to W^u(P_2; f) \cap G^s(P_3; f)$ [respectively $i_x^* : \mathcal{F}^*_{h_2(x)} \to W^u(P_2^*; g) \cap G^s(P_3^*; g)$] and construct $\bar{h}_3 = (i_x^*)^{-1} \cdot \bar{\bar{h}}_3 \cdot i_x$ which is an extension of $\bar{\bar{h}}_3$ to the leaf $\mathcal{F}_x$. As before, one considers $\bar{h}_3$ locally defined on $W^u(P_1; f) \cap G^s(P_3; f) \cap S_E(P_3)$ and defines $\tilde{h}_3 = g \cdot \bar{h}_3 \cdot f^{-1}$ (locally) on $W^u(P_1; f) \cap G^s(P_3; f) \cap S_I(P_3)$. Using again the IET to the local foliation $x \to \mathcal{F}_x$, x in a neighborhood of P_2 in $W^s_{\mathrm{loc}}(P_2; f) \cap W^u(P_1; f)$, one obtains a continuous (local) extension h_3 of $\bar{\bar{h}}_3$ coinciding with $\bar{h}_3$ on $W^u(P_1; f) \cap G^s(P_3; f) \cap S_E(P_3)$ and with $\tilde{h}_3$ on $W^u(P_1; f) \cap G^s(P_3; f) \cap S_I(P_3)$. Notice that $W^u(P_1; f) \cap W^s_{\mathrm{loc}}(P_3; f)$ and $W^u(P_1^*; g) \cap W^s_{\mathrm{loc}}(P_3^*; g)$ are C^1-close on compact sets. In order to extend h_3 (defined on the leaves of $W^u(P_1; f) \cap G^s(P_3; f)$ near $W^u(P_2; f) \cap G^s(P_3; f)$) to $W^u(P_1; f) \cap G^s(P_3)$, we extract a small tubular neighborhood of $W^u(P_2; f) \cap$

$G^s(P_3; f)$ in $\mathrm{Cl}[W^u(P_1; f) \cap W^s_{\mathrm{loc}}(P_3; f)]$ and apply again the IET for diffeomorphisms near the identity. In this way we can continuously extend h_3 to a full neighborhood of $W^u(P_2; f) \cap G^s(P_3; f)$ so that it satisfies the conjugacy equation $h_3 f = g h_3$ for points of $W^u(P_1; f) \cap G^s(P_3; f) \cap S_E(P_3)$.

We proceed, in an analogous way, with all possible sequences $(P_1, P_2', P_3) \in \mathrm{Per}(f)$ such that $\mathrm{beh}(P_1|P_2') = \mathrm{beh}(P_2'|P_3) = 1$. Consider, finally, the remaining sources $P_1' \in \mathrm{Per}(f)$ in the same conditions as P_1 and obtain a continuous h_3 defined on $G^s(P_3)$ with the equality $h_3 f = g h_3$ holding on $S_E(P_3)$ and then, a continuous h_3 defined on $A(f) \cap W^s(P_3)$, $h_3(P_3) = P_3^*$, with the desired conjugacy property $h_3 f = g h_3$.

The last step showed us, clearly, the full induction procedure. Assume we have constructed all maps h_k, satisfying $h_k(P_k) = P_k^*$ and $h_k f = g h_k$ on $A(f) \cap W^s(P_k)$ for all $P_k \in \mathrm{Per}(f)$ such that $\mathrm{beh}(F_i|P_k) \leq k - 1$, $k \geq 3$, where the F_i are all sources of $\mathrm{Per}(f)$; let $P_{k+1} \in \mathrm{Per}(f)$ be such that $\mathrm{beh}(F_i|P_{k+1}) \leq k$ for all sources $F_i \in \mathrm{Per}(f)$. Let $(F_1, P_2, \ldots, P_k, P_{k+1})$ be a sequence such that F_1 is a source and $\mathrm{beh}(F_1|P_2) = \mathrm{beh}(P_2|P_3) = \cdots = \mathrm{beh}(P_k|P_{k+1}) = 1$. We start the construction of h_{k+1} on $W_u(P_k; f) \cap G^s(P_{k+1})$, extend locally h_{k+1} to $W^u(P_{k+1}) \cap G^s(P_{k+1})$ and by a second induction procedure extend h_{k+1} to $W^u(P_{k+2}) \cap G^s(P_{k+1}), \ldots, W^u(F_1) \cap G^s(P_{k+1})$, as we did in the case $k = 2$. Do the same with all maximal sequences $(F_1, P_2', P_3', \ldots, P_{k+1})$ with F_1 and P_{k+1} fixed and, finally, with the remaining sources F_i to obtain h_{k+1} defined on $G^s(P_{k+1})$ verifying the equality $h_{k+1} f = g h_{k+1}$ defined on $S_E(P_{k+1})$. By forcing the conjugacy $h_{k+1} f = g h_{k+1}$ extend h_{k+1} to $A(f) \cap W^s(P_{k+1})$. The induction is complete and we reach the sinks. Since the disjoint union

$$\bigcup_{P \in \mathrm{Per}(f)} A(f) \cap W^s(P)$$

is equal to $A(f)$ the map $H = h_2 \cup h_3 \cup \ldots$ is well defined on $A(f)$, $H(P) = P^*$, and $H f(x) = g H(x)$ for all $x \in A(f)$.

The final step is to check the continuity of $H : A(f) \to A(g)$. Remark, first of all, that if H is continuous in $f(x)$ then H is continuous in $x \in A(f)$; in fact let $z_i \to x$, $z_i \in A(f)$; since f is continuous, $f(z_i) \to f(x)$ and the fact that H is continuous at $f(x)$ implies $H f(z_i) \to H f(x)$ that is, $g H(z_i) \to g H(x)$. But $H(z_i) \in A(g)$, $H(x) \in A(g)$ and g is reversible, then, $H(z_i) \to H(x)$. Given, now, $x \in A(f)$, it is clear that $x \in A(f) \cap W^s(P_k)$ for some $P_k \in \mathrm{Per}(f)$; it is sufficient to verify the continuity of H at the points x of a neighborhood of P_k in $A(f) \cap W^s(P_k)$. If P_k is a source or a sink the continuity is trivial. Assume P_k is a saddle and let $x_n \to x$, $x_n \in \mathcal{F}^u_{x_n}(P_k)$ and $x \in \mathcal{F}^u_x(P_k)$, $\mathcal{F}^u(P_k)$ being the global unstable foliation at P_k considered above. But, by the definition of $H = h_2 \cup h_3 \cup \ldots$ and by the construction of the maps h_k, we see that the set of accumulation points of $\{H(x_n)\}$ is contained in $W^s_{\mathrm{loc}}(P_k^*; g)$ and $H(x_n) \in \mathcal{F}^{u^*}_{h_k(x_n)}(P_k^*)$, $\mathcal{F}^{u^*}(P_k^*)$ being the global unstable foliation at P_k^*. Then $H(x_n) \to \mathcal{F}^{u^*}_{h_k(x)}(P_k^*) \cap W^s_{\mathrm{loc}}(P_k^*; g) = h_k(x) = H(x)$

proving the continuity of H. Similarly, H^{-1} is also continuous and the proof is complete.   ∎

Corollary 6.1.33. *Let B be a compact manifold and $\text{Diff}^r(B)$ the set of all C^r-diffeomorphisms of B, $r \geq 1$. Then the Morse–Smale diffeomorphisms of $\text{Diff}^r(B)$ are stable and form an open set.*

Proof: Remark that $\text{Diff}^r(B)$ satisfies the conditions to be a $KC^r(B,B) \subset C^r(B,B)$. In fact $A(f) = B$ for all $f \in \text{Diff}^r(B)$ and the reversibility is trivial. The result follows from Theorem 6.1.32.   ∎

Theorem 6.1.34. *Let $KC^r(B,B)$ be the subspace S, set of flow maps of all analytic RFDE $F \in \mathcal{X}^r$, $r \geq 1$, defined on an analytic compact manifold M. The Morse–Smale maps f of S are stable relatively to $A(f)$ and form an open set in S.*

Proof: Follows from Remark 6.1.14, Theorem 6.1.28 and Theorem 6.1.32.   ∎

The main results presented above relative to the infinite dimensional setting of Morse–Smale maps are due to Oliva [152] and [87]. A reference for a theory of local stable and unstable manifolds of a hyperbolic fixed point of a C^r-map is the seminal work done by Hirsch, Pugh and Shub in [102]. Proposition 6.1.6 (local λ-lemma) is due to Palis [160]. The proof of Prop. 6.1.27 is a simple generalization of Lemma 1.11 of [160]. The language of unstable foliations and compatible system of unstable foliations is due to Palis and Takens [162] where we can see also the statement and references for a proof of the Isotopy Extension Theorem (Lemma 6.1.31). Finally, Theorems 6.1.28 and 6.1.32 applied to the special case in which B is a compact manifold yield the proof for the stability of Morse–Smale diffeomorphisms (Corollary 6.1.33), originally established in [160] and [161]. Theorem 6.1.34 gives, in some sense, the answer to a fundamental question established in [72], [149] and [73].

6.2 Morse–Smale semiflows

We proceed now mentioning the continuous case. When a (continuous) semiflow is gradient, (α-limit and ω-limit sets of any bounded orbit are contained in the set of equilibrium points), the consideration of the (discrete) flow map at a certain fixed time (say $t = 1$) captures the essential properties of the semiflow; but, if there are T-periodic orbits ($T > 0$ is the minimum period) in a given semiflow, the discrete map will not capture this behavior. If we assume for a semiflow, as one does for flows, that the critical elements are only a finite number of equilibrium points and periodic orbits, each of which hyperbolic, all the unstable and local stable manifolds of those critical elements in general position, and the non-wandering set consists only of critical elements,

then we call the semiflow a Morse–Smale (MS) semiflow in analogy with the
MS maps. And, as it is mentioned by Hale and Lunel (see [86, p. 374]): *it is
reasonable to expect that a continuous version of the theorems for MS maps
is true, but no one has written a proof;* so, in the sequel we introduce, without proofs, the basic properties (openness and A-stability) of Morse–Smale
C^k-semiflows, that is, Morse–Smale semigroups of C^k-transformations acting
on a Banach manifold (for a complete presentation see Oliva [154]), as well as
we mention an example of a MS semiflow with an arbitrary given number of
periodic orbits, corresponding to a retarded functional differential equation
on the circle (see [154]).

To be more precise we start by recalling, now, the concept of semiflows
of C^k-transformations (C^k-semiflows). Let B be a closed smooth embedded
Banach submanifold of a Banach space $\mathcal{E}$. We will work with **semiflows of
C^k-transformations** of B, $k \geq 1$, that is, each transformation is a map that
leaves B invariant and belongs to $C^k(B, \mathcal{E})$, the Banach space of all $\mathcal{E}$-valued
C^k maps defined on B which are bounded together with their derivatives up
to the order k. The semiflows we will consider depend on parameters $f \in \mathcal{F}$,
$\mathcal{F}$ being a suitable Banach manifold. More precisely, let us recall the two next
definitions (see, for instance, Ruelle [176] and Fichmann and Oliva [56]):

Definition 6.2.1. *A C^k-semiflow, $k \geq 1$, defined by $f \in \mathcal{F}$, is a map
$T_f \colon \mathbb{R}_+ \times B \to B$ given by $T_f(t,x) := T(f,t,x)$ where $T \colon \mathcal{F} \times \mathbb{R}_+ \times B \to B$
is continuous and $\mathbb{R}_+ := \{t \in \mathbb{R} | t \geq 0\}$. If, moreover, $T_f(t) \colon B \to B$ is the
transformation $T_f(t)x := T(t, f, x)$, we assume to have:*

1. *$T_f(0)x = x$ and $T_f(t_1 + t_2) = T_f(t_1) \circ T_f(t_2)$ for all $f \in \mathcal{F}$, $x \in B$,
   $t_1, t_2 \in \mathbb{R}_+$;*
2. *For each fixed $t \geq 0$, the map T is C^k on $\mathcal{F} \times \{t\} \times B$, so $T_f(t) \in C^k(B, B)$.*

*The C^k-semiflow defined by $f \in \mathcal{F}$ has the compact property if for some $a \geq 0$
and every $t \in [a, \infty)$ the map $T_f(t) \colon B \to B$ is compact. If there exists $a > 0$
such that $T = T(f, t, x)$ is C^1 for $(f, t, x) \in \mathcal{F} \times (a, \infty) \times B$, one says that it
satisfies the smooth property.*

Definition 6.2.2. *A topological subspace $\tilde{\mathcal{F}}$ of a manifold $\mathcal{F}$ of parameters
of C^k-semiflows has the property $\mathcal{K}$ if:*

i) *$A(f)$ is compact for any $f \in \tilde{\mathcal{F}}$ (recall that $A(f)$ is the set of all $x \in B$
   such that x has a global and bounded extension with respect to T_f);*
ii) *The map $f \mapsto A(f)$ is upper semi-continuous, that is, given a neighbor-
    hood $\mathcal{V}$ of $A(f)$ in B, there exists a neighborhood $\mathcal{W}(f)$ of f in $\tilde{\mathcal{F}}$ such
    that $A(g) \subset \mathcal{V}$ for all $g \in \mathcal{W}(f)$;*
iii) *There exists a dense subset $\tilde{\mathcal{A}} \subset \tilde{\mathcal{F}}$ such that $g \in \tilde{\mathcal{A}}$ implies $T_g(t)|_{A(g)}$ is
     one-to-one for all $t \in \mathbb{R}_+$.*

When $\tilde{\mathcal{A}} = \tilde{\mathcal{F}}$ one says that $\tilde{\mathcal{F}}$ has the property $\mathcal{KC}$.

We have to remark that it would be nice to give verifiable conditions for which we obtain the required one-to-oneness hypotheses assumed in the last definition 6.2.2. For the case of delayed RFDE of type $\dot{x} = f(x(t), x(t-1))$, $x \in \mathbb{R}$, with f of class C^1 and everywhere non singular partial derivative of f with respect to the second argument, the implicit function theorem gives (locally) $x(t-1)$ as function of $\dot{x}$ and $x(t)$ proving the non existence of collision points.

Conjecture (Hale): On ω-limit sets $\omega(x)$, the maps $T_f(t)$ are one-to-one. We recommend the reader to Hale and Lunel ([86]- Section 12.5 p.379–382) where one can see contributions to the subject of one-to-oneness made by Nussbaum [146], Hale and Oliva [88], Sternberg [193] and Hale and Raugel [89].

Any C^k semiflow $T_f \colon \mathbb{R}_+ \times B \to B$ induces, by differentiation, the map $DT_f \colon \mathbb{R}_+ \times TB \to TB$; that is, for $t \in \mathbb{R}_+$ and $\psi \in T_x B$, one defines

$$DT_f(t, (x, \psi)) := (T_f(t)x, (DT_f(t))(x)\psi) \in T_{T_f(t)x}B.$$

DT_f is, in fact, a C^{k-1} semiflow, called the **linearization** of T_f. We denote by $A_b(DT_f)$ the set of all $(x, \psi) \in TB$ that have global bounded backwards extension (with respect to DT_f).

Definition 6.2.3. *The subspace $\widetilde{\mathcal{F}}$ of the manifold $\mathcal{F}$ has the property $\mathcal{K}C^1$ if it has the property $\mathcal{K}C$ and, besides the fact that $T_f(t)\,|_{A(f)}$ is one-to-one for all $t \in \mathbb{R}_+$ and $f \in \widetilde{\mathcal{F}}$, we also have that $DT_f(t)\,|_{A_b DT_f}$ is one-to-one, for all $t \in \mathbb{R}_+$ and $f \in \widetilde{\mathcal{F}}$.*

So, $\widetilde{\mathcal{F}}$ has the property $\mathcal{K}C^1$ if, and only if, $\widetilde{\mathcal{F}}$ has the property $\mathcal{K}C$ and for any $x \in A(f)$ and $(x, \psi_1), (x, \psi_2) \in A_b(DT_f)$ with $\psi_1 \neq \psi_2$, one has $(DT_f(t))(x)\psi_1 \neq (DT_f(t))(x)\psi_2$, for all $t \in \mathbb{R}_+$ and $f \in \widetilde{\mathcal{F}}$. We say that an element $f \in \mathcal{F}$ has the property $\mathcal{K}C$ (resp. $\mathcal{K}C^1$) if $\{f\} \subset \mathcal{F}$ has the property $\mathcal{K}C$ (resp. $\mathcal{K}C^1$).

In the next two examples we have one-to-oneness.

Example 6.2.4. (see Nussbaum [146], Sternberg [193], Fichmann and Oliva [55],[56]). Let M be a compact analytic manifold and $\mathcal{F}$ be the set of all C^k ($k \geq 1$) retarded functional differential equations (RFDE) defined on M with the same delay. The subspace $\widetilde{\mathcal{F}} \subset \mathcal{F}$ of all analytic RFDE is such that the corresponding semiflows have the property $\mathcal{K}C$. It is easy to check that $\widetilde{\mathcal{F}}$ has also the property $\mathcal{K}C^1$ and the C^k-semiflows of its elements have the compact property.

Example 6.2.5. Mallet-Paret (see [137]) considered a C^∞ scalar delay-differential equation $x' = -f(x(t), x(t-1))$ with a negative feedback condition in the delay that defines a C^k-semiflow on $B = C([-1,0], \mathbb{R})$, for any $k \geq 1$. Let $\mathcal{F}$ be a Banach manifold formed by maps f of this kind and $\widetilde{\mathcal{F}} \subset \mathcal{F}$ be any topological subspace. It follows that $\widetilde{\mathcal{F}}$ has the property $\mathcal{K}C^1$ and the corresponding semiflows have the compact property.

From now on we only consider semiflows satisfying the *smooth property*. We remark that the smooth property is not satisfied for general asymptotically smooth semigroups. In particular, semigroups generated by neutral FDE, damped hyperbolic equations and the beam equation are not included. It would be very desirable to have an extension of the results below to cover this case.

Hyperbolic fixed points and hyperbolic periodic orbits of a C^k-semiflow play an important role in the sequel. They have local stable and local unstable manifolds that persist under small perturbations of f (see [176], [72], [86], [56]). If $x \in B$ denotes a hyperbolic fixed point and $\gamma \subset B$ a hyperbolic periodic orbit of a C^k-semiflow T_f, f with the property $\mathcal{K}$, or, at least, $A(f)$ compact, one assumes, as in the previous section, that the corresponding local unstable manifolds $W^u_{loc}(x)$ and $W^u_{loc}(\gamma)$ have finite dimensions while the local stable manifolds $W^s_{loc}(x)$ and $W^s_{loc}(\gamma)$ have finite codimensions. One denotes by α a general critical element of T_f (we set $\alpha = \{x\}$ if x is a fixed point and $\alpha = \gamma$ when γ is a periodic orbit). So, if α is hyperbolic, one can talk about $W^u_{loc}(\alpha) = \bigcup_{p \in \alpha} W^u_{loc}(p)$ and $W^s_{loc}(\alpha) = \bigcup_{p \in \alpha} W^s_{loc}(p)$ (see [176, p. 91(c)], [163, p. 158]), where $W^u_{loc}(p)$ and $W^s_{loc}(p)$ are the local unstable and stable manifolds of $p \in \alpha$. The flow map $T_f(t)$ defines the invariant sets

$$W^u(\alpha) := \bigcup_{t \geq 0} T_f(t) W^u_{loc}(\alpha)$$

and

$$W^s(\alpha) \cap A(f) := A(f) \cap (\bigcup_{t \geq 0} (T_f(t))^{-1}(W^s_{loc}(\alpha))),$$

where

$$W^s(\alpha) := \bigcup_{t \geq 0} (T_f(t))^{-1}(W^s_{loc}(\alpha)).$$

In order to have a manifold structure for the invariant set $W^u(\alpha)$, the injectivity of $T_f(t)|_{A(f)}$ is not enough (see Example 3.2.18). However we have the following result:

Proposition 6.2.6. *Let $\mathcal{F}$ be a manifold of parameters of C^k semiflows and let us assume the existence of a topological subspace $\widetilde{\mathcal{F}} \subset \mathcal{F}$ that has the property $\mathcal{KC}^1$. Then, for any given $f \in \widetilde{\mathcal{F}}$ and α a hyperbolic critical element of the semiflow $T_f(t): B \to B$, the invariant set $W^u(\alpha)$ is a finite dimensional injectively immersed C^k submanifold of B.*

The manifold $W^u(\alpha)$, introduced under the hypotheses of the last proposition, is called the *global unstable manifold* of the hyperbolic element α.

The next fundamental statement for semiflows corresponds to Proposition 6.1.8 proved for maps. It is related with persistence of hyperbolic critical elements of a semiflow T_f under small perturbations of f.

More precisely, let $\mathcal{F}$ be a manifold of parameters of C^k semiflows and $\widetilde{\mathcal{F}} \subset \mathcal{F}$ be a subspace with the property $\mathcal{KC}$. From the results on persistence of hyperbolicity that appear in [176] (see also [56, section 6]), we can state the following

Proposition 6.2.7. *Let $f \in \widetilde{\mathcal{F}}$ and α be a hyperbolic critical element of T_f. Then, there exist a neighborhood $\mathcal{O}$ of α in B and a neighborhood $\mathcal{V}(f)$ of f in $\widetilde{\mathcal{F}}$ such that given $f' \in \mathcal{V}(f)$ there exists a unique homeomorphism $h = h(f')\colon \alpha \to h(\alpha) = \alpha' \subset \mathcal{O}$, close to the inclusion $i\colon \alpha \to B$ in the C^0-topology, and $\alpha' = h(\alpha)$ is a $T_{f'}$-invariant hyperbolic critical element of $T_{f'}$. Moreover, the correspondence $f' \in \mathcal{V}(f) \mapsto h(f')$ is continuous and the local stable and unstable manifolds $W^s_{loc}(\alpha')$ and $W^u_{loc}(\alpha')$ depend continuously on $f' \in \mathcal{V}(f)$ and so $dimW^u_{loc}(\alpha') = dimW^u_{loc}(\alpha)$ for all $f' \in \mathcal{V}(f)$. In particular, if α is a periodic orbit, $\mathcal{O}$ can be chosen endowed with the bundle structure $(\mathcal{O}, \alpha, p)$ of a tubular neighborhood of α with projection $p\colon \mathcal{O} \to \alpha$, $h = h(f')\colon \alpha \to \mathcal{O}$ being a continuous cross section of this bundle(i.e. $p \circ h = i$).*

Remark 6.2.8. It follows from above that when α is a periodic orbit, the map $h = h(f')$ is a homeomorphism from α onto α', for each $f' \in \mathcal{V}(f)$, preserving the sense of time (i.e. h is a α-*topological equivalence*).

We introduce now the definition of *Morse–Smale C^k-semiflows*

Definition 6.2.9. *Let $\mathcal{F}$ be a manifold of parameters of C^k semiflows and $\widetilde{\mathcal{F}} \subset \mathcal{F}$ a topological subspace with the property $\mathcal{KC}^1$. We say that $f \in \widetilde{\mathcal{F}}$ is* Morse–Smale *or that f defines a Morse–Smale semiflow T_f if:*

i) *T_f has a finite number of critical elements, all hyperbolic, and their union coincides with the non-wandering set $\Omega(f)$;*
ii) *if α, α' are critical elements, then the global unstable manifold $W^u(\alpha)$ and the local stable manifold $W^s_{loc}(\alpha')$ either do not meet or they intersect transversaly i.e $(x \in W^u(\alpha) \cap W^s_{loc}(\alpha')$ implies $T_xW^u(\alpha) + T_xW^s_{loc}(\alpha') = T_xB)$.*

The choices of the class $\widetilde{\mathcal{F}} \subset \mathcal{F}$ with property $\mathcal{KC}^1$ depend on the problems we are considering; for each such a choice we introduce the corresponding subset $MS \subset \widetilde{\mathcal{F}}$ of all elements $f \in \widetilde{\mathcal{F}}$ such that T_f is a Morse–Smale semiflow (for a series of examples of gradient-like Morse–Smale semiflows, see [100]). The main result of this section says that *"MS is open in $\widetilde{\mathcal{F}}$ and any $f \in MS$ is A-stable"*. A-stable, here, means structural stability with respect to the set $A(f)$, that is, there exists a neighborhood $\mathcal{V}(f)$ of f in $\widetilde{\mathcal{F}}$ such that to each $g \in \mathcal{V}(f)$ there corresponds a homeomorphism $h = h(g)\colon A(f) \to A(g)$ taking orbits in $A(f)$ onto orbits in $A(g)$ preserving the sense of time.

As for maps, the λ-*lemma* plays an important role in the proofs of this section (see Proposition 6.1.6).

Along the proof of the openness and stability of Morse–Smale semiflows we will deal, often, with two hyperbolic critical elements α and β of a C^k

semiflow $T_f \colon \mathbb{R}_+ \times B \to B$ defined by f, with the property $\mathcal{KC}^1$ and, correspondingly, are well defined the local stable manifold $W_{loc}^s(\beta)$ and the global unstable manifold $W^u(\alpha)$; by Proposition 6.2.6, $W^u(\alpha)$ is a finite dimensional injectively immersed C^k-submanifold of B.

Proposition 6.2.10. - *global λ-lemma (see [160, lemma 3.1]). Let α be a hyperbolic critical element of a C^1 semiflow $T_f \colon \mathbb{R}_+ \times B \to B$, f with the property $\mathcal{KC}^1$. Let N be an injectively immersed T_f-invariant submanifold of B having a point q of transversal intersection with the local stable manifold $W_{loc}^s(\alpha)$. If $\alpha = \{x\}$, so x is a fixed point, let B^u be an embedded open disc in $W_{loc}^u(x)$ centered at x; if, otherwise, α is a periodic orbit and $S_{\bar{q}}$ is a local cross section of α at $\bar{q} \in \alpha$, let B^u be a disc in $S_{\bar{q}} \cap W_{loc}^u(\alpha)$ centered at the point $\bar{q}$. Then, given $\epsilon > 0$, there exists a submanifold of N ϵ-close to B^u in the C^1-sense.*

Proof: If $\alpha = \{x\}$ we consider the C^1 map $g = T_f(1) \colon B \to B$ for which x is hyperbolic and apply λ-lemma taking $U = X = B$ and $D^u \subset N$. If α is a periodic orbit, we consider the Poincaré map $g \colon U \subset S_{\bar{q}} \to S_{\bar{q}}$ for which $\bar{q} \in \alpha$ is a hyperbolic fixed point, U being an open neighborhood of $\bar{q}$ in $S_{\bar{q}}$; in order to apply the λ-lemma to the Poincare' map g, we observe that $q \in N \cap W_{loc}^s(\alpha)$ and that N is invariant under T_f, so N has a point of transversal intersection with $W_{loc}^s(\bar{q}) = S_{\bar{q}} \cap W_{loc}^s(\alpha)$ in U and the result follows. ∎

We remark that the smoothness property was used in the proof in a significant way.

Let us state now the following proposition (for a proof see [154]):

Proposition 6.2.11. *Let $f \in MS$, then one has:*

1) *for any hyperbolic critical element α of f, the injectively immersed submanifold $W^u(\alpha)$ is, in fact, embedded in B. Also, for any $t \in \mathbb{R}$, $T_f(t)|_{W^u(\alpha)}$ is a C^k diffeomorphism of $W^u(\alpha)$;*

2) *the set of all critical elements has a partial order structure defined in the following way: if α_1, α_2 are hyperbolic critical elements we set $\alpha_1 \leq \alpha_2$ if, and only if, $W^u(\alpha_2) \cap W_{loc}^s(\alpha_1) \neq \emptyset$.*

Remark 6.2.12. The transitivity property of the partial order above shows also that if f is a Morse–Smale semiflow, $\Omega(f)$ does not have *cycles*. In fact, if we have a (finite) sequence of $n \geq 3$ hyperbolic critical elements in $\Omega(f)$ satisfying

$$\alpha_1 \leq \alpha_2 \leq \alpha_3 \leq \cdots \leq \alpha_n \leq \alpha_1,$$

that means, a cycle, one obtains that the set $W^u(\alpha_1) \cap W_{loc}^s(\alpha_1)$ contains (and is not equal to) α_1; so, we get a contradiction.

Let α be a hyperbolic critical element of a C^1 semiflow $T_f \colon \mathbb{R}_+ \times B \to B$, f with the property $\mathcal{KC}^1$. One says that α is a **source** if $W_{loc}^s(\alpha) \cap A(f) = \alpha$, is a **sink** if $W_{loc}^u(\alpha) \cap A(f) = \alpha$ and is a **saddle** otherwise.

Proposition 6.2.13. *Let* $f \in MS$. *Then, between the (hyperbolic) critical elements in* $\Omega(f)$, *at least one is a sink and at least one is a source. Moreover, the union of all unstable manifolds of all critical elements is equal to* $A(f)$.

We may have that $\Omega(f)$ does contain just one sink and just one source, both equal to the same hyperbolic critical element α. In this case $A(f) = \alpha$.

Definition 6.2.14. *The set* $Crit(f)$ *of all critical elements of* $f \in MS$ *with the partial order defined in Proposition 6.2.11 is called the* **phase diagram** *of the semiflow given by* f. *We set* $D(f) := (Crit(f); \leq)$.

Theorem 6.2.15. *The set* $MS \subset \widetilde{\mathcal{F}}$ *is open in* $\widetilde{\mathcal{F}}$. *Moreover, if* $f \in MS$, *then its phase diagram is stable (up to diagram isomorphism) under small perturbations of* f *in* $\widetilde{\mathcal{F}}$.

Definition 6.2.16. *A Morse–Smale* $f \in \widetilde{\mathcal{F}}$ *is A-stable if there exists a neighborhood* $\mathcal{V}(f)$ *of* f *in* $\widetilde{\mathcal{F}}$ *(so in* MS*) such that to each* $f' \in \mathcal{V}(f)$ *one can find a homeomorphism* $h = h(f')\colon A(f) \to A(f')$ *such that* h *takes orbits of* f *in* $A(f)$ *onto orbits of* f' *in* $A(f')$ *preserving the sense of time. We use to say that* h *is topological equivalence between* $A(f)$ *and* $A(f')$ *or a topological A-equivalence.*

The main statement in this section (for a proof see [154]) is the following

Theorem 6.2.17. *Any Morse–Smale* $f \in \widetilde{\mathcal{F}}$ *is A-stable.*

Corollary 6.2.18. *([160], [161]) Let* M *be a compact manifold and* $\mathcal{X}^k(M)$ *be the set of all* C^k- *vector fields,* $k \geq 1$. *Then the* C^k-*Morse–Smale vector fields are structurally stable and form an open set in* $\mathcal{X}^k(M)$.

Proof: We start by remarking that $\widetilde{\mathcal{F}} = \mathcal{F} = \mathcal{X}^k(M)$ satisfies the conditions in Definition 6.2.9 (with $B = M$) since $\widetilde{\mathcal{F}}$ has the property $\mathcal{KC}^1$. In fact $A(f) = M$ for all $f \in \mathcal{F}$ and the flow $T_f(t)$ verifies Definition 6.2.2. So, the result follows from Theorem 6.2.17. ∎

Corollary 6.2.19. *([154]) Let* $\widetilde{\mathcal{F}}$ *be the set of all analytic RFDE (see Example 6.2.4) defined on a compact analytic manifold* M. *The Morse–Smale RFDE in* $\widetilde{\mathcal{F}}$ *are stable relatively to* $A(f)$, $f \in \widetilde{\mathcal{F}}$, *and form an open set in* $\widetilde{\mathcal{F}}$.

6.3 An example

We will consider an RFDE on the circle depending on a parameter $M > 0$ with the property that, for any odd n_0, there is an $M(n_0) > 0$ such that its nonwandering set is the union of n_0 hyperbolic periodic orbits. Moreover, we intend to show that the RFDE is described by a Morse–Smale semiflow.

Start with a RFDE on $\mathbb{R}$, by considering a 2π-periodic C^∞ function $h: \mathbb{R} \to \mathbb{R}$, and the equation

$$\dot{\theta} = Mh(\theta(t) - \theta(t-1)), h(0) \neq 0, \qquad (6.1)$$

that is, a RFDE given by the map

$$\varphi \in C^0(I, \mathbb{R}) \mapsto Mh(\varphi(0) - \varphi(-1)) \in \mathbb{R} \sim T_{\varphi(0)}\mathbb{R}, \qquad (6.2)$$

where $I=[-1,0]$; here $\theta(t) = \theta(\varphi; t)$ means the solution of (6.1) determined by the initial condition $\varphi \in C^0(I, \mathbb{R})$ defined for $t \geq -1$ such that $\theta(\tau)=\varphi(\tau)$ for all $\tau \in I$.

To each global solution $\theta=\theta(t)$ of (6.1) one defines the function $u : \mathbb{R} \to \mathbb{R}$ by

$$u = u(t) := \theta(t) - \theta(t-1). \qquad (6.3)$$

It is easy to see that $u(t)$ satisfies the auxiliary RFDE on $\mathbb{R}$:

$$\dot{u}(t) = Mh(u(t)) - Mh(u(t-1)). \qquad (6.4)$$

If $g : \mathbb{R} \to \mathbb{R}$ is the function $g = Mh$, we see that equation (6.4) is included in a class of RFDE on $\mathbb{R}$ of the type

$$\dot{u} = g(u(t)) - g(u(t-1)). \qquad (6.5)$$

Cooke and Yorke [36] considered equations of type (6.5) with g of class C^1. They proved that any (not necessarily global) solution $u(t)$ of (6.5) defined by an initial condition $\sigma \in C(I, \mathbb{R})$ tends to a limit (finite or not) as $t \to \infty$. Since $u(\tau) = \sigma(\tau)$ for any $\tau \in I$ we have that

$$u(t) = \sigma(0) - \int_{-1}^{0} g(\sigma(\tau))d\tau + \int_{t-1}^{t} g(u(s))ds \qquad (6.6)$$

and, if g is bounded, the last equation (6.6) shows that any solution of (6.5) is also bounded and tends to a constant $c \in \mathbb{R}$, as $t \to \infty$. We also observe that any constant function $u(t) = c$ is a solution of (6.4). Another remark is that the variational equation of (6.4) associated to a global solution $u = u(t)$ is

$$\dot{y} = Mh'(u(t))y(t) - Mh'(u(t-1))y(t-1). \qquad (6.7)$$

Also, the variational equation of (6.1) associated to a global solution $\theta = \theta(t)$ is

$$\dot{Y} = Mh'(\theta(t) - \theta(t-1))Y(t) - Mh'(\theta(t) - \theta(t-1))Y(t-1). \qquad (6.8)$$

The covering map of the circle:

$$p : \theta \in \mathbb{R} \mapsto e^{i\theta} \in S^1 \subset \mathbb{R}^2,$$

$i = (-1)^{1/2}$ and the 2π- periodicity of h show that equation (6.1) induces a RFDE on the compact manifold S^1. In fact, to each $\psi \in C^0(I, S^1)$ one considers $\psi(0) \in S^1$ and let $\varphi(0) \in \mathbb{R}$ be any real number (determined up to 2π) such that $e^{i\varphi(0)} = \psi(0)$; call $\varphi \in C^0(I, \mathbb{R})$ be the (unique) continuous lifting of ψ such that $e^{i\varphi(\tau)} = \psi(\tau)$ for all $\tau \in I$. The RFDE on S^1 is then defined by the C^∞ map

$$f : \psi \in C^0(I, S^1) \mapsto f(\psi) = i\psi(0)Mh(\varphi(0) - \varphi(-1)) \in T_{\psi(0)}S^1. \qquad (6.9)$$

Observe that, as a complex number, the vector $i\psi(0)$ is tangent to S^1 at the point $\psi(0)$ and the real number $h(\varphi(0) - \varphi(-1))$ is well determined because the choice of $\varphi(0)$ was given up to 2π. For later purposes, one remarks, also, that f is bounded in $C^0(I, S^1)$, together with its Fréchet derivative.

Let us describe some results concerning equation (6.1) and the RFDE defined by the C^∞ map f in (6.9):

i) $h(0) \neq 0$ implies that equation (6.1) does not have constant solutions.

ii) Any linear function $\theta(t) = \omega t + \alpha$ (so with constant velocity $\dot{\theta}(t) = \omega$) is a solution of (6.1) if, and only if, ω satisfies

$$\omega \doteq Mh(\omega). \qquad (6.10)$$

Moreover, it is easy to see that, choosing M properly, equation (6.10) admits, in the analytic case, a finite (but arbitrary) number of roots. In any case, for each ω satisfying (6.10), there corresponds a periodic solution of the RFDE defined by (6.9) with minimum period $T = 2\pi(\omega)^{-1}$. This last assertion follows from the fact that

$$\theta(t + 2\pi(\omega)^{-1}) = \theta(t) + 2\pi,$$

for all $t \in \mathbb{R}$.

iii) Any global solution $\theta(t)$ of (6.1) that defines a T-periodic orbit of the RFDE corresponding to f given by (6.9), has constant velocity. In fact, the existence of such a T-periodic closed orbit defined by $\theta(t)$ implies that

$$\theta(t + T) = \theta(t) + 2k\pi, \ k \in Z \ \forall t \in \mathbb{R};$$

from (6.3) we have that $u(t) = \theta(t) - \theta(t - 1)$ and

$$u(t + T) = \theta(t + T) - \theta(t + T - 1) \ \forall t \in \mathbb{R},$$

so

$$u(t + T) = \theta(t) + 2k\pi - \theta(t - 1) - 2k\pi = u(t)$$

and then $u(t)$ is a T-periodic solution of (6.4). But the existence of a limit as $t \to +\infty$ for a periodic function implies that $u(t)$ is necessarily a constant function, say A, that is, $u(t) = A = \theta(t) - \theta(t - 1)$; the general solution of this difference equation is the sum of a particular solution (say

$At)$ with a 1-periodic function $\bar{\theta}(t)$, that is $\theta(\bar{t}) = \bar{\theta}(t-1)$ for all $t \in \mathbb{R}$. But the sum $\theta(t) = At + \bar{\theta}(t)$ has to satisfy equation (6.1), then

$$\dot{\theta}(t) = A + \dot{\bar{\theta}}(t) = Mh(At + \bar{\theta}(t) - A(t-1) - \bar{\theta}(t-1))$$

so $\dot{\theta}(t) = Mh(A)$ and then $\theta(t)$ has constant velocity.

iv) If h is real analytic, the flow map at the time one is injective on $A(f)$ and so the non-wandering set $\Omega(f)$ of the RFDE (6.9) is an invariant set.

In fact, even in the case in which h is only C^∞ and $\omega = Mh(\omega)$ implies $h'(\omega) \neq 0$, the nonwandering set $\Omega(f)$ is invariant and equal to the union of all periodic orbits of the RFDE (6.9). This is a non trivial fact and a complete proof appears in [154]:

Proposition 6.3.1. *Let $M > 0$ be such that the roots of $\omega = Mh(\omega)$ satisfy $h'(\omega) \neq 0$. Then the nonwandering set $\Omega(f)$ of the RFDE (6.9) is invariant and equal to the union of all periodic orbits of f.*

We remark that in this last Proposition 6.3.1 the hypothesis saying that the roots of $\omega = Mh(\omega)$ satisfy $h'(\omega) \neq 0$, is fundamental in the C^∞ category, because it constitutes a sufficient condition for the non existence of collision points on global orbits. On the other hand, since in the analytic case there are no such collision points, it follows, from above, the following:

Proposition 6.3.2. *In the analytic case the non-wandering set $\Omega(f)$ is invariant and equal to the finite union of all periodic orbits of f.*

We state now the main result of this section:

Theorem 6.3.3. *Given an arbitrary odd integer n_0, there exists an interval of positive values for the constant M such that if equation (6.9) is analytic, it defines an RFDE on the circle and its non-wandering set is formed by n_0 hyperbolic periodic orbits and no critical points. Moreover, the corresponding semiflow is Morse–Smale.*

Proof: A complete proof is in [154] and here we mention only the main steps. In fact one needs to check transversality between unstable and local stable manifolds of the hyperbolic periodic orbits (for the discussion on the hyperbolicity of the orbits we also refer the reader to [154]); in fact, it will be enough to show that there are no saddle connections. The way to do this is to show that if $\theta(t)$ corresponds to a saddle connection between two saddle orbits of constant velocities ω_{r_1} and ω_{r_2}, the existence of the associated solution $\bar{u}(t) := \theta(t) - \theta(t-1)$ of equation (6.4) will lead to a contradiction. First of all we recall that the hyperbolic periodic orbits of velocities ω_{r_1} and ω_{r_2} correspond to fixed points $\bar{c}_1 = \omega_{r_1}$ and $\bar{c}_2 = \omega_{r_2}$ of equation (6.4) and the trajectory of the solution $\bar{u}(t)$ belongs, simultaneously, to the unstable manifold of $\bar{c}_1 = \omega_{r_1}$ and to the local stable manifold of $\bar{c}_2 = \omega_{r_2}$. The contradiction appears after an analysis of the changes of signs of the non zero solutions $y = y(t)$ of the variational equation of (6.4) associated to the global solution $\bar{u}(t)$ (see (6.7)):

$$\dot{y} = Mh'(\bar{u}(t))y(t) - Mh'(\bar{u}(t-1))y(t-1). \tag{6.11}$$

Remark that $\lim_{t \to -\infty} \bar{u}(t) = \bar{c}_1$ and $\lim_{t \to \infty} \bar{u}(t) = \bar{c}_2$, and that (6.10) implies $\bar{c}_i = \omega_{r_{1i}} = Mh(\omega_{r_i})$, $i = 1, 2$. From this and the fact that $h'(\omega_{r_i}) \neq 0$, $i = 1, 2$, we observe, by continuity, that there are numbers $a, b \in \mathbb{R}$ such that $h'(\bar{u}(t-1)) \neq 0$ for all $t \in (-\infty, a] \cup [b, +\infty)$.

We are interested on the set Ψ of all global solutions y of (6.11) that are bounded on $(-\infty, 0]$. By a theorem of Nussbaum (see [146]) these solutions y are real analytic as well as the solution $\bar{u}(t)$ of the non linear equation (6.4). One defines the translation flow $\Psi \times \mathbb{R} \to \Psi$ denoted $y.t$, by setting

$$(y.t)(\xi) = y(t + \xi), \ \xi \in \mathbb{R}. \tag{6.12}$$

Also, let $V : \Psi - \{0\} \to \{1, 3, 5, \dots\}$ be a discrete functional which measures the number of zeros of $y \in \Psi$, $y \neq 0$, in a unit interval $(\sigma - 1, \sigma]$ when $y(\sigma) = 0$. For this we consider $\sigma := \inf\{s \geq 0 | y(s) = 0\}$ and define $V(y)$ as the number of zeros (counting multiplicity) of $y(t)$ in $(\sigma - 1, \sigma]$ or equal to 1 if σ does not exist. One can show that if $y \in \Psi - \{0\}$ (so $V(y)$ is an odd integer) then $V(y.t)$ is a non-increasing function of $t \in (-\infty, a] \cup [b, +\infty)$. The linearization of equation (6.4) at the critical points $\bar{c}_i = \omega_{r_i} = \lim_{t \to \infty} \bar{u}(t)$ and the analysis of the corresponding characteristic equation imply, assuming the existence of a saddle connection, that equation (6.4) in a neighborhood of the (non hyperbolic) critical point $\bar{c}_2$ presents an invariant C^k one-dimensional unstable manifold corresponding to the unique positive root and an invariant C^k center stable manifold defined by the remaining roots; the theorem for the existence of the center stable manifold (see [176] p.32), depends on the nature of these roots and on the fact that we have, in this case, the C^k extension property that is the possibility of using the cut-off technique (see [157]). But, since the root $\lambda = 0$ is simple and the center manifold is formed by fixed points of the time one (compact) flow map, one can show that the center stable manifold of $\bar{c}_2$ has, locally, a C^k fibration $\pi : U \to B$ where U is an open neighborhood of $\bar{c}_2$ in the center stable manifold and B is an open ball centered in $\bar{c}_2$ and contained in the center manifold, such that for each $x \in B$, $\pi^{-1}(x)$ is the local stable manifold of x. The connecting orbit of $\bar{u}(t)$ tends to $\bar{c}_2$ passing through its local stable manifold so the solutions $y(t)$ of the linearized equation will present infinitely many oscillations because all the roots corresponding to that local stable manifold are complex conjugate (see [98]). A contradiction appears from the fact that these solutions $y(t)$ cannot eventually have infinitely many oscillations because the number of changes of sign is finite and does not increase when we approach $\bar{c}_2$. This avoids saddle connections and then proves transversality.   ∎

7 One-to-Oneness, Persistence, and Hyperbolicity

The persistence of a (compact) normally hyperbolic manifold and the existence of its stable and unstable manifolds, corresponding to the action of C^r maps or semiflows, $r \geq 1$, are well known facts (see, for instance, [102] and [176]). The more general case of (compact) hyperbolic sets that are invariant under (not necessarily injective) maps was also carefully studied (see Ruelle [176], Shub [189]and Palis and Takens [163]). The persistence and smoothness of (compact) hyperbolic invariant manifolds for RFDE were considered in detail in Magalhães [132] based on skew-product semiflows defined for RFDE, locally around hyperbolic invariant manifolds, and their spectral properties, associated with exponential dichotomies, following the lines of work developed by Sacker and Sell for flows([186], [187]).

In Sections 7.1 and 7.2 of this Chapter 7, we recall some preliminaries as well as the definition and examples of hyperbolic sets. In sections 7.3 and 7.4, we present (omitting a few technical proofs) the work done in [56] by Fichmann and Oliva who considered the case of semiflows in infinite dimensions studying, mainly, the persistence of a compact hyperbolic invariant set (not necessarily a manifold) without singularities, treating the possible presence of collisions.

The hyperbolicity of invariant sets we will talk about in these first four sections is, in a certain sense, uniform, and, only in Section 7.5, we deal with the so called nonuniform hyperbolicity, by discussing, in this more general kind of hyperbolicity, the notions of stable and unstable manifolds in the setting of finite dimension. Section 7.5 was written by Luis Barreira; thanks to him for the suggestive collaboration.

As we already observed, there are special and interesting phenomena that do not make sense in the setting of flows acting on finite dimensional manifolds but appear, often, in infinite dimensional systems, showing that the global theory is still in its infancy. For example, there are cases of two global trajectories of a semiflow, i.e. both defined for all $t \in \mathbb{R}$, with a collision point and, of course, after the collision they keep evolving together; that can happen even in a C^∞ system and with two C^∞ global solutions as we saw in Example 3.2.18.

Assume that a compact set, invariant under a certain flow, is hyperbolic, do not have fixed points and admits a collision point in its interior. There is persistence of hyperbolicity, so it is possible, in some cases, to create a new hyperbolic (compact) and invariant set, without collisions, and as close as we want to the compact invariant set given initially. Due to the existence of collision we do not have, in general, a topological equivalence between the two sets which is, in fact, weakened to a topological semi-equivalence.

If we are dealing with a certain class of semigroups, one can see that hyperbolicity of a compact invariant set implies hyperbolicity of a fixed point of a suitable operator, but the converse needs some additional hypotheses, as we will see later.

We already saw, in Chapter 6, Section 6.2, the concept of *semiflows of C^k-transformations of $\mathcal{B}$*, also known as *semigroups of C^k-transformations*, $\mathcal{B}$ being a closed embedded Banach submanifold of a Banach space $\mathcal{E}$. A transformation is a map (not necessarily one-to-one) that leaves $\mathcal{B}$ invariant and belongs to $C^k(\mathcal{B}, \mathcal{E})$, the Banach space of all $\mathcal{E}$-valued C^k maps defined on $\mathcal{B}$ which are bounded together with their derivatives up to the order $k \geq 1$.

The semiflows we consider depend on a parameter $f \in \mathcal{F}$, $\mathcal{F}$ being a suitable Banach manifold.

A point $x_0 \in \mathcal{B}$ has a *global extension with respect to T_f* if there exists a continuous function $u : \mathbb{R} \to \mathcal{B}$ such that $u(0) = x_0$ and $T_f(t)[u(\tau)] = u(t+\tau)$ for all $t \geq 0$ and $\tau \in \mathbb{R}$. When there is such an extension u through x_0 (not necessarily unique) it makes sense to talk about the *global orbit $\gamma_f^u(x_0)$ of x_0 corresponding to u* by setting

$$\gamma_f^u(x_0) := \bigcup_{\tau \in \mathbb{R}} u(\tau).$$

It is also easy to check that, if x_0 has a global extension u with respect to T_f, then

$$T_f(t)\gamma_f^u(x_0) \subset \gamma_f^u(x_0), \text{ for all } t \in \mathbb{R}_+.$$

If $x_0 \in \mathcal{B}$ admits two global extensions u and v with respect to T_f, we may have $T_f(\beta)x_1 = T_f(\beta)x_2 = x$ for some $\beta > 0$ and $x_1 \in \gamma_f^u(x_0)$, $x_2 \in \gamma_f^v(x_0)$, $x_1 \neq x_2$. When β is the smallest number with that property, x is called a *collision point* (see Example 3.2.18).

Recall that a set $\Lambda \subset \mathcal{B}$ is said to be T_f-*invariant* if any $x_0 \in \Lambda$ has a global extension u with respect to T_f and $\gamma_f^u(x_0) \subset \Lambda$.

Let us consider the following T_f-invariant set:

$$A(f) := \{x \in \mathcal{B} | x \text{ has a global and bounded extension with respect to } T_f\}.$$

We assume, from now on in this chapter, that $\mathcal{F}$ satisfies the *smooth property* and has the *property $\mathcal{K}$* (see Section 6.2).

7.1 The semiflow of an RFDE on a compact manifold M

Set $\mathcal{B} := C^0([-r, 0], M)$. We already stated in Section 6.2 the following results for the class $\mathcal{F}$ of all C^k RFDE on M:

Proposition 7.1.1. *Let $f \in \mathcal{F}$ be a C^k $(k \geq 1)$ RFDE on a compact C^∞ manifold M. Then $T_f : \mathbb{R}_+ \times \mathcal{B} \to \mathcal{B}$ is a semiflow of C^k-transformations that satisfies the smooth property. Moreover $A(f) \subset \mathcal{B}$ is a compact connected global attractor and $T_f(t) : \mathcal{B} \to \mathcal{B}$ is a compact C^k-map for all $t \geq r$.*

Proposition 7.1.2. *The manifold of parameters $\mathcal{F}$ is a Banach space and, when M is analytic, it has the property $\mathcal{K}$.*

7.2 Hyperbolic invariant sets

Let us come back to the general class of semiflows T_f of C^k-transformations $(k \geq 1)$ acting on $\mathcal{B}$, parametrized by $f \in \mathcal{F}$ with *property $\mathcal{K}$* (may have collisions), and satisfying the *smooth property*.

Definition 7.2.1. *A compact T_f-invariant set Λ is hyperbolic if for any $p \in \Lambda$ we have a splitting of $T_p\mathcal{B}$ in closed subspaces, that is,*

$$T_p\mathcal{B} = E_p^+ \oplus E_p^- \oplus Y_p, \quad \dim E_p^- < \infty, \quad Y_p = \frac{d}{dt}(T_f(t)p)|_{t=0};$$

the splitting varies continuously in $p \in \Lambda$, that is, the corresponding canonical projections depend continuously on $p \in \Lambda$. The subbundles E^+, E^-, Y are invariant on Λ and such that $DT_f(t)E^- = E^-$ and $DT_f(t)E^+ \subset E^+$ for all $t \in \mathbb{R}^+$ (here D means the partial derivative of T_f with respect to $x \in \mathcal{B}$). Moreover, there exist $a, c > 0$ such that for all $p \in \Lambda$:

1. *$|(DT_f(t)p)\xi| \leq a|\xi|e^{-ct}, \forall \xi \in E_p^+, \forall t \geq 0$;*
2. *there exists $[DT_f(t)p|_{E_p^-}]^{-1}$ and $|(DT_f(t)p)^{-1}\eta| \leq a|\eta|e^{-ct}, \forall \eta \in E_p^-$, $\forall t \geq 0$.*

The hypothesis $\dim E_p^- < \infty$ is a technical condition that we used strongly in the proofs of the main results.

Examples of hyperbolic sets

1. $\Lambda = \{p\}$, p a hyperbolic fixed point of T_f, $f \in \mathcal{F}$.
2. A hyperbolic periodic orbit Λ (in the sense of hyperbolicity of its Poincaré map) of T_f, $f \in \mathcal{F}$.
3. Any hyperbolic invariant set of the flow generated by a C^k-vector-field on a compact manifold M, $k \geq 1$.
4. Compact manifolds with Anosov flows (see [191], [176]).

7.3 Hyperbolic sets as hyperbolic fixed points

For a sake of completeness, we recall the following definition. If E_1, E_2 are Banach spaces and $V \subset E_1$ is an open set, we say that $\varphi : V \to E_2$ is a $C^{(k,\gamma)}$

map (or a *Hölder C^k map with Hölder constant γ*), if φ is C^k and there exists $A > 0$ such that

$$\| D_x^k \varphi - D_y^k \varphi \| \leq A \| x - y \|^\gamma, \quad for \ \ all \ \ x, y \in V.$$

Also, when $0 < \gamma < 1$, one denotes by $C^{k+\gamma}$ semiflow a $C^{(k,\gamma)}$ semiflow.

Assume (from now on in this and in the next section) that we are dealing with $C^{1+\gamma}$ semiflows for each bounded open subset of $\mathcal{B}$.

A hyperbolic fixed point is a hyperbolic set and we will see that, conversely, a hyperbolic set always corresponds to a hyperbolic fixed point of a suitable map defined on an infinite dimensional Banach manifold.

One starts with a *first lifting* (see [163] Appendix 1, p.166 and [176],p.94, where the authors considered endomorphisms) by defining $\hat{\mathcal{B}} := BC(\mathbb{R}, \mathcal{B})$ to be the set of all bounded continuous mappings from $\mathbb{R}$ to $\mathcal{B} \subset \mathcal{E}$. $\hat{\mathcal{B}}$ is a subset of $BC(\mathbb{R}, \mathcal{E})$, the space of all bounded continuous maps from $\mathbb{R}$ to $\mathcal{E}$ endowed with the topology of the uniform convergence on compact parts of $\mathbb{R}$. Let $\pi : \hat{\mathcal{B}} \to \mathcal{B}$ be the C^0-projection $\pi(u) = u(0)$, and define the lifting

$$\hat{T}_f(t) : \hat{\mathcal{B}} \to \hat{\mathcal{B}}$$

by the equality

$$\hat{T}_f(t)u := T_f(t) \circ u, \quad \text{for all } t \geq 0 \text{ and } u \in \hat{\mathcal{B}}.$$

Given a compact T_f-invariant set $\Lambda \subset \mathcal{B}$ (so $\Lambda \subset A(f)$), one defines:

$$\hat{\Lambda} = \{u \in \hat{\mathcal{B}} | u \text{ is a global extension in } \Lambda \text{ with respect to } T_f \text{ of } u(0) \in \Lambda\}.$$

Then $[\hat{T}_f(t)u](\tau) = [T_f(t)]u(\tau) = u(t + \tau)$ belongs to Λ for all $t \geq 0$ and $\tau \in \mathbb{R}$. Also $\hat{T}_f(t)\hat{\Lambda} = \hat{\Lambda}$. The existence of collision points in Λ means that the projection $\pi|_{\hat{\Lambda}}$ is not one-to-one.

Proposition 7.3.1. $\hat{\Lambda}$ *is a compact $\hat{T}_f$-invariant subset of $\hat{\mathcal{B}}$.*

Proof: The fact that $T_f(t, x)$ is C^1 for $(t, x) \in (r, \infty) \times \mathcal{B}$, $r > 0$, and that $[T_f(t)]u(\tau) = u(t + \tau)$, $t \geq 0$, $\tau \in \mathbb{R}$, $u \in \hat{\Lambda}$, imply that u is C^1 on $\mathbb{R}$; moreover the derivative $(T_f(t)p)'_{t=0}$ is continuous in $p \in \Lambda$. Then $\hat{\Lambda}$ is a family of equicontinuous functions defined on the (separable) space $\mathbb{R}$ with values in $\Lambda \subset \mathcal{B}$; that follows from the uniform bound of $|(T_f(t)p)'_{t=0}|$ on the compact set Λ. Also $\{u(\tau) | u \in \hat{\Lambda}\}$ has compact closure. So we apply Arzelà's theorem to show that $\hat{\Lambda}$ is compact (see [173]). ∎

Corollary 7.3.2. $\widehat{A(f)}$ *is a compact $\hat{T}_f$-invariant subset of $\hat{\mathcal{B}}$.*

Proof: The space $\mathcal{F}$ has the property $\mathcal{K}$ so $A(f)$ is compact and invariant. ∎

Corollary 7.3.3. *The maps $\hat{T}_f(t)$, $t \geq 0$, are one-to-one on $\widehat{A(f)}$. Then, after considering the inverses of $\hat{T}_f(t)$, we can define on $\widehat{A(f)}$ a one parameter group ($t \in \mathbb{R}$) of C^0-transformations. Moreover, equality $\pi \circ \hat{T}_f(t) = T_f(t) \circ \pi$ holds on $\widehat{A(f)}$, for each $t \geq 0$.*

Proceed now with a *second lifting*. Assume that Λ *is a compact T_f-invariant set without fixed points*, that is, for all $p \in \Lambda$ one has $\dim Y_p = \dim[\frac{d}{dt}(T_f(t)p)|_{t=0}] = 1$. Let us consider the closed submanifold $\mathcal{B}_{\hat{\Lambda}} = C(\hat{\Lambda}, \mathcal{B})$ of $\mathcal{E}_{\hat{\Lambda}} = C(\hat{\Lambda}, \mathcal{E})$, the Banach space of all continuous functions from $\hat{\Lambda}$ to $\mathcal{E}$, and also the closed submanifold $\mathcal{B}_{\hat{\Lambda}}^0 = \Gamma^0(\hat{\Lambda}, \mathcal{B})$, of $\mathcal{E}_{\hat{\Lambda}}^0 = \Gamma^0(\hat{\Lambda}, \mathcal{E})$, the Banach space of all bounded functions from $\hat{\Lambda}$ to $\mathcal{E}$ (both Banach spaces endowed with the sup norm). We have $\mathcal{B}_{\hat{\Lambda}} \subset \mathcal{B}_{\hat{\Lambda}}^0$ and $\mathcal{E}_{\hat{\Lambda}} \subset \mathcal{E}_{\hat{\Lambda}}^0$.

Set

$$T_{f,f'}(t)g := T_{f'}(t) \circ g \circ \hat{T}_f(-t)|_{\hat{\Lambda}}, \quad t \geq 0, \quad f' \in \mathcal{F}, g \in \mathcal{B}_{\hat{\Lambda}}^0.$$

The maps $T_{f,f'}(t)$, $t \geq 0$, define a semigroup of transformations acting on $\mathcal{B}_{\hat{\Lambda}}^0$. We see that $\mathcal{B}_{\hat{\Lambda}}$ is positively invariant under this semigroup. The $C^{1+\gamma}$ hypothesis on T_f assumed above implies that this last semigroup is in fact C^1.

Let us denote, from now on, $\pi|_{\hat{\Lambda}}$ as π; note that $\pi \in \mathcal{B}_{\hat{\Lambda}}$ and is a fixed point of $T_{f,f}(t)$, $t \geq 0$. Take a continuous decomposition of $T_\Lambda \mathcal{B} = \bigcup_{p \in \Lambda}(p, T_p\mathcal{B})$ in closed subspaces:

$$T_p\mathcal{B} = S_p + Y_p, \quad p \in \Lambda, \quad Y_p = [(T_f(t)p)'_{t=0}]$$

(if Λ is hyperbolic one chooses $S_p = E_p^+ \oplus E_p^-$). We have $T_\Lambda \mathcal{B} = S_\Lambda \oplus Y_\Lambda$ (S_Λ and Y_Λ are defined in the same way as $T_\Lambda \mathcal{B}$). For each $p \in \Lambda$ define, using an exponential map (see [121]), a manifold $\mathcal{S}_p$, transversal to the flow at p, such that $\mathcal{S}_p \cap \mathcal{S}_{p'} = \emptyset$ if $p' = T_f(t)p$ for $t \in [0, 8]$, the tangent space to $\mathcal{S}_p$ at the point p is S_p and also for each $u \in \hat{\Lambda}$, there exists a neighborhood O_u of u in $\hat{\mathcal{B}}$ and $\varepsilon_u > 0$ such that

$$u' \in O_u \Rightarrow B_{u'(0)}(\varepsilon_u) \subset \bigcup_{t \in [-4,4]} \mathcal{S}_{u'(t)}$$

(here $B_p(\varepsilon)$ means the ball in $\mathcal{B}$ centered at $p \in \Lambda$ with radius $\varepsilon > 0$). Let

$$\tilde{C} = \{g \in \mathcal{B}_{\hat{\Lambda}} \mid g(u) \in \mathcal{S}_{u(0)}, \, \forall u \in \hat{\Lambda}\} \text{ and}$$
$$\tilde{\Gamma} = \{g \in \mathcal{B}_{\hat{\Lambda}}^0 \mid g(u) \in \mathcal{S}_{u(0)}, \, \forall u \in \hat{\Lambda}\};$$

$\pi \in \tilde{C} \subset \tilde{\Gamma}$ and they are closed submanifolds of $\mathcal{B}_{\hat{\Lambda}}$ and $\mathcal{B}_{\hat{\Lambda}}^0$, respectively. The tangent spaces at π of $\tilde{C}$, $\mathcal{B}_{\hat{\Lambda}}$, $\tilde{\Gamma}$ and $\mathcal{B}_{\hat{\Lambda}}^0$ are:

$$T_\pi \tilde{C} = \{\eta \in C(\hat{\Lambda}, S_\Lambda) \mid \eta(u) \in S_{u(0)}, \ \forall u \in \hat{\Lambda}\},$$

$$T_\pi \mathcal{B}_{\hat{\Lambda}} = \{\eta \in C(\hat{\Lambda}, T_\Lambda \mathcal{B}) \mid \eta(u) \in T_{u(0)}\mathcal{B}, \ \forall u \in \hat{\Lambda}\},$$

$$T_\pi \tilde{\Gamma} = \{\eta \in \Gamma^0(\hat{\Lambda}, S_\Lambda) \mid \eta(u) \in S_{u(0)}, \ \forall u \in \hat{\Lambda}\},$$

$$T_\pi \mathcal{B}_{\hat{\Lambda}}^0 = \{\eta \in \Gamma^0(\hat{\Lambda}, T_\Lambda \mathcal{B}) \mid \eta(u) \in T_{u(0)}\mathcal{B}, \ \forall u \in \hat{\Lambda}\},$$

Lemma 7.3.4. *There exist a neighborhood $\mathcal{U}$ of f in $\mathcal{F}$ and a neighborhood V of π in $\tilde{\Gamma}$ such that, given $\sigma \in]0, 2]$, $f' \in \mathcal{U}$, $g \in V$ and $u \in \hat{\Lambda}$, there exists $\tilde{t} = \tilde{t}(\sigma, f', g, u) > 0$ such that*

$$T_{f'}(\tilde{t})[g(\hat{T}_f(-\sigma)u)] \in S_{u(0)} \ and \ T_{f'}(s)[g(\hat{T}_f(-\sigma)u)] \notin S_{u(0)} \ if \ 0 \le s < \tilde{t} \ .$$

Moreover, $\tilde{t}(\sigma, f, \pi, u) = \sigma$ for all $u \in \hat{\Lambda}$, $\sigma \in]0, 2]$; also, there exists $\frac{\partial \tilde{t}}{\partial g}(\sigma, f, \pi, u)$ and we have:

$$\frac{\partial \tilde{t}}{\partial g}(\sigma, f, \pi, u) = 0 \ for \ all \ u \in \hat{\Lambda} \ \Leftrightarrow \ S_\Lambda \ is \ DT_f(\sigma)\text{-}invariant, \sigma \in]0, 2].$$

Proof: (see [56])   ∎

It is possible to introduce now a *family of return maps* that turns out to be one of the main tools for the proofs of the main results of this section. These maps we denote by $\tilde{T}_{f,f'}(t)$ and are induced by $T_{f,f'}(t)$. So $\tilde{T}_{f,f'}(t) : V \subset \tilde{\Gamma} \to \tilde{\Gamma}$, $t \in [0, 2]$, $f' \in \mathcal{U}$ is the map defined by

$$\left[\tilde{T}_{f,f'}(t)g\right]u := [T_{f'}(\tilde{t}) \circ g \circ \hat{T}_f(-t)](u) \in S_{u(0)},$$

$g \in V$, $u \in \hat{\Lambda}$, where $\tilde{t} = \tilde{t}(t, f', g, u)$ is given by Lemma 7.3.4 above, when $t \in]0, 2]$ and $\tilde{t}(0, f', g, u) = 0$ for all $f' \in \mathcal{U}$, $g \in V$, $u \in \hat{\Lambda}$.

Remark 7.3.5.   1. Lemma 7.3.4 shows that the map $\tilde{T}_{f,f'}(t)$ is well defined and that $\tilde{T}_{ff'}(t)|_{V \cap \tilde{C}} : V \cap \tilde{C} \to \tilde{C}$.
2. π is a fixed point of $\tilde{T}_{ff}(t)$, $\forall t \in [0, 2]$.
3. With $\mathcal{U}$ and V small enough, $\tilde{t}$ is arbitrarily close to t.

Theorem 7.3.6. *Let $\Lambda \in \mathcal{B}$ be a compact T_f-invariant set, without fixed points. If Λ is hyperbolic, then π is a hyperbolic fixed point of $\tilde{T}_{f,f}(t)$, $t \in]0, 2]$.*

Proof: For $t \in]0, 2]$, consider the map

$$\tilde{F}_t := D\tilde{T}_{f,f}(t)[\pi] : T_\pi \tilde{\Gamma} \to T_\pi \tilde{\Gamma}$$

given by

$$(\tilde{F}_t \eta)(u) = DT_f(t)[u(-t)](\eta(\hat{T}_f(-t)u)),$$

because $\frac{\partial \tilde{t}}{\partial g}(t, f, \pi, u) = 0$, $t \in]0, 2]$. For $n \in \mathbb{N}^*$ we have

$$(\tilde{F}_t^n \eta)(u) = DT_f(nt)[u(-nt)](\eta(\hat{T}_f(-nt)u)).$$

Since Λ is hyperbolic we choose $S_\Lambda = E_\Lambda^+ \oplus E_\Lambda^-$ and

$$T_\pi \tilde{\Gamma} = \{\eta \in \Gamma^0(\hat{\Lambda}, E_\Lambda^+ \oplus E_\Lambda^-) \mid \eta(u) \in E_{u(0)}^+ \oplus E_{u(0)}^-\}$$
$$= \{\eta \in \Gamma^0(\hat{\Lambda}, E_\Lambda^+) \mid \eta(u) \in E_{u(0)}^+\} \oplus \{\eta \in \Gamma^0(\hat{\Lambda}, E_\Lambda^-) \mid \eta(u) \in E_{u(0)}^-\}$$
$$:= \mathcal{E}^+ \oplus \mathcal{E}^-.$$

For $\eta \in \mathcal{E}^+$ and $n \in \mathbb{N}^*$, one has

$$\|\tilde{F}_t^n \eta\| = \sup_{u \in \hat{\Lambda}} \|DT_f(nt)[u(-nt)](\eta(\hat{T}_f(-nt)u))\|$$
$$= \sup_{v \in \hat{\Lambda}} \|DT_f(nt)[v(0)](\eta(v))\| \leq a\|\eta\|e^{-cnt}, \text{ because } \eta(v) \in E_{v(0)}^+.$$

For $\eta \in \mathcal{E}^-$ and $n \in \mathbb{N}^*$ one can write

$$(\tilde{F}_t^n \eta)(u) = DT_f(nt)[v(0)]\eta(v), \quad \eta(v) \in E_{v(0)}^-, \quad v = \hat{T}_f(-nt)u,$$

so

$$(DT_f(nt)[v(0)])^{-1}[(\tilde{F}_t^n \eta)(u)] = \eta(v),$$

or

$$(DT_f(nt)[v(0)])^{-1}((\tilde{F}_t^n \eta)(\hat{T}_f(nt)v)) = \eta(v).$$

Then there exists

$$\left[(\tilde{F}_t^n)^{-1}(\xi)\right](v) = (DT_f(nt)[v(0)])^{-1}(\xi(\hat{T}_f(nt)v)), \quad \xi \in \mathcal{E}^-,$$

and

$$\|(\tilde{F}_t^n)^{-1}(\xi)\| = \sup_{v \in \hat{\Lambda}} \|(DT_f(nt)[v(0)])^{-1}(\xi(\hat{T}_f(nt)v))\| \leq a\|\xi\|e^{-cnt},$$

so $T_\pi \tilde{\Gamma}$ is the direct sum of the stable and unstable subspaces of $\tilde{F}_t$. ∎

Consider now the converse to Theorem 7.3.6 but, assuming that there are no collision points in Λ and, also, with an additional hypothesis that it will be described in the sequel.

Recall that we have $T_\Lambda \mathcal{B} = S_\Lambda \oplus Y_\Lambda$ with projections $\rho_p : T_p \mathcal{B} \to S_p$ and $I_p - \rho_p : T_p \mathcal{B} \to Y_p$, that vary continuously with $p \in \Lambda$ (I_p is the identity of $T_p \mathcal{B}$). Recall that we have also the transversal sections S_Λ with $T_p S_p = S_p$, the sets $\tilde{C} \subset \tilde{\Gamma}$ and the return maps $\tilde{T}_{f,f'}(t)$, $t \in [0, 2]$.

Assume that π is a hyperbolic fixed point of $\tilde{T}_{f,f}(t_0)$ for some $t_0 \in]0, 2]$ and $T_\pi \tilde{\Gamma} = \mathcal{E}^+ \oplus \mathcal{E}^-$, $\mathcal{E}^+$ and $\mathcal{E}^-$ being the closed stable and unstable subspaces of $\tilde{F}_{t_0}$. Note that

$$[\tilde{F}_{t_0}\eta](u) = DT_f(t_0)[u(-t_0)](\eta(\hat{T}_f(-t_0)u))$$

$$+ \frac{\partial T_f(t_0)}{\partial t}[u(-t_0)]\frac{\partial \tilde{t}}{\partial g}(t_0, f, \pi, u)(\eta)$$

$$= \rho_{u(0)}DT_f(t_0)[u(-t_0)](\eta(\hat{T}_f(-t_0)u)).$$

For each $u \in \hat{\Lambda}$, one considers the evaluation map $\theta_u : \tilde{\Gamma} \to S_\Lambda$ given by $\theta_u(g) = g(u) \in S_{u(0)}$, with derivative $D\theta_u(\pi) : T_\pi \tilde{\Gamma} \to S_{u(0)}$, the linear evaluation $\eta \to \eta(u)$. Since we are assuming that there are no collision points in Λ, for each $p \in \Lambda$ there exists a unique $u \in \hat{\Lambda}$ such that $u(0) = p$. Define $E_p^+ = D\theta_u[\pi](\mathcal{E}^+)$ and $E_p^- = D\theta_u[\pi](\mathcal{E}^-)$.

Lemma 7.3.7. E_p^+ and E_p^- are closed subspaces of S_p and $S_p = E_p^+ \oplus E_p^-$, for all $p \in \Lambda$.

Proof: Fix $p \in \Lambda$ and $u \in \hat{\Lambda}$ with $u(0) = p$ (u is unique because there are no collision points). For each $\xi \in S_p$, let $\eta \in T_\pi \tilde{\Gamma}$ be such that $\eta(u) = \xi$. Then $\xi = \eta^+(u) + \eta^-(u)$, $\eta^\pm \in \mathcal{E}^\pm$, so $S_p = E_p^+ + E_p^-$. To see that it is a direct sum, take $\xi \in E_p^+ \cap E_p^-$ so $\xi = \eta^+(u) = \eta^-(u)$ for some $\eta^+ \in \mathcal{E}^+$ and $\eta^- \in \mathcal{E}^-$. Take $\eta_{u,\xi} \in T_\pi \tilde{\Gamma}$ given by $\eta_{u,\xi}(u) = \xi$ and $\eta_{u,\xi}(v) = 0$ for $v \neq u$, and show that $\tilde{F}_{t_0}^n(\eta_{u,\xi}) \to 0$ as $n \to \pm\infty$. This follows from the following two facts:

I) $\eta(u) = z(u) \Rightarrow (\tilde{F}_{t_0}\eta)(\hat{T}_f(t_0)u) = (\tilde{F}_{t_0}z)(\hat{T}_f(t_0)u)$ because

$$\left[\tilde{F}_{t_0}(\eta - z)\right](\hat{T}_f(t_0)u) = \frac{\partial T_f(t_0)}{\partial t}[u(0)]\frac{\partial \tilde{t}}{\partial g}(t_0, f, \pi, \hat{T}_f(t_0)u)[\eta - z]$$

$$\in S_{u(t_0)} \cap Y_{u(t_0)} = \{0\},$$

and

II) by induction $\eta(u) = z(u) \Rightarrow (\tilde{F}_{t_0}^n\eta)(\hat{T}_f(nt_0)u) = (\tilde{F}_{t_0}^n z)(\hat{T}_f(nt_0)u)$.

Then

$$(\tilde{F}_{t_0}^n\eta_{u,\xi})(v) = (\tilde{F}_{t_0}^n\eta^+)(v) = (\tilde{F}_{t_0}^n\eta^-)(v) \text{ for } v = \hat{T}_f(nt_0)u$$

and

$$(\tilde{F}_{t_0}^n\eta_{u,\xi})(v) = 0 \text{ for } v \neq \hat{T}_f(nt_0)u,$$

so $\sup_{v \in \hat{\Lambda}} \|(\tilde{F}_{t_0}^n\eta_{u,\xi})(v)\| \to 0$ as $n \to \pm\infty$, that implies $\eta_{u,\xi} \in \mathcal{E}^+ \cap \mathcal{E}^- = \{0\}$, that is, $\xi = \eta_{u,\xi}(u) = 0 \in S_p$. From this isomorphism between $E_p^\pm$ and $\{\eta_{u,\xi} \in T_\pi \tilde{\Gamma} \mid \xi \in E_p^\pm\} \subset \mathcal{E}^\pm$, one concludes that E_p^+ and E_p^- are closed sets. ∎

Remark 7.3.8. The bundles E_Λ^+ and E_Λ^- may be not invariant under $DT_f(s)$, $s > 0$; but it is possible to find a continuous family (in $p \in \Lambda$) of continuous linear maps with domains in these bundles and images in Y_Λ whose graphs are $DT_f(s)$-invariant (see [176]).

Define these two families:

$$C_\Lambda^\pm = \left\{ A \in C(\Lambda, \bigcup_{p \in \Lambda} \mathcal{L}(E_p^\pm, Y_p)) \mid A_p \in \mathcal{L}(E_p^\pm, Y_p), p \in \Lambda \right\}.$$

C_Λ^+ and C_Λ^- are Banach spaces endowed with the norm

$$\|A\| = \sup_{p \in \Lambda} \|A_p\| = \sup_{p \in \Lambda} \sup_{\xi \in E_p^\pm : \|\xi\| = 1} \|A_p(\xi)\|.$$

Look for $A^\pm \in C_\Lambda^\pm$ such that $\mathrm{graph} A_p^\pm = \{(\xi + A_p^\pm \xi) \in T_p B \mid \xi \in E_p^\pm\}$ is $DT_f(s)$-invariant, that means, $DT_f(s)[\mathrm{graph} A_p^\pm] \subset [\mathrm{graph} A_{T_f(s)p}^\pm]$, $\forall p \in \Lambda$, $\forall s > 0$, or, equivalently, $\forall p \in \Lambda$, $\forall s > 0$, $\forall \xi \in E_p^\pm$ we want

$$(I_{T_f(s)p} - \rho_{T_f(s)p}) DT_f(s)[p](\xi + A_p^\pm \xi)$$
$$= A_{T_f(s)p}^\pm \left[\rho_{T_f(s)p} DT_f(s)[p](\xi + A_p^\pm \xi) \right].$$

Since $\rho_{T_f(s)p} DT_f(s)[p] A_p^\pm \xi = 0$, the last equality can be written as

$$DT_f(s)[p] A_p^\pm \xi = A_{T_f(s)p}^\pm \rho_{T_f(s)p} DT_f(s)[p]\xi - (I_{T_f(s)p} - \rho_{T_f(s)p}) DT_f(s)[p]\xi$$

or

$$A_p^\pm = DT_f(s)[p]|_{Y_p}^{-1} \circ A_{T_f(s)p}^\pm \circ \rho_{T_f(s)p} \circ DT_f(s)[p]$$
$$- DT_f(s)[p]|_{Y_p}^{-1} \circ (I_{T_f(s)p} - \rho_{T_f(s)p}) \circ DT_f(s)[p]$$

because $DT_f(s)[p]|_{Y_p}$ is a linear isomorphism of Y_p onto $Y_{T_f(s)p}$. So $\mathrm{graph} A_p^\pm$ is $DT_f(s)$-invariant if, and only if, $A^\pm \in C_\Lambda^\pm$ is a fixed point of the operator

$$T_s^\pm : C_\Lambda^\pm \to C_\Lambda^\pm \text{ given by } T_s^\pm(A) = \Phi_s^\pm A + \mathcal{K}_s^\pm$$

with

$$\Phi_s^\pm \in \mathcal{L}(C_\Lambda^\pm) \mid (\phi_s^\pm A)[p] = DT_f(s)[p]|_{Y_p}^{-1} \circ A_{T_f(s)p} \circ \rho_{T_f(s)p} \circ DT_f(s)[p]|_{E_p^\pm}$$

and

$$\mathcal{K}_s^\pm \in C_\Lambda^\pm \mid \mathcal{K}_s^\pm(p) = -DT_f(s)[p]|_{Y_p}^{-1} \circ (I_{T_f(s)p} - \rho_{T_f(s)p}) \circ DT_f(s)[p]|_{E_p^\pm}.$$

With one more result (Proposition 7.3.9) we can see the statement of Theorem 7.3.10, the converse of Theorem 7.3.6. Observe that one can also define $\tilde{F}_t : T_\pi \tilde{\Gamma} \to T_\pi \tilde{\Gamma}$ for $t \geq 0$, with the formula

$$(\tilde{F}_t \eta)(u) = \rho_{u(0)} DT_f(t)[u(-t)](\eta(\hat{T}_f(-t)u)),$$

and see that $\{\tilde{F}_t\}_{t \geq 0}$ is a semigroup of continuous linear transformations of $T_\pi \tilde{\Gamma}$.

Proposition 7.3.9. *Assume that $\Lambda \subset \mathcal{B}$ is a compact T_f-invariant set without fixed points and assume also that there are no collision points in Λ. Suppose that π is a hyperbolic fixed point of $\tilde{T}_{f,f}(t_0)$ for some $t_0 \in]0,2]$, and $T_\pi \tilde{\Gamma} = \mathcal{E}^+ \oplus \mathcal{E}^-$, the direct sum of the closed stable and unstable subspaces of $\tilde{F}_{t_0}$, is such that $\dim D\theta_u(\pi)(\mathcal{E}^-) < \infty$ for all $u \in \hat{\Lambda}$. Then $\tilde{F}_t|_{\mathcal{E}^-}$ is invertible, $\mathcal{E}^+$ and $\mathcal{E}^-$ are the stable and unstable subspaces of $\tilde{F}_t$ for all $t > 0$ and are characterized as*

$$\mathcal{E}^+ = \{\eta \in T_\pi \tilde{\Gamma} \mid \tilde{F}_t \eta \to 0 \text{ as } t \to \infty\}$$
$$\mathcal{E}^- = \{\eta \in T_\pi \tilde{\Gamma} \mid \tilde{F}_t^{-1} \eta \to 0 \text{ as } t \to \infty\};$$

moreover, there exist constants K and $\alpha > 0$ such that

$$\|\tilde{F}_t \eta\| \le K e^{-\alpha t} \|\eta\| \text{ for } \eta \in \mathcal{E}^+, \ t \ge 0,$$

and

$$\|\tilde{F}_t^{-1} \eta\| \le K e^{-\alpha t} \|\eta\| \text{ for } \eta \in \mathcal{E}^-, \ t \ge 0.$$

Proof: (see [56] for the complete proof where it is used the technical hypothesis $\dim E_p^- < \infty$ of definition 7.2.1). ∎

Let us come back to the concepts and notations introduced in Remark 7.3.8 where we considered the Banach spaces $\mathcal{C}_\Lambda^\pm$ and the operators $\mathcal{T}_s^\pm$ defined as $\mathcal{T}_s^\pm(A) = \Phi_s^\pm A + \mathcal{K}_s^\pm$. One obtains the following theorem (for a proof see [56]):

Theorem 7.3.10. *Assume that $\Lambda \subset \mathcal{B}$ is a compact T_f-invariant set without fixed points and assume also that there are no collision points in Λ. Suppose that π is a hyperbolic fixed point of $\tilde{T}_{f,f}(t_0)$ for some $t_0 \in]0,2]$ and that there exists $s_0 > 0$ such that $1 \in R(\Phi_s^\pm)$, $\forall 0 < s \le s_0$ ($R(T)$ denotes the resolvent of the operator T), that is, $(I - \Phi_s^\pm)^{-1} \in \mathcal{L}(\mathcal{C}_\Lambda^\pm)$ for all $0 < s \le s_0$ (where I is the identity of $\mathcal{C}_\Lambda^\pm$). Then Λ is a hyperbolic set.*

7.4 Persistence of hyperbolicity and perturbations with one-to-oneness

Let $\mathcal{A}$ be the dense subset of $\mathcal{F}$ introduced in the definition of property $\mathcal{K}$ (see Section 6.2).

Recall Theorem 7.3.6 where a compact T_f-invariant set $\Lambda \subset \mathcal{B}$ is given, without fixed points; if Λ is hyperbolic then $\pi \in \tilde{\mathcal{C}}$ is a hyperbolic fixed point of $\tilde{T}_{f,f}(t)$, $t \in]0,2]$. Now, given $\varepsilon > 0$, from the properties of $T : \mathcal{F} \times \mathbb{R}_+ \times \mathcal{B} \to \mathcal{B}$ and compactness of Λ, one can take $f' \in \mathcal{F}$, close to f, such that $\tilde{T}_{f,f}(t)$ and $\tilde{T}_{f,f'}(t)$ are $\varepsilon - C^1$ close, $\forall t \in [0,2]$.

Lemma 7.4.1. *Choosing $\varepsilon > 0$, properly, the map $\tilde{T}_{f,f'}(t)$ has a unique hyperbolic fixed point $h = h(f, f')$, close to π in $\tilde{C}$ (and also in $\tilde{\Gamma}$), that does not depend on $t \in]0, 2]$.*

Proof: The fact that $\tilde{T}_{f,f'}(t)$ is close to $\tilde{T}_{f,f}(t)$ (in $\tilde{C}$ or $\tilde{\Gamma}$) implies the existence of a unique fixed point $h_t = h(t, f, f')$ of $\tilde{T}_{f,f'}(t)$, $t \in]0, 2]$, with h_t close to π (in $\tilde{C}$ or in $\tilde{\Gamma}$, then $h_t \in \tilde{C}$). The function $t \in]0, 2] \to h(t, f, f') \in \tilde{C}$ is continuous, because, as operators of $\tilde{C}$, $\tilde{T}_{f,f'}(s) \to \tilde{T}_{f,f'}(t)$ as $s \to t$. Let $h = h_2$ (for $t = 2$) and let $r_{m,j} = \frac{2j}{2^m}$, $m \in \mathbb{N}$, $j = 1, \ldots, 2^m$. From $\tilde{t}(t, f', h_t, u) + \tilde{t}(t, f', h_t, \hat{T}_f(-t)u) = \tilde{t}(2t, f', h_t, u)$, $\forall t \in]0, 1]$, $u \in \hat{\Lambda}$, we see that $h_{r_{m,1}}$ and $h_{r_{m,j}}$ are hyperbolic fixed points of $\tilde{T}_{f,f'}(r_{m,j})$, $\forall m \in \mathbb{N}$, $j = 1, \ldots, 2^m$ and are close to π, then they coincide. From this and the continuity in t of h_t, one concludes that $h_t = h$, $\forall t \in]0, 2]$.   ∎

One also has $T_\pi \tilde{\Gamma} = \mathcal{E}_\pi^+ \oplus \mathcal{E}_\pi^-$ and $T_h \tilde{\Gamma} = \mathcal{E}_h^+ \oplus \mathcal{E}_h^-$ for the closed stable and unstable subspaces relative to $D\tilde{T}_{f,f}(t)[\pi]$ and $D\tilde{T}_{f,f'}(t)[h]$, $\forall t \in]0, 2]$, respectively. As above, one can also show that the subspaces $\mathcal{E}_h^\pm$ do not depend on $t \in]0, 2]$. And, given $\varepsilon > 0$, one can choose f' sufficiently close to f in order to get h close to π such that $\mathcal{E}_\pi^\pm$ and $\mathcal{E}_h^\pm$ be ε-close, respectively.

Since h is a fixed point of $\tilde{T}_{f,f'}(t)$, that is,

$$(*): \quad T_{f'}(\tilde{t}(t, f', h, u))h(\hat{T}_f(-t)u) = h(u), \quad \forall u \in \hat{\Lambda}, \ \forall t \in]0, 2],$$

we see that $\tilde{t}(t, f', h, u) \geq 0$ is well defined for all $t \geq 0$ such that $(*)$ holds . Note that $\tilde{t}(\cdot, f', h, u)$ is continuous in $t \geq 0$ and $\tilde{t}(t, f', h, u) \to \infty$ as $t \to \infty$ for each fixed $u \in \hat{\Lambda}$.

Definition 7.4.2. *Set $\Lambda' := h(\hat{\Lambda}) \subset \mathcal{B}$.*

Proposition 7.4.3. *$\Lambda' = T_{f'}(t)\Lambda'$ for all $t \geq 0$, that is, Λ' is $T_{f'}$-invariant.*

As we know, if $T_{f'}(t)$ is one-to-one on Λ', then $\tilde{t}(t, f', h, u) \in \mathbb{R}$ is well defined for $t \in \mathbb{R}$ such that $T_{f'}(\tilde{t}(t, f', h, u))h(\hat{T}_f(-t)u) = h(u)$ for all $u \in \hat{\Lambda}$. Note that, in this case, $\tilde{t}(\cdot, f', h, u)$ is continuous in $t \in \mathbb{R}$ and $\tilde{t}(t, f', h, u) \to -\infty$ as $t \to -\infty$ for each fixed $u \in \hat{\Lambda}$. The way to obtain $T_{f'}(t)$ one to one on $A(f')$, and so on Λ', is to assume that $\mathcal{F}$ has the property $\mathcal{K}$ and the smooth property.

Proposition 7.4.4. *Let Λ be a compact T_f-invariant hyperbolic set without fixed points. Assume that the space of parameters $\mathcal{F}$ of the semiflow T_f has the property $\mathcal{K}$ and satisfies the smooth property. Then one can find f' sufficiently close to f in $\mathcal{A} \subset \mathcal{F}$ such that $h = h_{f,f'}$ is injective and arbitrarily close to π in $\tilde{C}$.*

Proof: Given $\varepsilon > 0$, take $f' \in \mathcal{A}$, close enough to $f \in \mathcal{F}$ such that

$$\text{dist}(h(u), \pi(u)) < \varepsilon/2 \text{ for all } u \in \hat{\Lambda}$$

and this implies that

$$\text{dist}(h(\hat{T}_f(t)u), u(t)) < \varepsilon/2, \ \forall u \in \hat{\Lambda} \text{ and } \forall t \in \mathbb{R}.$$

Assume $u_1, u_2 \in \hat{\Lambda}$, $u_1 \neq u_2$, such that $h(u_1) = h(u_2)$. We have that

$$\text{dist}(u_1(0), u_2(0)) < \varepsilon,$$

then u_1 and u_2 are not in the same $\hat{T}_f$-orbit because $h(u_1) \in \mathcal{S}_{u_1(0)}$, $h(u_2) \in \mathcal{S}_{u_2(0)}$, and these two sections do not intersect for distinct and near points of the same T_f-orbit. Now,

$$\begin{aligned} h(\hat{T}_f(t)u_1) &= T_{f'}(\tilde{t}(t, f', h, \hat{T}_f(t)u_1)h(u_1) \\ &= T_{f'}(\tilde{t}(s(t), f', h, \hat{T}_f(s(t))u_2))h(u_2) \\ &= h(\hat{T}_f(s(t))u_2) \end{aligned}$$

where $s(t) \in \mathbb{R}$ is such that $\tilde{t}(t, f', h, \hat{T}_f(t)u_1) = \tilde{t}(s(t), f', h, \hat{T}_f(s(t))u_2)$ for all $t \in \mathbb{R}$; $s(t)$ is well defined because $h(\hat{T}_f(t)u_1)$ and $h(\hat{T}_f(t)u_2)$ are on the same $T_{f'}$-orbit, for all $t \in \mathbb{R}$. Also, $s(t)$ is continuous in t and $s(t) \to \pm\infty$ as $t \to \pm\infty$. So, $\text{dist}(h(\hat{T}_f(t)u_1), u_1(t)) < \varepsilon/2$ and $\text{dist}(h(\hat{T}_f(s(t))u_2), u_2(s(t))) < \varepsilon/2$ imply that $\text{dist}(u_1(t), u_2(s(t))) < \varepsilon$ for all $t \in \mathbb{R}$. Since $T_f(t)$ is expansive on the hyperbolic set Λ (see [176], p.99), taking $\varepsilon > 0$ sufficiently small, one obtains $u_1 = u_2$ (contradiction), so h is injective.   ∎

Under the hypotheses of the last proposition one sees that $h : \hat{\Lambda} \to \Lambda' = h(\hat{\Lambda})$ is a homeomorphism of $\hat{\Lambda}$ onto Λ'. Also, we consider $\hat{T}_{f'}(t) : \hat{B} \to \hat{B}$, $t \in \mathbb{R}$, $\hat{\Lambda}' \subset \hat{B}$ and $\pi' : \hat{\Lambda}' \to \Lambda'$, repeating all the previous arguments with f' replacing f. Since there are no collisions on Λ' one obtains that π' is a homeomorphism between the compact sets $\hat{\Lambda}'$ and Λ'. Define $\hat{h} : \hat{\Lambda} \to \hat{\Lambda}'$ by $\hat{h} = (\pi')^{-1} \circ h$. For each $p' \in \Lambda'$, there exists a unique $u \in \hat{\Lambda}$ and $u' \in \hat{\Lambda}'$ such that $p' = u'(0) = h(u)$ and $u' = \hat{h}(u)$. One can consider the transversal sections $\mathcal{S}'_{p'} = \mathcal{S}_{u(0)}$ and the splitting $T_{p'}B = S'_{p'} \oplus Y'_{p'}$, as before; here $S'_{p'} = T_{h(u)}\mathcal{S}_{u(0)} = T_{p'}S'_{p'}$. We see that $\tilde{C}' \subset \tilde{\Gamma}'$, $\tilde{t}'(t, f'', g', u')$ (for $f'' \in \mathcal{F}$, $g' \in \tilde{\Gamma}'$, $u' \in \hat{\Lambda}'$ and $t \in]0, 2]$) and $\tilde{T}_{f'f''}(t)$ are well defined, and all the previous results hold taking f' instead of f.

Coming back to f, the following lemma can be stated:

Lemma 7.4.5. *For* $t \in]0, 2]$, *the operator* $\tilde{T}_{f,f'}(\tau_t) : V \subset \tilde{\Gamma} \to \tilde{\Gamma}$ *(V is a suitable neighborhood of* π *with* $h \in V$ *) defined by*

$$(\tilde{T}_{f,f'}(\tau_t)g)(u) = T_{f'}(\tilde{t}(\tau_t(u), f', g, u)) \left[g(\hat{T}_f(-\tau_t(u))u) \right]$$

(where $\tau_t(u)$ *is such that* $\tilde{t}(\tau_t(u), f', h, u) = t$, $\forall u \in \hat{\Lambda}$*) has* h *as a hyperbolic fixed point.*

Proof: Clearly h is a fixed point of $\tilde{T}_{ff'}(\tau_t)$. To see that h is hyperbolic we have to show that $\mathcal{E}_h^+$ and $\mathcal{E}_h^-$ ($T_h\tilde{\Gamma} = \mathcal{E}_h^+ \oplus \mathcal{E}_h^-$ as before) are the stable and unstable subspaces of $D\tilde{T}_{f,f'}(\tau_t)[h]$. Since

$$(D\tilde{T}_{f,f'}(\tau_t)[h]\eta)(u) = \rho_{u(0)}DT_{f'}(t)[h(\hat{T}_f(-\tau_t(u))u)](\eta(\hat{T}_f(-\tau_t(u))u)),$$

with a direct computation one verifies that

$$(D\tilde{T}_{f,f'}(\tau_t)[h])^n(\eta) \to 0 \text{ as } n \to \infty, \text{ for } \eta \in \mathcal{E}_h^+$$

(because $(D\tilde{T}_{f,f'}(\sigma)[h])^n(\eta) \to 0$ as $n \to \infty, \forall \sigma \in]0,2]$), that $D\tilde{T}_{f,f'}(\tau_t)[h]|_{\mathcal{E}_h^-}$ is a linear isomorphism and that

$$(D\tilde{T}_{f,f'}(\tau_t)[h])^{-n}(\eta) \to 0 \text{ as } n \to \infty, \text{ for } \eta \in \mathcal{E}_h^-.$$

∎

Define now $\Psi : \tilde{\Gamma} \to \tilde{\Gamma}'$ by $\Psi(g) = g \circ (\hat{h})^{-1}$. One sees that Ψ is a homeomorphism onto $\tilde{\Gamma}'$ and it is quite easy to show the commutativity of the diagram:

$$
\begin{array}{ccc}
V \subset \tilde{\Gamma} & \xrightarrow{\tilde{T}_{f,f'}(\tau_t)} & \tilde{\Gamma} \\
\Psi \downarrow & & \downarrow \Psi \\
\Psi(V) \subset \tilde{\Gamma}' & \xrightarrow{\tilde{T}_{f',f'}(t)} & \tilde{\Gamma}'
\end{array}
$$

For $t \in]0,2]$, if one of the two next equivalent facts is true: h is a hyperbolic fixed point of $\tilde{T}_{ff'}(\tau_t)$, or $\pi' = \Psi(h)$ is a hyperbolic fixed point of $\tilde{T}_{f',f'}(t)$, then $T_h\tilde{\Gamma} = \mathcal{E}_h^+ \oplus \mathcal{E}_h^-$, $T_{\pi'}\tilde{\Gamma}' = \mathcal{E}_{\pi'}^+ \oplus \mathcal{E}_{\pi'}^-$, where $\mathcal{E}_{\pi'}^{\pm} = D\Psi[h]\mathcal{E}_h^{\pm}$.

The previous definitions of the Banach spaces $\mathcal{C}_\Lambda^{\pm}$ and of the operators $\Phi_s^{\pm} \in \mathcal{L}(\mathcal{C}_\Lambda^{\pm})$, $s \geq 0$, make sense for Λ compact, T_f-invariant and hyperbolic, even if there are collision points in Λ. Still may define the Banach spaces

$$\mathcal{C}_{\hat{\Lambda}}^{\pm} = \{\hat{A} \in C(\hat{\Lambda}, \bigcup_{p\in\Lambda} \mathcal{L}(E_p^{\pm}, Y_p)) \mid \hat{A}_u \in \mathcal{L}(E_{u(0)}^{\pm}, Y_{u(0)}), \ u \in \hat{\Lambda}\}$$

with the sup norm $\|\hat{A}\| = \sup_{u\in\hat{\Lambda}} \|\hat{A}_u\|$; we also have the natural maps

$$\mathcal{Z}^{\pm} : \mathcal{C}_\Lambda^{\pm} \to \mathcal{C}_{\hat{\Lambda}}^{\pm}, [\mathcal{Z}^{\pm}(A)]_u = A_{u(0)}, \ \forall u \in \hat{\Lambda}.$$

These maps $\mathcal{Z}^{\pm}$ are linear isomorphisms onto their images

$$\mathcal{Z}^{\pm}(\mathcal{C}_\Lambda^{\pm}) = \{\hat{A} \in \mathcal{C}_{\hat{\Lambda}}^{\pm} \mid u_1(0) = u_2(0) \text{ implies } \hat{A}_{u_1} = \hat{A}_{u_2}\}.$$

Define, finally, $\hat{\Phi}_s^{\pm} \in \mathcal{L}(\mathcal{C}_{\hat{\Lambda}}^{\pm})$ by the formulae

$$[\hat{\Phi}_s^{\pm}(\hat{A})]_u = DT_f(s)[u(0)]|_{Y_{u(0)}}^{-1} \hat{A}_{\hat{T}_f(s)u} \rho_{u(s)} DT_f(s)[u(0)]|_{E_{u(0)}^{\pm}}, \ u \in \hat{\Lambda}, \ s \geq 0.$$

One sees that $\mathcal{Z}^{\pm}(\mathcal{C}_{\Lambda}^{\pm})$ is invariant under $\hat{\varPhi}_s^{\pm}$ and

$$\hat{\varPhi}_s^{\pm}|_{\mathcal{Z}^{\pm}(\mathcal{C}_{\Lambda}^{\pm})} = \mathcal{Z}^{\pm} \circ \varPhi_s^{\pm} \circ (\mathcal{Z}^{\pm})^{-1}.$$

Now, the main result of the present section is stated as follows.

Theorem 7.4.6. *Let Λ be a compact, T_f-invariant, hyperbolic set without fixed points. Assume that the space of parameters $\mathcal{F}$ of the semiflow T_f has the property $\mathcal{K}$ and satisfies the smooth property. Then we can find $f' \in \mathcal{A} \subset \mathcal{F}$, sufficiently close to f, such that $\Lambda' = h(\hat{\Lambda})$ is arbitrarily close to Λ and, moreover, Λ' is compact, $T_{f'}$-invariant, hyperbolic and without collision points.*

Proof: Taking $f' \in \mathcal{A}$ sufficiently close to $f \in \mathcal{F}$, we get $h \in \tilde{C}$ arbitrarily close to π, which is a hyperbolic fixed point of $\tilde{T}_{f,f'}(t)$, $t \in]0, 2]$, and injective. From Proposition 7.4.4 and Lemma 7.4.5, h is a hyperbolic fixed point of $\tilde{T}_{f,f'}(\tau_t)$ in $\tilde{\Gamma}$, then π' is a hyperbolic fixed point of $\tilde{T}_{f',f'}(t)$ in $\tilde{\Gamma}'$. So, $\Lambda' = h(\hat{\Lambda})$ is compact, $T_{f'}$-invariant and without collision points with respect to $T_{f'}$. Only need to prove hyperbolicity of Λ'. Taking f', Λ', etc., instead of f, Λ, etc., in the previous definitions, one considers the spaces $\mathcal{C}_{\Lambda'}^{'\pm}$, $\mathcal{C}_{\hat{\Lambda}'}^{'\pm}$ and the operators $\varPhi_s^{'\pm} \in \mathcal{L}(\mathcal{C}_{\Lambda'}^{'\pm})$, $\hat{\varPhi}_s^{'\pm} \in \mathcal{L}(\mathcal{C}_{\hat{\Lambda}'}^{'\pm})$, $s \geq 0$, and $\mathcal{Z}^{'\pm} \in \mathcal{L}(\mathcal{C}_{\Lambda'}^{'\pm}, \mathcal{C}_{\hat{\Lambda}'}^{'\pm})$. Since there are no collision points in Λ', $\mathcal{Z}^{'\pm}$ are isomorphisms between $\mathcal{C}_{\Lambda'}^{'\pm}$ and $\mathcal{C}_{\hat{\Lambda}'}^{'\pm}$ and $\hat{\varPhi}_s^{'\pm} = \mathcal{Z}^{'\pm} \circ \varPhi_s^{'\pm} \circ (\mathcal{Z}^{'\pm})^{-1}$, $s \geq 0$. Now $\mathcal{E}_h^{\pm}$ and $\mathcal{E}_{\pi}^{\pm}$ are isomorphic (they are arbitrarily close), then $\mathcal{E}_{\pi'}^{\pm}$ and $\mathcal{E}_{\pi}^{\pm}$ are also isomorphic. But, as we already saw, for each $p' \in \Lambda'$, there exist unique $u' \in \hat{\Lambda}'$ and $u \in \hat{\Lambda}$ such that $p' = u'(0) = h(u)$, $u' = \hat{h}(u)$. Set $p = u(0)$. The spaces $E_p^{\pm} = D\theta_u[\pi]\mathcal{E}_{\pi}^{\pm}$ and $E_{p'}^{'\pm} = D\theta_{u'}'[\pi']\mathcal{E}_{\pi'}^{\pm}$ are isomorphic, that is, there exist $z_{p'}^{\pm} : E_{p'}^{'\pm} \to E_p^{\pm}$ as linear isomorphisms. So, one obtains the natural isomorphisms $I^{\pm}$ between $\mathcal{C}_{\hat{\Lambda}}^{\pm}$ and $\mathcal{C}_{\hat{\Lambda}'}^{'\pm}$ given by $(I^{\pm}\hat{A})_{u'}\xi' = y_{u'}[\hat{A}_u\xi]$ where $u = (\hat{h})^{-1}u'$, $\xi = z_{u'(0)}^{\pm}(\xi')$ and $y_{u'}$ is the natural isomorphism between $Y_{u(0)}$ and $Y'_{u'(0)}$, for all $u' \in \hat{\Lambda}'$, $\xi' \in E_{u'(0)}^{'\pm}$. Fix now $s_0 > 0$; if f' is sufficiently close to f, the operators $\hat{\varPhi}_s^{\pm}$ and $I^{\pm} \circ \hat{\varPhi}_s^{'\pm} \circ (I^{\pm})^{-1} \in \mathcal{L}(\mathcal{C}_{\hat{\Lambda}}^{\pm})$ are arbitrarily close for $0 < s \leq s_0$. Since $\| DT_f(s)[u(0)]|_{Y_{u(0)}}^{-1} \|$ is uniformly bounded for $s > 0$ and $u \in \hat{\Lambda}$, above and below, one sees that $(\hat{\varPhi}_s^+)^n$ and $(\hat{\varPhi}_s^-)^{-n}$ are contractions for $s > 0$ and some suitable power $n \in \mathbb{N}$. From that $1 \in R(\hat{\varPhi}_s^{\pm})$ which implies that $1 \in R(I^{\pm} \circ \hat{\varPhi}_s^{'\pm} \circ (I^{\pm})^{-1})$ for $0 < s \leq s_0$; from this it follows that $1 \in R(\hat{\varPhi}_s^{'\pm})$ and so, finally, $1 \in R(\varPhi_s^{'\pm})$, $0 < s \leq s_0$. Using the result of Theorem 7.3.10 for f', Λ', π', etc., replacing f, Λ, π, etc., one arrives to the hyperbolicity of Λ'.   ∎

Final remarks

1. When there are no collision points in Λ, one obtains a topological equivalence between Λ and Λ' given by the homeomorphism $h \circ (\pi)^{-1}$ with inverse $\pi \circ h^{-1}$. If, otherwise, Λ has collision points, π is not invertible and the continuous map $\pi \circ h^{-1}$ is a semi-equivalence, only.
2. When the hyperbolic T_f-invariant set Λ is equal to $A(f)$, we have that, in fact, T_f is Anosov like $(A(f)$ may not be a manifold) and Theorem 7.4.6 proves openness and $A(f)$-stability of such semiflows, in $\mathcal{A}$, so generalizes the famous result of Anosov for C^1-flows on compact manifolds (see [191]).

7.5 Nonuniform hyperbolicity and invariant manifolds

We want to emphasize some of the geometric consequences of nonuniform hyperbolicity in infinite dimensional spaces, by recalling, first of all, what happens in finite dimension and, finally, showing briefly to the reader, that there does exist the possibility of analyzing those consequences in infinite dimension. Particularly, this is the case of parabolic semilinear equations and retarded functional differential equations. We shall deal, mainly, with results concerning the existence of families of local stable and unstable manifolds.

The results are similar for maps and semiflows, and as such, we shall restrict ourselves to the case of maps for which the formulation is a bit simpler.

7.5.1 Nonuniform hyperbolicity

We first consider the case of dynamics in finite dimensional spaces. It is well known that the occurrence of hyperbolicity in a smooth (finite dimensional) manifold guarantees a very rich structure. In particular it guarantees the existence of families of stable and unstable manifolds, with the celebrated Hadamard–Perron theorem. The results described here apply in fact to a weaker form of hyperbolicity, namely *nonuniform hyperbolicity*. It is believed that nonuniform hyperbolicity is much more common than the classical (uniform) hyperbolicity.

Let $f\colon M \to M$ be a diffeomorphism in a finite dimensional manifold. The trajectory of a point $x \in M$ under f is called *nonuniformly hyperbolic* if there is a decomposition $T_{f^n x}M = E^s(f^n x) \oplus E^u(f^n x)$ for each $n \in \mathbb{Z}$, numbers $\lambda \in (0,1)$ and $\varepsilon > 0$, and a positive function C defined in the trajectory of x such that if $k, n \in \mathbb{Z}$ then:

1. $d_x f^k E^s(x) = E^s(f^k x)$ and $d_x f^k E^u(x) = E^u(f^k x)$;
2. if $v \in E^s(f^k x)$ and $n > 0$ then $\|d_{f^k x} f^n v\| \le C(f^k x)\lambda^n e^{\varepsilon n}\|v\|$;
3. if $v \in E^u(f^k x)$ and $n < 0$ then $\|d_{f^k x} f^n v\| \le C(f^k x)\lambda^{|n|} e^{\varepsilon |n|}\|v\|$;
4. $\angle(E^u(f^k x), E^s(f^k x)) \ge [C(f^k x)]^{-1}$;
5. $C(f^{k+n} x) \le e^{\varepsilon |k|} C(f^n x)$.

Recall that the angle between two subspaces E and F such that $E \cap F = \{0\}$ is defined by $\angle(E, F) = \inf\{\angle(u, v) : u \in E \setminus \{0\}, v \in F \setminus \{0\}\}$. The notion of nonuniform hyperbolicity is rooted in seminal work of Pesin in the mid-seventies (see [166]). We shall shortly see that nonuniform hyperbolicity is a rather natural concept.

The following statement was established by Pesin in [166], and constitutes a version of the Hadamard–Perron theorem for nonuniform hyperbolicity.

Theorem 7.5.1. *Let $f : M \to M$ be a $C^{1+\alpha}$ diffeomorphism on a compact manifold, for some $\alpha > 0$, and $x \in M$ a point whose trajectory is nonuniformly hyperbolic. Then there exist C^1 manifolds $V^s(x)$ and $V^u(x)$ containing x, and a function D defined in the trajectory of x such that:*

1. *$T_x V^s(x) = E^s(x)$ and $T_x V^u(x) = E^u(x)$;*
2. *if $y \in V^s(x)$ and $n > 0$ then $d(f^n x, f^n y) \le D(x) \lambda^n e^{\varepsilon n} d(x, y)$;*
3. *if $y \in V^u(x)$ and $n < 0$ then $d(f^n x, f^n y) \le D(x) \lambda^{|n|} e^{\varepsilon |n|} d(x, y)$;*
4. *$D(f^{k+n} x) \le e^{10 \varepsilon |k|} D(f^n x)$ for every $k, n \in \mathbb{Z}$.*

Besides the original approach, other proofs are due to Ruelle [174], Fathi, Herman and Yoccoz [53], and Pugh and Shub [170]. In [13] Barreira and Pesin made a detailed exposition following closely the approach in [166].

7.5.2 Regular points

For each $x \in M$ and $v \in T_x M$ we define the Lyapunov exponent

$$\chi(x, v) = \lim_{n \to +\infty} \frac{1}{n} \log \|d_x f^n v\|.$$

It follows from the abstract theory of Lyapunov exponents (see [13] for a detailed description) that for each $x \in M$ there exist $s(x) \in \mathbb{N}$, numbers $\chi_1(x) < \cdots < \chi_{s(x)}(x)$, and subspaces

$$E_i(x) = \{v \in T_x M : \chi(x, v) \le \chi_i(x)\}$$

such that if $v \in E_i(x) \setminus E_{i-1}(x)$ then $\chi(x, v) = \chi_i(x)$.

We now explain how nonuniform hyperbolicity may appear in a natural way. In order to do this we first introduce the concept of regularity for the diffeomorphism f. We say that the point $x \in M$ is *Lyapunov regular* or simply *regular* if there exists a decomposition $T_x M = \bigoplus_{i=1}^{s(x)} H_i(x)$ such that:

1. if $i = 1, \ldots, s(x)$ and $v \in H_i(x) \setminus \{0\}$ then

$$\lim_{n \to \pm\infty} \frac{1}{n} \log \|d_x f^n v\| = \chi_i(x); \tag{7.1}$$

2.

$$\lim_{n \to \pm\infty} \frac{1}{n} \log |\det d_x f^n| = \sum_{i=1}^{s(x)} \chi_i(x) \dim H_i(x). \tag{7.2}$$

If x is regular then $E_k(x) = \bigoplus_{i=1}^{k} H_i(x)$ for each $k = 1, \ldots, s(x)$. One can easily verify that when a point is regular every point in its trajectory is also regular. The concept of regularity requires a lot from the asymptotic behavior of the trajectory of x, and in particular requires the compatibility of the behavior into the past and into the future. However, we shall see that it is rather ubiquitous under fairly general assumptions.

Recall that a measure μ in M is f-invariant if $\mu(f^{-1}A) = \mu(A)$ for every measurable set $A \subset M$. We now formulate the celebrated Oseledets multiplicative ergodic theorem [158] in the case of a diffeomorphism acting on a compact manifold.

Theorem 7.5.2. *If $f\colon M \to M$ is a C^1 diffeomorphism on a compact manifold, and μ is an f-invariant finite measure in M, then the set of regular points has full μ-measure.*

More generally we can formulate a version of Theorem 7.5.2 for non-compact manifolds provided that $\log^+ \|df\|$ and $\log^+ \|df^{-1}\|$ are μ-integrable. We emphasize that the notion of regularity does not require any invariant measure. See [12] for a detailed exposition of the multiplicative ergodic theorem (in particular containing the proofs of Oseledets [158] and Raghunathan [171]).

It follows from Theorem 7.5.2 that at least from the point of view of ergodic theory, the set of regular points is rather large. It should however be noted that the set of non-regular points may also be large, from other points of view. Namely, it follows from work of Barreira and Schmeling [14] that for a *generic* C^1 diffeomorphism on a surface with a (saddle-type) hyperbolic set, the set of non-regular points has positive Hausdorff dimension (although the complement, i.e., the set of regular points has full measure with respect to *any* invariant measure).

7.5.3 Hyperbolic measures and nonuniform hyperbolicity

We say that an f-invariant measure μ is *hyperbolic* if for μ-almost every point $x \in M$ we have $\chi(x,v) \neq 0$ for every $v \in T_x M$. We want to show that diffeomorphisms with a hyperbolic measure possess many nonuniformly hyperbolic trajectories.

For simplicity (and without loss of generality) we assume here that μ is ergodic, i.e., that if $A \subset M$ if a measurable set with $f^{-1}A = A$ then either $\mu(A) = 0$ or $\mu(M \setminus A) = 0$. Since the functions $x \mapsto s(x)$, $x \mapsto \chi_i(x)$ and $x \mapsto \dim E_i(x)$ are f-invariant and measurable, it follows from the ergodicity of μ that there exist constants s^μ, χ_i^μ, and k_i^μ such that

$$s(x) = s^\mu, \quad \chi_i(x) = \chi_i^\mu, \quad \dim E_i(x) = k_i^\mu \tag{7.3}$$

for μ-almost every $x \in M$ and every $i = 1, \ldots, s^\mu$.

Let now $\Lambda_\mu \subset M$ be the set of regular points $x \in M$ satisfying (7.3). It was established by Pesin in [166] that: if $f: M \to M$ is a C^1 diffeomorphism on a compact manifold, preserving an ergodic hyperbolic finite measure μ in M, then every point in Λ_μ has a nonuniformly hyperbolic trajectory. Furthermore, if $x \in \Lambda_\mu$ then the spaces $E^s(x)$ and $E^u(x)$ appearing in the definition of nonuniformly hyperbolic trajectory are given by

$$E^s(x) = \bigoplus_{i:\chi_i(x)<0} H_i(x) \quad \text{and} \quad E^u(x) = \bigoplus_{i:\chi_i(x)>0} H_i(x).$$

The following is now an immediate consequence of Theorem 7.5.2.

Theorem 7.5.3. *If $f: M \to M$ is a C^1 diffeomorphism on a compact manifold, then the set of points whose trajectory is nonuniformly hyperbolic has full measure with respect to any ergodic hyperbolic f-invariant finite measure in M.*

This result shows that nonuniform hyperbolicity is rather common when there exist hyperbolic measures. Combining Theorems 7.5.1 and 7.5.3, we conclude that almost every point in the support of an ergodic hyperbolic invariant finite measure has associated stable and unstable manifolds.

Under the assumptions of Theorem 7.5.3 the constants λ and ε in the definition of a nonuniformly hyperbolic trajectory are replaced by constants $\lambda(x)$ and $\varepsilon(x)$ that depend measurably on x.

The study of the stable and unstable manifolds in the nonuniformly hyperbolic theory is much more delicate than the corresponding study in the classical (uniform) theory. We refer the reader to [12, 13, 110] for a detailed discussion, including the study of the so-called absolute continuity of the stable and unstable manifolds and of the ergodic properties of diffeomorphisms preserving hyperbolic measures.

It is appropriate to recall here the role played by regularity in the stability theory of non-autonomous ordinary differential equations. Let

$$x' = A(t)x \tag{7.4}$$

be a differential equation in $\mathbb{C}^n$ such that $A(t)$ depends continuously on t, and $\sup\{\|A(t)\| : t \in \mathbb{R}\} < \infty$. It is well known that if all Lyapunov exponents

$$\chi(x_0) = \limsup_{t \to +\infty} \frac{1}{n} \log \|x(t)\|,$$

where $x(t)$ denotes the solution of (7.4) with $x(0) = x_0$, are negative then the zero solution of (7.4) is asymptotically stable. However there may still exist arbitrarily small perturbations $h(t, x)$ with $h(t, 0) = 0$ such that the zero solution of

$$x' = A(t)x + h(t, x) \tag{7.5}$$

is not asymptotically stable (see [13] for an example going back to Perron). On the other hand, it can be shown that if the linear system in (7.4) is *regular* (in this case $v \mapsto A(t)v$ replaces $v \mapsto d_x f^n v$ in (7.1) and (7.2), and we let $t \to \pm\infty$), and $\chi(x_0) < 0$ for every $x_0 \in \mathbb{C}^n$, then for any sufficiently small perturbation $h(t, x)$ with $h(t, 0) = 0$ zero is an asymptotically stable solution of the perturbed system (7.5). This is a continuous time counterpart of the statements formulated above (where, for regular points, we are able to show that the behavior along the linear subspaces $E^s(x)$ and $E^u(x)$ induces similar behaviors along stable and unstable manifolds $V^s(x)$ and $V^u(x)$). The proofs are however of different nature.

7.5.4 The infinite dimensional case

We finish by remarking that it is possible and relevant to describe counterparts of the above results for dynamical systems in infinite dimensional spaces. It should however be noted that the theory is not yet as developed as the theory in finite dimensional spaces. Ruelle [175] was the first to obtain related results in Hilbert spaces. Later on Mañé [130] considered transformations in Banach spaces under some compactness and reversibility assumptions including the case of differentiable maps with compact derivative at each point. The results of Mañé were extended by Thieullen in [199] for a class transformations satisfying a certain asymptotic compactness. Mañé, in [130], also mentioned the possibility of applying these techniques in order to clarify some special phenomena that appear in the setting of infinite dimensional maps and flows relative to the case of semilinear parabolic equations and retarded functional differential equations.

8 Realization of Vector Fields and Normal Forms

8.1 Realization of vector fields on center manifolds

The meaning of *realization* comes from the following observation. Since the flow for a RFDE evolves in an infinite dimensional space, for any given integers N and n, it is perhaps to be expected that, for any system of ordinary differential equations of dimension N, there is an RFDE of dimension n such that the flow for the RFDE can be mapped onto the flow for the ODE. Of course, if $n \geq N$, this is the case. If $n < N$, we will observe below that there are flows defined by an N dimensional ODE which cannot be realized by an RFDE in $\mathbb{R}^n$. We verify this observation by considering the restrictions imposed on the ODE which determines the flow on a center manifold.

In the modeling of particular situations by differential equations involving delays, one wants to have the equation to be as simple as possible as well as to depend upon the least number of delays. Therefore, the above concepts should be considered in the context not only of the dimension n of the RFDE, but also with the vector field in the RFDE restricted by the models. It is therefore important to investigate the restrictions imposed on the types of flows that can occur in RFDE. Not too much is known, but there are some local results.

To be more precise, we need to recall some well known facts about linear RFDE (see, for example, Hale and Verduyn-Lunel [86] for more details and the proofs). Consider the linear equation

$$\dot{x}(t) = Lx_t \equiv \int_{-\delta}^{0} d\eta(\theta)\, x(t+\theta), \tag{8.1}$$

where $\eta(\theta)$ is an $n \times n$ matrix function of θ of bounded variation.

If we let $T_L(t), t \geq 0$, be the linear semigroup on C generated by the solutions of (8.1), then the infinitesimal generator $A_L : \mathcal{D}(A_L) \to C$ of $T_L(t), t \geq 0$, is given by

$$\mathcal{D}(A_L) = \{\varphi \in C : \frac{d\varphi}{d\theta} \in C \quad and \quad \frac{d\varphi(0)}{d\theta} = L\varphi\}$$

$$A_L\varphi(\theta) = \frac{d\varphi(\theta)}{d\theta}, \quad \theta \in [-\delta, 0]. \tag{8.2}$$

The spectrum $\sigma(A_L)$ of A_L plays an important role in the discussion of the asymptotic behavior of $T(t)$ as $t \to \infty$.

After a few elementary calculations, it can be shown that $\sigma(A_L) = \sigma P(A_L)$, where $\sigma P(A_L)$ is the point spectrum of A_L and

$$\sigma P(A_L) = \{\lambda : \det \Delta(\lambda) = 0\},$$

$$\Delta(\lambda) = \lambda I - \int_{-\delta}^{0} e^{\lambda \theta} d\eta(\theta). \tag{8.3}$$

The elements λ of $\sigma P(A_L)$ are called the *eigenvalues* of (8.1) and coincide with those values of λ for which there is a nonzero n-vector c_λ such there is a solution of (8.1) given by $c_\lambda \exp(\lambda t)$. It is easy to verify that the resolvent of A_L is compact and, as a consequence, the generalized eigenspace of any eigenvalue λ is finite dimensional. Furthermore, for any real number ρ, there are only a finite number of eigenvalues of (8.1) with real parts $\geq \rho$.

Since $T_L(t)$ is a compact linear operator for $t \geq \delta$, it follows that the spectrum $\sigma(T_L(t)) = \{0\} \cup \sigma P(T(t))$, where $\sigma P(T(t))$ designates the point spectrum. It also is well known that $\sigma P(T(t)) = \exp(\sigma P(A_L)t)$.

Let $\Lambda = \{\lambda_1, \ldots, \lambda_p\}$ be a finite set of eigenvalues of (8.1) and let $\Phi_\Lambda = \mathrm{col}\,(\varphi_1, \ldots, \varphi_m)$ be a basis for the solutions of (8.1) corresponding to generalized eigenspaces of the elements of Λ. The elements of Φ_Λ span a linear subspace P_Λ of dimension m which is invariant under the semigroup $T_L(t), t \geq 0$. Since P_Λ is invariant and Φ_Λ is a basis for P_Λ, we know that there is an $m \times m$ matrix B_Λ such that $T_L(t)\Phi_\Lambda = \Phi_\Lambda \exp(B_\Lambda t)$; that is, the flow on the subspace $P_\Lambda \subset C$ for (8.1) is equivalent to the linear ordinary differential equation on $\mathbb{R}^m$:

$$\dot{y} = B_\Lambda y. \tag{8.4}$$

The set of eigenvalues of B_Λ coincides with the set Λ and the multiplicity of an eigenvalue λ is the same as the algebraic multiplicity of λ as an element of the point spectrum of A_L. There also is a linear subspace Q_Λ of C, $P_\Lambda \cap Q_\Lambda = \{0\}$, which is invariant under $T_L(t)$ and $C = P_\Lambda \oplus Q_\Lambda$. The spectrum of $T_L(t)$ restricted to Q_Λ is all of the spectrum of $T_L(t)$ except the elements corresponding to the eigenvalues of the set Λ.

The decomposition $C = P_\Lambda \oplus Q_\Lambda$ defines a projection operator which projects C onto P_Λ along Q_Λ. It is possible to give an explicit representation of this projection by using the adjoint equation of (8.1). In fact, if

$$L\varphi = \int_{-\delta}^{0} d\eta(\theta)\varphi(\theta)d\theta, \tag{8.5}$$

then the adjoint equation is

$$\dot{w}(t) = -\int_{-\delta}^{0} w(t-\theta)d\eta(\theta), \tag{8.6}$$

where $w \in \mathbb{R}^{n*}$, the space of row n-vectors. Solutions of this equation are defined for negative t with initial functions ψ in the space $C([0, \delta], \mathbb{R}^{n*})$. If $w(t, \psi)$, $t \leq \delta$, is the solution of (8.6) through ψ at $t = 0$, and we define $w^t(\cdot, \psi) \in C([0, \delta], \mathbb{R})$ by the relation $w^t(\cdot, \psi)(\zeta) = w(t + \zeta, \psi)$, $\zeta \in [0, \delta]$ and $\tilde{T}(t)\psi = w^t(\cdot, \psi)$, then $\tilde{T}(t), t \leq 0$, is a C^0-semigroup on $C([0, \delta], \mathbb{R}^{n*})$. Associated with (8.1) and (8.6) is the bilinear form

$$(\psi, \phi) = \psi(0) \cdot \varphi(0) - \int_{-\delta}^{0} \int_{0}^{\theta} \psi(\xi - \theta) d\eta(\theta)\varphi(\xi)d\xi,$$

$$w \in C([0, \delta], \mathbb{R}^{n*}), \varphi \in C. \quad (8.7)$$

This form has the property that, if $x(t)$ is a solution of (8.1) and $w(t)$ is a solution of (8.6) on a common interval of existence, then (w^t, x_t) is independent of t.

For each $\lambda_j \in \Lambda$, $-\lambda_j$ is an eigenvalue of (8.6) with the same multiplicity. Let $\Psi_\Lambda = \mathrm{col}\,(\psi_1, \ldots, \psi_m)$ be a basis for the generalized eigenfunctions of the eigenvalues $-\lambda_1, \ldots, -\lambda_p$. If we use the bilinear form (8.7) and define the $m \times m$ matrix $(\Psi_\Lambda, \Phi_\Lambda)$ by specifying that its i, j element is $(\Psi_\Lambda, \Phi_\Lambda)_{ij} = (\psi_i, \varphi_j)$, then it can be shown that $(\Psi_\Lambda, \Phi_\Lambda)$ is nonsingular. As a consequence, we may assume without loss of generality that

$$(\Psi_\Lambda, \Phi_\Lambda) = I. \quad (8.8)$$

With this remark, we see that we can decompose elements of C in a unique way as follows:

$$\varphi = \varphi^{P_\Lambda} + \varphi^{Q_\Lambda}, \quad \varphi^{P_\Lambda} = \Phi_\Lambda(\Psi_\Lambda, \varphi), \quad (\Psi_\Lambda, \varphi^{Q_\Lambda}) = 0, \quad (8.9)$$

where, of course, $\varphi^{P_\Lambda} \in P_\Lambda$ and $\varphi^{Q_\Lambda} \in Q_\Lambda$.

Suppose that $g \in C(\mathbb{R}, \mathbb{R}^n)$ is a given function and consider the nonhomogeneous linear equation

$$\dot{x}(t) = Lx_t + g(t). \quad (8.10)$$

It is not difficult to show that any solution of (6.10) can be represented by the variation of constants formula

$$x_t = T_L(t)\varphi + \int_{0}^{t} T_L(t - s)X_0 g(s)\, ds, \quad (8.11)$$

where the function X_0 is defined by $X_0(\theta) = 0, -\delta \leq \theta < 0$, $X_0(0) = I$. There is no difficulty in showing that a solution of (8.1) exists with initial value X_0. The integral in (8.11) is to be interpreted as a regular integral for $x_t(\theta)$ for each value of $\theta \in [-\delta, 0]$.

In order to keep the notation to a minimum, let us now suppose that $n = 1$; that is, (8.12) is a *scalar* RFDE. Suppose that Λ consists of all of the eigenvalues of the linear equation (8.1) on the imaginary axis and consider the perturbed linear system

$$\dot{x}(t) = Lx_t + f(x_t) \tag{8.12}$$

where f is a C^∞-function from C to $\mathbb{R}$ and is small.

If we use the decomposition (8.9) for the solution of (8.12), then $x_t = \Phi y(t) + x_t^Q$, $y(t) = (\Psi, x_t)$. Using the variation of constants formula , it can be shown that

$$\dot{y} = By + bf(\Phi y + x_t^Q), \quad b = \Psi(0)$$

$$x_t^Q = T_L(t)\varphi^Q + \int_0^t T_L(t-s)X_0^Q f(\Phi y(s) + x_s^Q)ds. \tag{8.13}$$

Recall that the function X_0 is defined by $X_0(\theta) = 0, -\delta \le \theta < 0, X_0(0) = 1$, $X_0 = X_0^P + X_0^Q$ where $X_0^P = (\Psi, X_0) = \Psi(0)$. We remark again that the integral in (8.13) is to be interpreted as a regular integral for $x_t^Q(\theta)$ for each value of $\theta \in [-\delta, 0]$.

For each given f, we can apply the theory of invariant manifolds near the origin to equations (8.13) to determine a center manifold $CM(f)$ of (8.12) given by

$$CM(f) = \{\varphi \in C : \varphi = \Phi y + h(f, y)\} \tag{8.14}$$

where $y \in \mathbb{R}^m$ varies in a neighborhood of zero and $h(f, y) \in Q$, $h(0, y) = 0$, is C^r in y if f is a C^r-function. The flow of (8.12) on $CM(f)$ is given by $x_t = \Phi y(t) + h(f, y(t))$, where $y(t)$ is a solution of the ordinary differential equation

$$\dot{y} = By + bf(\Phi y + h(f, y)) \equiv By + bY(f, y), \tag{8.15}$$

where b is the m-vector given in (8.13).

To obtain other properties of (8.15), we use the concept of complete controllability. The pair (B, b) is said to be *completely controllable* if the matrix $[b, Bb, \ldots, B^{m-1}b]$ is nonsingular. To show complete controllability for our system, it is convenient to have B in Jordan canonical form, $B = \mathrm{diag}(D_1, \ldots, D_q)$, where $D_j = \lambda_j I + R_j$ has dimension m_j and R_j is nilpotent, $j = 1, 2, \ldots, q$. If we partition b as

$$b = \begin{pmatrix} b^{(1)} \\ \vdots \\ b^{(q)} \end{pmatrix}, \quad b^{(j)} = \begin{pmatrix} b_1^{(j)} \\ \vdots \\ b_{m_j}^{(j)} \end{pmatrix}$$

and use the fact that $-\dot{\Psi}(s) = B\Psi(s)$ for all s and Ψ is a basis, we observe that $b_{m_j}^{(j)} \ne 0, j = 1, 2, \ldots q$.

With these remarks and a few elementary computations, we see that (B, b) is completely controllable.

Since (B, b) is completely controllable, it is known that there is a nonsingular matrix C such that

$$CBC^{-1} = \begin{pmatrix} 0 & 1 & 0 & \cdots & 0 \\ 0 & 0 & 1 & \cdots & 0 \\ \vdots & \vdots & \vdots & \cdots & \vdots \\ -a_m & -a_{m-1} & -a_{m-2} & \cdots & -a_1 \end{pmatrix} \qquad Cb = \begin{pmatrix} 0 \\ \vdots \\ 0 \\ 1 \end{pmatrix}. \qquad (8.16)$$

Therefore, if we make the change of variables $y = C^{-1}z$, in (8.15), then $z = (z_1, \ldots, z_m)$ is a solution of

$$\dot{z} = CBC^{-1}z + Cb Y\,(f, C^{-1}z). \qquad (8.17)$$

From the form of CBC^{-1}, we have $z = (z_1, z_1^{(1)}, \ldots, z_1^{(m-1)})$. From (8.16), equation (8.17) is equivalent to an mth-order scalar ODE,

$$z_1^{(m)} + a_1 z_1^{(m-1)} + \ldots a_m z_1 = Y(f, C^{-1}z). \qquad (8.18)$$

These observations already yield some interesting information about RFDE. Even though our scalar RFDE is an evolutionary equation in the infinite dimensional space C, the flow on a center manifold of dimension m is restricted to the form (8.18). In particular, if $m > 1$, there are systems of ODE of dimension m which cannot be realized as the flow on a center manifold of dimension m. For example, for $m = 2$, the flow on a center manifold is given by a second order equation (8.18). Therefore, in the $(z_1, \dot{z}_1)$-plane, all equilibrium points of the flow on the center manifold are on the z_1 axis.

The basic problem is to determine the range of the mapping $f \in C^r(C, \mathbb{R}) \mapsto Y(f, \cdot) \in C^r(\mathbb{R}^m, \mathbb{R})$; that is, describe those scalar ODE of order m of the form

$$z_1^{(m)} + a_1 z_1^{(m-1)} + \ldots a_m z_1 = g(z) \qquad (8.19)$$

with $g \in C^r(\mathbb{R}^m, \mathbb{R})$, $r \geq 1$, which can be realized on the center manifold.

To accomplish this, we need the following result.

Lemma 8.1.1. *Let $\Phi = (\varphi_1, \ldots, \varphi_m)$ be a set of linearly independent functions on C. Then there exist constants $0 < \delta_j \leq \delta$, $1 \leq j \leq m - 1$, such that the $m \times m$ matrix*

$$E_m = \begin{pmatrix} \Phi(0) \\ \Phi(-\delta_1) \\ \vdots \\ \Phi(-\delta_{m-1}) \end{pmatrix}$$

is nonsingular.

Proof: For any given $\delta_j \in [-\delta, 0]$, $1 \leq j \leq m - 1$, let $\Xi_i(\delta_1, \ldots, \delta_{m-1}) = \text{col}\,(\varphi_i(0), \varphi_i(-\delta_1), \ldots, \varphi_i(-\delta_{m-1}))$, $1 \leq i \leq m$. The matrix E_m to be nonsingular is equivalent to showing that the vectors $\Xi_i(\delta_1, \ldots, \delta_{m-1})$, $1 \leq i \leq m$, are linearly independent in $\mathbb{R}^m$. We prove this latter fact by induction. It is clearly true for $m = 1$. Suppose that the statement is true for k; that is, $\Xi_i(\delta_1, \ldots, \delta_{k-1})$, $1 \leq i \leq k$, are linearly independent. If the statement is not true for $k + 1$, then there exist functions $\alpha_j(t)$, $0 \leq j \leq k$, such that

$$\Sigma_{j=1}^k \alpha_j(t) \Xi_j(\delta_1, \ldots, \delta_{k-1}, t) = \Xi_{k+1}(\delta_1, \ldots, \delta_{k-1}, t)$$

Since the upper $k \times k$ minor of this matrix has rank k, it follows that the α_j do not depend upon t. The last row in this matrix asserts that $\Sigma_{j=1}^k \alpha_k \varphi_j(t) = \varphi_{k+1}(t)$. This contradicts the fact that the $\varphi_j(t)$ are linearly independent. Thus, there must exist a t_{k+1} such that the assertion is true for $k + 1$. This proves the lemma.   ∎

Theorem 8.1.2. *Suppose that (8.12) is a scalar equation and that $Y(f, y)$ is defined as in (8.15), C as in (8.16) and $0 < \delta_1 < \cdots < \delta_{m-1}$ are such that the matrix E_m in Lemma 8.1.1 is nonsingular. For any $F \in C^r(\mathbb{R}^m, \mathbb{R})$, define $f_F \in C^r(C, \mathbb{R})$ by the relation*

$$f_F(\varphi) = F(\varphi(0), \varphi(-\delta_1), \ldots, \varphi(-\delta_{m-1})). \tag{8.20}$$

Then there are positive constants η_r, ν_r such that, for any $g \in C^r(\mathbb{R}^m, \mathbb{R})$, $\|g\|_r < \eta_r$, there exists an $F \in C^r(\mathbb{R}^m, \mathbb{R})$ with $\|F\|_r < \nu_r$ such that

$$Y(f_F, C^{-1}z) = g(z); \tag{8.21}$$

that is, in a neighborhood of zero, an arbitrary flow on a center manifold of the form (8.19) can be realized by the differential difference equation

$$\dot{x}(t) = F(x(t), x(t - \delta_1), \ldots, x(t - \delta_{m-1})) \tag{8.22}$$

with the $m - 1$ delays $\delta_1, \ldots, \delta_{m-1}$.

The first result on the realization of vector fields on center manifolds for $n = 1$ is due to Hale [76] where it was shown that the jet of every vector field could be realized (it was falsely claimed that every vector field could be realized). Faria and Magalhães [51] extended the result to n dimensions using normal forms. Rybakowski [184] used the Nash-Moser iteration technique to realize all vector fields in C^{m+15} by functions $f \in C^m$ if $m \geq 17$. Faria and Magalhães, also in [51], established that these technical conditions on regularity were not needed by showing that a necessary and sufficient condition for all vector fields

$$\dot{y} = By + G(y), \quad G \in C^r,$$

to be realized, in a neighborhood of the origin, by functions $f \in C^r$, $r \geq 1$, is that the range of the restriction of G to such a neighborhood be contained in the column space of $\Psi(0)$. The results in the text were proved by Prizzi [169] using some of the ideas of Poláčik and Rybakowski [167] in their work on realization of vector fields on center manifolds for parabolic equations. Hale [77] extended the result on jets to NFDE with a stable D operator. The proof below will yield the equivalent complete realization theorem for such NFDE.

Proof: We give the ideas of the proof without giving all of the technical estimates. To motivate it, we first recall the manner in which one proves the existence of a center manifold. In order to transfer the local problem to a global one, the vector field f is extended in a small strip containing the span of the columns of Φ in such a way that f and its derivatives up through order r are small in the strip. In terms of the variable z in (8.17), the center manifold described by a function $H(f, z)$ taking z to an element in $Q \subset C$, the complementary subspace of C to $P = [\Phi]$, satisfies the equation

$$H(f, z_0) = \int_{-\infty}^{\infty} [d_s K(-s)] C^{-1} b f(\Phi C^{-1} z(s, z_0) + H(f, z(s, z_0))) ds \quad (8.23)$$

for a certain kernel $K(t)$ for which there are positive constants k, α such that

$$|d_s K(t)| \le k e^{-\alpha|t|}, \quad t \in \mathbb{R}$$

and $z(t, z_0)$, $z(0, z_0) = z_0$ is the solution of the equation

$$\dot{z}(t) = CBC^{-1} z(t) + Cbf(\Phi C^{-1} z(t) + H(f, z(t))). \quad (8.24)$$

We make a few remarks about the kernel $K(t)$. The space Q can be decomposed as $Q = Q_+ \oplus Q_-$ with Q_+, Q_- being invariant under $T_L(t)$ and there are positive constants k, α such that

$$\|T_L(t)\|_{\mathcal{L}(Q_+, C)} \le k e^{\alpha t}, \quad t \le 0,$$

$$\|T_L(t)\|_{\mathcal{L}(Q_-, C)} \le k e^{-\alpha t}, \quad t \ge 0.$$

If we decompose X_0^Q as $X_0^Q = X_0^{Q+} + X_0^{Q-}$, then

$$K(t) C^{-1} b = -T_L(t) X_0^{Q+} \quad \text{when} \quad t > 0,$$

$$K(t) C^{-1} b = T_L(t) X_0^{Q-} \quad \text{when} \quad t < 0.$$

If we consider the right hand side of (8.23) as a map on a special class of functions H with z satisfying (8.24) for this H, then one shows that a center manifold is shown to exist by showing that there is a fixed point of this map.

Using this observation, the proof of the Theorem 8.1.2 will be complete if we show that, for any $g \in C^r(\mathbb{R}^m, \mathbb{R})$, there are functions $F \in C^r(\mathbb{R}^m, \mathbb{R})$ and $\Gamma \in C^r(\mathbb{R}^n, Q)$ such that

$$f_F(\Phi C^{-1} z + \Gamma(z)) = g(z)\lambda \quad (8.25)$$

$$\Gamma(z_0) = \int_{-\infty}^{\infty} [d_s K(-s)] f_F(\Phi C^{-1} z(s, z_0) + \Gamma(z(s, z_0))) ds \qquad (8.26)$$

where $z(t, z_0)$, $z(0, z_0) = z_0$, is the solution of the equation

$$\dot{z}(t) = CBC^{-1} z(t) + Cb f_F(\Phi C^{-1} z(t) + \Gamma(z(t))). \qquad (8.27)$$

Since $E_m C^{-1}$ has an inverse CE_m^{-1}, we may deduce from the definition of f_F that relation (8.25) is equivalent to

$$g(z) =$$
$$= F[E_m C^{-1}(z + CE_m^{-1} \mathrm{col}(\Gamma(z)(0), \Gamma(z)(-\delta_1), \dots, \Gamma(z)(-\delta_{m-1})))]. \qquad (8.28)$$

As a first step in showing that (8.28) has a solution F, Γ, we observe that the Implicit Function Theorem implies that there is a $\nu_r > 0$ such that, for any $\Gamma \in C^r(\mathbb{R}^n, Q)$ with $\|\Gamma\| < \nu_r$, there is a unique function $V(\Gamma, z) \in C^r(\mathbb{R}^m, \mathbb{R}^m)$, $V(0, z) = z$, which satisfies the equation

$$V(\Gamma, z + CE_m^{-1} \mathrm{col}(\Gamma(z)(0), \Gamma(z)(-\delta_1), \dots, \Gamma(z)(-\delta_{m-1}))) = z$$

for z sufficiently small. If we define

$$F(w) = g(V(\Gamma, CE_m^{-1} w)), \quad w \in \mathbb{R}^m,$$

then (8.25) is satisfied.

To conclude the proof, we must choose the function Γ in such a way that (8.26) is satisfied. Since (8.25) is satisfied, the right hand side of (8.26) is independent of Γ and can be used to define Γ by the formula

$$\Gamma(z_0) = \int_{-\infty}^{\infty} [d_s K(-s)] g(z(s, z_0)) ds.$$

Of course, some technical estimates are needed to show that all of the above steps are justified for η_r and ν_r sufficiently small. Performing these estimates will complete the proof of the theorem.   ∎

Theorem 8.1.2 can be generalized to n-dimensional RFDE on $C = C([-\delta, 0], \mathbb{R}^n)$. As one can imagine, much more notation is needed to describe the special characteristics of the vector fields on center manifolds (the analogue for $n = 1$ of an mth order scalar ODE). However, it is easy to state the result in a qualitative way and the reader should consult the previously mentioned references for the detailed construction. The complete proof follows the main ideas as above.

Theorem 8.1.3. *If a given linear RFDE in $C([-\delta, 0], \mathbb{R}^n)$,*

$$\dot{x}(t) = Lx_t,$$

has exactly m eigenvalues on the imaginary axis, then there are integers q,
m_j, $1 \leq j \leq q$, such that the flow on each center manifold of

$$\dot{x}(t) = Lx_t + f(x_t)$$

is determined by q functions $Y_j(f, \cdot) \in C^r(\mathbb{R}^{m_j}, \mathbb{R})$, $1 \leq j \leq q$. Furthermore,
there exist an integer p and positive constants $0 < -\delta_1 < -\delta_2 < \cdots < -\delta_{p-1}$
such that, for any given function $g_j \in C^r(\mathbb{R}^{m_j}, \mathbb{R})$, $1 \leq j \leq q$, there is a
function $F \in C^r(\mathbb{R}^m, \mathbb{R}^n)$ such that

$$Y_j(f_F, \cdot) = g_j, \quad 1 \leq j \leq q, \tag{8.29}$$

where $f_F(\varphi) = F(\varphi(0), \varphi(-\delta_1), \dots, \varphi(-\delta_{p-1}))$. Of course, all functions have
sufficiently small norms.

When considering above the realization on a center manifold of a given ODE
in $\mathbb{R}^m$: $\dot{y} = By + G(y)$, with $G \in C^r(r \geq 1)$, the $m \times m$ matrix B is obtained,
by the construction preceding (8.4), from the linear operator L which gives
the linear part of the RFDE on $C = C([-\delta, 0], \mathbb{R}^n)$ considered for realization.
The question of which conditions must be satisfied so that any given matrix
B could be obtained by RFDE was considered by Faria and Magalhães in [51]
who established that a necessary and sufficient condition is that n be larger
or equal to the largest number of Jordan blocks associated with each one of
the eigenvalues of B. The proof of this result involves the use of a result of
Pandolfi [164] on feedback stabilization of RFDE.

Let us now consider a specific situation where the functional differential
equation is a scalar, the vector field depends only upon one delay, vanishes
at the origin and the linearization near the origin has a double eigenvalue
zero, the analogue of the famous *Bogdanov–Takens singularity* for ODE. In
this case, we can write the equation as

$$\dot{x}(t) = x(t) - x(t-1) +$$
$$+ \frac{1}{2}[A_{20}x^2(t) + A_{11}x(t)x(t-1) + A_{02}x^2(t-1)] + k(x(t), x(t-1)), \tag{8.30}$$

where k vanishes at the origin together with derivatives up through order
two.

The first objective is to determine the vector field on the center manifold
up through the second order terms. In the next section, the theory of normal
forms near an equilibrium point is presented. It permits one to constructively
obtain this vector field on a center manifold up to any order.

Performing the normal form computations (see details in this Chapter
in Subsection 8.4.1, in particular Example 8.4.2), we obtain the differential
equation on the center manifold of (8.30) up through terms of second order
as

$$\dot{y}_1 = y_2$$
$$\dot{y}_2 = ay_1^2 + by_1y_2 \tag{8.31}$$

where

$$a = A_{20} + A_{11} + A_{02}, \quad b = \frac{1}{3}(2A_{20} - A_{11} - 4A_{02}). \tag{8.32}$$

To determine the nature of the flow of a perturbation of (8.31) near the origin, keeping the origin as an equilibrium point, it is known that it is sufficient to consider a two parameter family of linear perturbations of the form

$$\dot{y}_1 = y_2$$
$$\dot{y}_2 = \lambda_1 y_1 + \lambda_2 y_2 + a y_1^2 + b y_1 y_2 \tag{8.33}$$

provided that $a \neq 0$ and $b \neq 0$.

The complete bifurcation diagrams for (8.33) are known.

We can obtain the perturbations involving λ_1, λ_2 by considering a linear perturbation

$$\nu_1 x(t) - \nu_2 x(t-1)$$

obtaining

$$\lambda_1 = -2(\nu_2 - \nu_1), \quad \lambda_2 = \frac{2}{3}(\nu_2 - \nu_1) + 2\nu_2.$$

From (8.32), we observe that we can obtain all values of a, b by varying the coefficients of the second order terms in the delay differential equation. Therefore, we can obtain all flows associated with the Bogdanov–Takens singularity for ODE.

If the nonlinear terms in (8.30) depend only upon $x(t-1)$; that is, $A_{20} = 0 = A_{11}$ (which sometimes is the case in modeling), then, from (8.32), we see that a and b have opposite sign. Therefore, in this class of FDE, we cannot obtain flows which occur for (8.33) when $ab > 0$. This result is not surprising after the computations have been performed, but a priori, since the problem is infinite dimensional, one would think that we should obtain all flows for the ODE.

Related with the above question of realization of vector fields on center manifolds of RFDE by ODE, one can think of looking for restrictions on the possible local flows around singularities of an RFDE. For $n = 1$, from Theorem 8.1.2 we know that, for a given $m \times m$ matrix B, any arbitrary m^{th}-order scalar ODE of the form

$$z_1^{(m)} + a_1 z_1^{(m-1)} + \ldots a_m z_1 = g(z)$$

where the coefficients $a_1, \ldots, a_m$ are as in (8.16), is realized as the equation on the center manifold of a scalar perturbed equation (8.12) with nonlinearities depending on only a finite number of delays. For higher dimensions, a similar result is valid, where the number of delays in (8.20) necessary to assure the realization above is equal to $m - q$, m being the dimension of the center manifold and $q = rank\,\Phi(0)$ (see Faria and Magalhães [51] and Prizzi [169]). For the scalar case, in order to find restrictions on the possible flows, we should then consider scalar fields f_F defined by (8.20), where either f_F does not depend on $\varphi(0)$ or m is less than the dimension of the center manifold.

The theory of normal forms can be very useful in finding such restrictions, as the last example showed. For the Bogdanov–Takens singularity of (8.30), the normal form construction allows us to deduce that the equation on its center manifold of the origin is given by (8.31), up to second order terms. This result will be shown in Section 8.2 of the present chapter. Therefore, if one wants to obtain restrictions on the possible flows for the unfolding of the zero singularity of (8.30), significant restrictions on the values of a, b in (8.32) should be achieved. This means that, instead of considering (8.33) with all possible second order terms, one has to take into account only the terms in $x_1^2, x_1 x_2$ in the second equation, since all the others are irrelevant, in the sense that they can be annihilated and do not appear in the normal form. This illustrates the use of the theory of normal forms for RFDE and motivates its study.

Normal forms have been extensively applied to finite-dimensional ODE since the works of Birkhoff, Lyapunov and Poincaré. The main idea of this technique is to transform a nonlinear differential equation into an equation with a simpler algebraic expression, called a *normal form*, by changes of variables that eliminate all irrelevant terms from the equation, but keep the qualitative properties of the flow. Also for RFDE and other differential equations in infinite-dimensional spaces, normal forms have been considered for the ODE associated with the restriction of the flow to the center manifold of a singularity. In applications, this is of particular interest, since the qualitative behavior of solutions can be described by the flow on center manifolds.

The usual approach for constructing normal forms for RFDE, as well as for ODE in $\mathbb{R}^n$, is to assume that the center manifold reduction has been applied to the original equation, and then to compute the normal form for the finite-dimensional ODE describing the flow on this manifold. Even for the case of ODE, this method is not very efficient because it requires two steps. Moreover, this procedure does not provide a general setting to be naturally extended to infinite-dimensional spaces.

The normal form theory for RFDE presented here, first developed in [49] and [50], starts from a different perspective. It relies on the existence of center manifolds and on the classic formal adjoint theory for linear RFDE in [86], which is used to set a suitable system of coordinates near an equilibrium point. The goal is to perform a sequence of transformations of variables such that, at each step j, the change of variables effects simultaneously a projection of the original RFDE into the center manifold and removes the *non-resonant terms* of order j from the ODE on it. We point out that this idea has already been used for ODE in $\mathbb{R}^n$ (see [30], Chap. 12 and references therein). Not only is the method more efficient, in the sense that it does not require the computation of the center manifold beforehand, but also the coefficients of the normal form for the flow within the center manifold appear *explicitly* given in terms of the original RFDE.

These normal forms are also applicable to determine the flow on other finite-dimensional invariant manifolds tangent to an invariant space for the infinitesimal generator of the linearized equation at a singularity, if certain non-resonance conditions hold. We remark that situations with parameters are of fundamental interest, since the normal form theory developed here is particularly powerful in the study of bifurcation problems.

8.2 Normal forms for RFDE in finite dimensional spaces

This and the remaining sections 8.3 to 8.8 of the present chapter were written by Teresa Faria who also solved, in section 8.8 a question proposed by Hale. The present section is a concise presentation of the normal form theory developed for RFDE in [49] and [50].

We start with autonomous semilinear RFDE in $\mathbb{R}^n$ and with an equilibrium point at the origin, in the form

$$\dot{u}(t) = L(u_t) + F(u_t) \qquad (t \geq 0), \tag{8.34}$$

where $r > 0$, $C := C([-r, 0]; \mathbb{R}^n)$ is the Banach space of continuous mappings from $[-r, 0]$ to $\mathbb{R}^n$ equipped with the sup norm, $u_t \in C$ is defined by $u_t(\theta) = u(t + \theta)$ for $t \in [-r, 0]$, $L : C \to \mathbb{R}^n$ is a bounded linear operator and F is a C^k function $(k \geq 2)$, with $F(0) = 0, DF(0) = 0$. All the results presented here can be directly applied to equations whose solutions have values in $\mathbb{C}^n$, i.e., for $C = C([-r, 0]; \mathbb{C}^n)$. Consider also the linearized equation at zero,

$$\dot{u}(t) = L(u_t), \tag{8.35}$$

and suppose that η is the $n \times n$ matrix-valued function of bounded variation on $[-r, 0]$ such that $L(\varphi) = \int_{-r}^{0} d\eta(\theta)\varphi(\theta)$. As above, we still refer the reader to [86] for standard notation and results on RFDE. So, the infinitesimal generator $A : C \to C$ for the C_0-semigroup defined by the solutions of (8.35), is given by

$$A\varphi = \dot{\varphi}, \quad D(A) = \{\varphi \in C : \dot{\varphi} \in C, \dot{\varphi}(0) = L(\varphi)\},$$

and has only point spectrum, $\sigma(A) = \sigma_P(A)$. Furthermore, for any $\alpha \in \mathbb{R}$ the set $\{\lambda \in \sigma(A) : Re\lambda \geq \alpha\}$ is finite. For $\lambda \in \mathbb{C}$, we note that $\lambda \in \sigma(A)$ if and only if λ is a *characteristic value* for (8.35), that is, if λ satisfies the *characteristic equation*

$$det\,\Delta(\lambda) = 0, \quad \Delta(\lambda) := \lambda I - L(e^{\lambda \cdot} I), \tag{8.36}$$

Throughout the next sections of this chapter, for the sake of simplicity, we abuse the notation and often write $L(\varphi(\theta))$ instead of $L(\varphi)$, for $\varphi \in C$.

Let $\mathbb{R}^{n*}$ be the space of row n-vectors and $C^* := C([0, r]; \mathbb{R}^{n*})$. The *formal duality* is the bilinear form $(\cdot, \cdot)$ from $C^* \times C$ to the scalar field given by

$$(\alpha, \varphi) = \alpha(0)\varphi(0) - \int_{-r}^{0} \int_{0}^{\theta} \alpha(\xi - \theta) d\eta(\theta) \varphi(\xi) d\xi, \quad \alpha \in C^*, \varphi \in C.$$

If $\lambda \in \sigma(A)$, we denote by $\mathcal{M}_\lambda(A)$ the generalized eigenspace associated with λ. For a fixed set $\Lambda = \{\lambda_1, \ldots, \lambda_s\} \subset \sigma(A)$, let P be the generalized eigenspace associated with the eigenvalues in Λ, $P = \mathcal{M}_{\lambda_1}(A) \oplus \cdots \oplus \mathcal{M}_{\lambda_s}(A)$, P^* its dual space, and consider bases Φ, Ψ for P, P^*. The phase space C is decomposed by Λ as $C = P \oplus Q$, where $Q = \{\varphi \in C : (\Psi, \varphi) = 0\}$. For dual bases Φ, Ψ such that $(\Psi, \Phi) = I_p$, $p = \dim P = \sum_{i=1}^{s} p_i$, $p_i = \dim \mathcal{M}_{\lambda_i}(A)$, there exists a $p \times p$ real matrix B with $\sigma(B) = \Lambda$, that satisfies simultaneously

$$\dot{\Phi} = \Phi B \quad \text{and} \quad -\dot{\Psi} = B\Psi. \tag{8.37}$$

For most of the material in this section, we refer to [49], [50] for more information and proofs. To develop a normal form theory for RFDE, first it is necessary to enlarge the phase space C in such a way that Equation (8.34) can be written as an abstract ODE. An adequate phase space to accomplish this is the space BC,

$$BC := \{\psi : [-r, 0] \to \mathbb{R}^n \mid \psi \text{ is continuous on } [-r, 0), \exists \lim_{\theta \to 0^-} \psi(\theta) \in \mathbb{R}^n\},$$

with the sup norm. The elements of BC have the form $\psi = \varphi + X_0 \alpha, \varphi \in C, \alpha \in \mathbb{R}^n$, where

$$X_0(\theta) = \begin{cases} 0, & -r \leq \theta < 0 \\ I, & \theta = 0, \quad (I \text{ is the } n \times n \text{ identity matrix}), \end{cases}$$

so that BC is identified with $C \times \mathbb{R}^n$, with the norm $|\varphi + X_0 \alpha| = |\varphi|_C + |\alpha|_{\mathbb{R}^n}$.

In BC we define an extension of the infinitesimal generator A, denoted by $\tilde{A}$,

$$\tilde{A} : C^1 \subset BC \to BC, \quad \tilde{A}\varphi = \dot{\varphi} + X_0[L(\varphi) - \dot{\varphi}(0)], \tag{8.38}$$

where $D(\tilde{A}) = C^1 := \{\varphi \in C \mid \dot{\varphi} \in C\}$. Setting $v(t) = u_t \in C$, from (8.34) we have

$$\frac{dv}{dt}(0) = L(v) + F(v), \quad \frac{dv}{dt}(\theta) = \frac{dv}{d\theta}(\theta) \text{ for } \theta \in [-r, 0),$$

or simply

$$\frac{dv}{dt} = \tilde{A}v + X_0 F(v), \tag{8.39}$$

which is the abstract ODE in BC associated with (8.34). We also define an extension of the canonical projection $C = P \oplus Q \to P$ as

$$\pi : BC \to P, \quad \pi(\varphi + X_0 \alpha) = \Phi[(\Psi, \varphi) + \Psi(0)\alpha]. \tag{8.40}$$

By using integration by parts and (8.1.4), one can prove that

Lemma 8.2.1. π *is a continuous projection onto* P, *which commutes with* $\tilde{A}$ *in* C^1.

Decomposition $C = P \oplus Q$ and the above lemma allow us to decompose BC as a topological direct sum,

$$BC = P \oplus Ker\,(\pi), \tag{8.41}$$

where the subspace Q is contained in the null space of π. Therefore, Equation (8.39) is decomposed as a system of abstract ODE in $\mathbb{R}^p \times Ker\,(\pi) \equiv BC$, as follows. From (8.41), we write $v(t) \in C^1$ as $v(t) = \Phi x(t) + y(t)$, with $x(t) = (\Psi, v(t)) \in \mathbb{R}^p, y(t) = (I - \pi)v(t) \in Ker\,(\pi) \cap C^1 = Q \cap C^1 = \{\varphi \in Q : \dot{\varphi} \in C\} := Q^1$. Thus, $v(t)$ is a solution of (8.39) if and only if

$$\Phi\frac{dx}{dt}(t) + \frac{dy}{dt}(t) = \tilde{A}\Phi x(t) + (I - \pi)\tilde{A}y(t)$$

$$+\Phi\Psi(0)F(\Phi x(t) + y(t)) + (I - \pi)X_0F(\Phi x(t) + y(t)).$$

Since $\tilde{A}\Phi = \Phi B$ and $\tilde{A}\pi = \pi\tilde{A}$ in C^1, the above equation is equivalent to the system in $\mathbb{R}^p \times Ker\,(\pi)$

$$\begin{cases} \dot{x} = Bx + \Psi(0)F(\Phi x + y) \\ \dot{y} = A_{Q^1}y + (I - \pi)X_0F(\Phi x + y), \quad x \in \mathbb{R}^p, y \in Q^1 \end{cases} \tag{8.42}$$

(here the dot denotes the derivative with respect to t), where A_{Q^1} is the restriction of $\tilde{A}$ to Q^1 interpreted as an operator acting in the Banach space $Ker\,(\pi)$, i.e.,

$$A_{Q^1} : Q^1 \subset Ker\,(\pi) \rightarrow Ker\,(\pi), \quad A_{Q^1}\varphi = \tilde{A}\varphi, \quad \text{for } \varphi \in Q^1.$$

The spectrum of A_{Q^1} will be an important tool for the construction of normal forms. This is the reason why it is crucial to restrict the range of $\tilde{A}|_{Q^1}$, by considering A_{Q^1} in the space $Ker\,(\pi)$, rather than the full space BC. From the definition of $\tilde{A}$, it is obvious that $\sigma_P(\tilde{A}) = \sigma(A)$. On the other hand, $\tilde{A}$ is a closed operator and $\tilde{A} - \lambda I$ is surjective for $\lambda \in \rho(A)$, because the map $\varphi \mapsto L(\varphi) - \dot{\varphi}(0)$ is also. Thus, $\sigma(\tilde{A}) = \sigma_P(\tilde{A}) = \sigma(A)$. However, the spectrum of A_{Q^1} is smaller.

Lemma 8.2.2. $\sigma(A_{Q^1}) = \sigma_P(A_{Q^1}) = \sigma(A) \setminus \Lambda$.

Proof: . The proof is done in several steps. First, one can prove that (see [49] for details) $\sigma_P(A_{Q^1}) = \sigma(A) \setminus \Lambda$ and $\sigma(A_{Q^1}) \subset \sigma(\tilde{A})$. Since $\sigma(\tilde{A}) = \sigma(A)$, it is now sufficient to show that $\lambda \in \Lambda$ implies that $Im\,(A_{Q^1} - \lambda I) = Ker\,(\pi)$. Let $\lambda \in \Lambda$ and consider $f \in Ker\,(\pi)$. As $f = (I - \pi)f$, $\tilde{A}$ commutes with π in its domain and $C^1 \cap Ker\,(\pi) = Q^1$, then $f \in Im\,(A_{Q^1} - \lambda I)$ if and only if $f \in Im\,(\tilde{A} - \lambda I)$. Hence to justify that $Im\,(A_{Q^1} - \lambda I) = Ker\,(\pi)$ it is sufficient to show that for each $f = \phi + X_0\alpha \in BC$ with

$$(\Psi, \phi) + \Psi(0)\alpha = 0, \tag{8.43}$$

there exists $h \in C^1$ such that $(\tilde{A} - \lambda I)h = \phi + X_0\alpha$, which is equivalent to

$$\begin{cases} \dot{h} - \lambda h = \phi \\ L(h) - \dot{h}(0) = \alpha. \end{cases}$$

The solution of the first equation is $h(\theta) = e^{\lambda\theta}b + \int_0^\theta e^{\lambda(\theta-\xi)}\phi(\xi)d\xi$, where $b = h(0)$. Moreover we have $\dot{h}(0) = \lambda b + \phi(0)$. By substituting these expressions into the second equation, we conclude that there is $h \in C^1$ satisfying this system if and only if there is $b \in \mathbb{R}^n$ such that

$$\Delta(\lambda)b = -L\left(\int_0^\theta e^{\lambda(\theta-\xi)}\phi(\xi)d\xi\right) + \phi(0) + \alpha. \tag{8.44}$$

Let $\lambda = \lambda_i$ for some $i \in \{1, 2, \cdots, s\}$ and let $\{\psi_1^{\lambda_i}, \cdots, \psi_k^{\lambda_i}\}$ ($k \le p_i$) be a basis of $Ker\,(A^* - \lambda_i I)$, where A^* is the formal adjoint of A. It follows from [86] that

$$\psi_j^{\lambda_i}(s) = e^{-\lambda_i s}x_j^*, \quad s \in [0, r], \quad j = 1, 2, \cdots, k,$$

where $\{x_1^*, \cdots, x_k^*\}$ is a basis of $N := \{x^* \in \mathbb{R}^{n*} : x^*\Delta(\lambda) = 0\}$. Now (8.43) clearly implies that for $j = 1, \cdots, k$,

$$0 = (\psi_j^{\lambda_i}, \phi) + \psi_j^{\lambda_i}(0)\alpha = x_j^*\phi(0) - x_j^*L\left(\int_0^\theta e^{\lambda_i(\theta-\xi)}\phi(\xi)d\xi\right) + x_j^*\alpha.$$

That is, $-L\left(\int_0^\theta e^{\lambda_i(\cdot-\xi)}\phi(\xi)d\xi\right) + \phi(0) + \alpha \in N^\perp = Im\,(\Delta(\lambda))$, i.e., there is $b \in \mathbb{R}^n$ solution of (8.44). ∎

For the sake of applications, we are particularly interested in obtaining normal forms for equations giving the flow on center manifolds. Therefore, we turn our attention to the case where Λ is the set of eigenvalues of A on the imaginary axis. However, we shall analyze situations corresponding to other choices of Λ. In the following, we use formal series, although in applications only a few terms of these series will be computed. The subject of the convergence of these series is irrelevant for our purposes and is not addressed here.

Suppose $\Lambda = \{\lambda \in \sigma(A) : Re\,\lambda = 0\} \ne \emptyset$. The center manifold theorem assures that there is an p-dimensional invariant manifold of (8.34) tangent to the center space P of (8.35) at zero. In the center manifold, which has the form $\{\varphi \in C : \varphi = \Phi x + h(x), x \in V\}$ for $V \subset \mathbb{R}^p$ a neighborhood of zero and $h : V \to Q$ C^k-smooth such that $h(0) = 0, h'(0) = 0$, the solutions of (8.34) are the solutions of the ODE

$$\dot{x} = Bx + \Psi(0)F(\Phi x + h(x)). \tag{8.45}$$

We want now to linearize the function h defining the center manifold, and at the same time simplify (8.45), that is, we want to transform (8.45) into an ODE that should be in normal form.

Consider Equation (8.34) written in the form (8.42) and expand F in Taylor series as

$$F(v) = \sum_{j \geq 2} \frac{1}{j!} F_j(v), \ v \in C,$$

where F_j is jth Fréchet derivative of F. Equation (8.42) becomes

$$\begin{cases} \dot{x} = Bx + \sum_{j \geq 2} \frac{1}{j!} f_j^1(x, y) \\ \dot{y} = A_{Q^1} y + \sum_{j \geq 2} \frac{1}{j!} f_j^2(x, y), \end{cases} \qquad (8.46)$$

with $f_j := (f_j^1, f_j^2), j \geq 2$, defined by

$$f_j^1(x, y) = \Psi(0) F_j(\Phi x + y), \quad f_j^2(x, y) = (I - \pi) X_0 F_j(\Phi x + y). \qquad (8.47)$$

As for autonomous ODE in $\mathbb{R}^n$, normal forms are obtained by a recursive process of changes of variables. At each step the terms of order $j \geq 2$ are computed from the terms of the same order and the terms of lower orders already computed in previous steps, by a change of variables that has the form

$$(x, y) = (\bar{x}, \bar{y}) + \frac{1}{j!} (U_j^1(\bar{x}), U_j^2(\bar{x})), \qquad (8.48)$$

where $x, \bar{x} \in \mathbb{R}^p$, $y, \bar{y} \in Q^1$ and $U_j^1 : \mathbb{R}^p \to \mathbb{R}^p, U_j^2 : \mathbb{R}^p \to Q^1$ are homogeneous polynomials of degree j in $\bar{x}$. For each j, the aim is to choose (8.48) in such a way that all the *non-resonant* terms of degree j vanish in the transformed equation.

We describe now the algorithm for computing such normal forms. Suppose that the general changes of variables (8.48), have already been performed for $2, 3, \ldots, j - 1$. Denote by $\tilde{f}_j = (\tilde{f}_j^1, \tilde{f}_j^2)$ the terms of last order j in (x, y) obtained after these transformations, and effect then (8.48). This recursive process transforms (8.46) into

$$\begin{cases} \dot{\bar{x}} = B\bar{x} + \sum_{j \geq 2} \frac{1}{j!} g_j^1(\bar{x}, \bar{y}) \\ \dot{\bar{y}} = A_{Q^1} \bar{y} + \sum_{j \geq 2} \frac{1}{j!} g_j^2(\bar{x}, \bar{y}), \end{cases} \qquad (8.49)$$

where $g_j := (g_j^1, g_j^2)$ are the new terms of order j given by

$$g_j^1(x, y) = \tilde{f}_j^1(x, y) - [DU_j^1(x)Bx - BU_j^1(x)]$$

$$g_j^2(x, y) = \tilde{f}_j^2(x, y) - [DU_j^2(x)Bx - A_{Q^1}(U_j^2(x))], \ j \geq 2.$$

We introduce some notation: for $j, p \in \mathbb{N}$ and Y a normed space, let $V_j^p(Y)$ denote the space of homogeneous polynomials of degree j in p variables, $x = (x_1, \ldots, x_p)$, with coefficients in Y, $V_j^p(Y) = \{\sum_{|q|=j} c_q x^q : q \in \mathbb{N}_0^p, c_q \in$

$Y\}$, with the norm $|\sum_{|q|=j} c_q x^q| = \sum_{|q|=j} |c_q|_Y$. Define also the operators $M_j = (M_j^1, M_j^2), j \geq 2$, by

$$M_j^1 : V_j^p(\mathbb{R}^p) \to V_j^p(\mathbb{R}^p),$$
$$(M_j^1 h_1)(x) = Dh_1(x)Bx - Bh_1(x);$$
$$M_j^2 : V_j^p(Q^1) \subset V_j^p(Ker(\pi)) \to V_j^p(Ker(\pi)),$$
$$(M_j^2 h_2)(x) = D_x h_2(x)Bx - A_{Q^1}(h_2(x)). \qquad (8.50)$$

Setting $U_j = (U_j^1, U_j^2)$, it is clear that

$$g_j = \tilde{f}_j - M_j U_j. \qquad (8.51)$$

Note that the operators M_j^1 are defined by the Lie brackets that appear in computing normal forms for finite-dimensional ODE. The infinite-dimensional part in the transformation formulas is handled through the operators M_j^2. Clearly, the ranges of M_j^1, M_j^2 contain exactly the terms that can be removed from the equation. They are determined (in general not in a unique way) by the choices of complementary spaces for $Im\,(M_j)$. Similarly, the functions U_j depend on the choices of complementary spaces for $Ker\,(M_j)$. Naturally, the situation $Im\,(M_j^2) = V_j^p(Ker\,(\pi)), j \geq 2$, is of particular interest, since it allows us to choose U_j^2 such that $g_j^2(x,0) = \tilde{f}_j^2(x,0) - (M_j^2 U_j^2)(x) = 0$, so that the equation $y = h(x)$ defining the center manifold, where h is as in (1.12), is transformed into the equation $y = 0$ (up to a certain order k, if F is C^k-smooth). Hence, it is important to characterize the spectrum of $M_j^2, j \geq 2$.

Lemma 8.2.3. *Let Λ be a nonempty finite subset of $\sigma(A)$. The linear operators $M_j^2, j \geq 2$, are closed and their spectra are*

$$\sigma(M_j^2) = \sigma_P(M_j^2) = \{(q, \bar{\lambda}) - \mu : \mu \in \sigma(A_{Q^1}), q \in \mathbb{N}_0^p, |q| = j\},$$

where $\bar{\lambda} = (\lambda_1, \dots \lambda_p)$, $\lambda_1, \dots, \lambda_p$ are the elements of Λ, counting multiplicities, and $(q, \bar{\lambda}) = q_1\lambda_1 + \cdots + q_p\lambda_p$, $|q| = q_1 + \cdots + q_p$, for $q = (q_1, \dots, q_p)$.

Proof: . By using arguments as for finite dimensional ODE in [30], pp. 408-410], we obtain

$$\sigma_P(M_j^2) = \{(q, \bar{\lambda}) - \mu : \mu \in \sigma(A_{Q^1}), q \in \mathbb{N}_0^p, |q| = j\}.$$

The proof of $\sigma(M_j^2) = \sigma_P(M_j^2)$ can be found in [49]. In both cases, the proofs are algebraic and include an inductive reasoning. ∎

Theorem 8.2.4. *Let $\Lambda = \{\lambda \in \sigma(A) : Re\,\lambda = 0\} \neq \emptyset$ and consider the space BC decomposed by Λ, $BC \equiv \mathbb{R}^p \times Ker\,(\pi)$. Then, there exists a formal change of variables $(x, y) = (\bar{x}, \bar{y}) + O(|\bar{x}|^2)$, such that*

i) Equation (8.46) is transformed into Equation (8.49), where $g_j^2(\bar{x}, 0) = 0, j \geq 2$;

ii) a local center manifold for Equation (8.34) at zero satisfies $\bar{y} = 0$; furthermore, the flow on it is given by the ODE

$$\dot{\bar{x}} = B\bar{x} + \sum_{j \geq 2} \frac{1}{j!} g_j^1(\bar{x}, 0), \quad \bar{x} \in \mathbb{R}^p, \tag{8.52}$$

which is in normal form (in the usual sense of normal forms for ODE).

Proof: From Lemma 8.2.2, $\sigma(A_{Q^1}) = \sigma(A) \setminus \Lambda$. Then, for $\mu \in \sigma(A_{Q^1}), q \in \mathbb{N}_0^p, |q| = j$, we have $Re\,[(q, \bar{\lambda}) - \mu] = -Re\,\mu \neq 0$, and Lemma 8.2.3 implies that $0 \in \rho(M_j^2)$ and $Im\,(M_j^2) = V_j^p(Ker\,(\pi)), j \geq 2$. It is then possible to choose U_j^2 so that $\tilde{f}_j^2(x, 0) = (M_j^2 U_j^2)(x)$, and (i) follows from (8.51). Clearly, for (8.49) a local center manifold is now given by $\bar{y} = 0$, and (8.52) describes the flow on it. For adequate choices of $U_j^1, j \geq 2$, this ODE in $\mathbb{R}^p$ is in normal form, since the operators M_j^1 defined in (8.50) coincide with those operators defined for computing normal forms for ODE in $\mathbb{R}^p$ (see [30], [69]). ∎

Suppose now that another nonempty finite subset Λ of $\sigma(A)$ is chosen, and consider the decomposition (8.41) of BC by Λ. Assume that there exists a local invariant manifold $\mathcal{M}_{\Lambda, F}$ for (8.34) tangent to P at zero. For instance, if $\Lambda = \{\lambda \in \sigma(A) : Re\,\lambda \geq 0\} \neq \emptyset$, P is the center-unstable space for the linear equation $\dot{u}(t) = L(u_t)$ and $\mathcal{M}_{\Lambda, F}$ is the center-unstable manifold for (8.34) at zero. In general, provided the existence and regularity of $\mathcal{M}_{\Lambda, F}$, we obtain a similar result to the one stated above for the case of center manifolds, if some additional non-resonance conditions are assumed.

Definition 8.2.5. *Let Λ be a nonempty finite subset of $\sigma(A)$. Equation (8.46) (or Equation (8.34)) is said to satisfy the* **non-resonance conditions relative to Λ** *if*

$$(q, \bar{\lambda}) \neq \mu, \quad \text{for all } \mu \in \sigma(A) \setminus \Lambda, q \in \mathbb{N}_0^p, |q| \geq 2, \tag{8.53}$$

where $\bar{\lambda}$ and $(q, \bar{\lambda})$ are as in Lemma 8.2.3.

From Lemmas 8.2.2 and 8.2.3, if (8.53) holds then $0 \in \rho(M_j^2)$ and $Im\,(M_j^2) = V_j^p(Ker\,(\pi))$, for all $j \geq 2$, and we can state the following:

Theorem 8.2.6. *If (8.53) is satisfied, the statements in Theorem 8.2.4 are valid for other invariant manifolds associated with other nonempty finite subsets Λ of $\sigma(A)$, assuming that these manifolds exist. In particular, they are valid for the case of center-unstable and unstable manifolds.*

For $\Lambda = \{\lambda \in \sigma(A) : Re\,\lambda = 0\} \neq \emptyset$ as before, or in a more general setting for Λ such that (8.53) holds, we give now the definition of normal forms relative to Λ.

Definition 8.2.7. *If the non-resonance conditions (8.53) are satisfied, Equation (8.49) is said to be a* **normal form** *for Equation (8.46) (or Equation (8.34))* **relative to** Λ *if* $g_j = (g_j^1, g_j^2)$ *are defined by (8.51), with* $U_j^2(x) = (M_j^2)^{-1} \tilde{f}_j^2(x, 0)$ *and* U_j^1 *($j \geq 2$) are chosen in such a way that the ODE (8.52) is in normal form.*

Remark 8.2.8. Equation (8.52) is in normal form if U_j^1 are chosen according to the method of normal forms for finite dimensional ODE. This means that $U_j^1(x) = (M_j^1)^{-1} P_j^1 \tilde{f}_j^1(x, 0), j \geq 2$, where P_j^1 is the projection of $V_j^p(\mathbb{R}^p)$ onto $Im\,(M_j^1)$ and $(M_j^1)^{-1}$ is a right inverse of M_j^1, with P_j^1 and $(M_j^1)^{-1}$ depending on the choices of complementary spaces for $Im\,(M_j^1)$ and $Ker\,(M_j^1)$ in $V_j^p(\mathbb{R}^p)$, respectively (see [30], [49]).

Remark 8.2.9. Consider Equation (8.34), with $F \in C^k$ for some $k \geq 2$, and assume that the non-resonance conditions (8.53) are fulfilled, but only for $|q| = j$, $2 \leq j \leq k$ (instead of $|q| \geq 2$). Using the algorithm described above, steps of order j, $2 \leq j \leq k$, can be performed through changes of variables of the form (8.48). We obtain then a *normal form relative to Λ up to k-order terms*:

$$\begin{cases} \dot{\bar{x}} = B\bar{x} + \sum_{j=2}^{k} \frac{1}{j!} g_j^1(\bar{x}, \bar{y}) + h.o.t. \\ \dot{\bar{y}} = A_{Q^1}\bar{y} + \sum_{j=2}^{k} \frac{1}{j!} g_j^2(\bar{x}, \bar{y}) + h.o.t., \end{cases}$$

where $h.o.t$ stands for higher order terms. The first equation at $\bar{y} = 0$ gives the normal form up to k-order terms on the invariant manifold associated with Λ, if it exists, while the invariant manifold is given by an equation in the form $\bar{y} = o(|\bar{x}|^k)$.

Remark 8.2.10. The terms $g_j^1(x, 0)$ in (8.52) are recursively given in terms of the coefficients of the original FDE (8.34), according to the following scheme (see Remark 8.2.8 for notation): First step ($j = 2$) : $\tilde{f}_2^1 = f_2^1$; $g_2^1(x, 0) = (I - P_2^1) f_2^1(x, 0)$. Second step ($j = 3$) : $U_2^1(x) = (M_2^1)^{-1} P_2^1 f_2^1(x, 0)$; $U_2^2(x) = (M_2^2)^{-1} f_2^2(x, 0)$;

$$\tilde{f}_3^1(x, 0) = f_3^1(x, 0)$$
$$+ \frac{3}{2}[(D_x f_2^1)(x, 0) U_2^1(x) + (D_y f_2^1)(x, 0) U_2^2(x) - (D_x U_2^1)(x) g_2^1(x, 0)];$$
$$g_3^1(x, 0) = (I - P_3^1) \tilde{f}_3^1(x, 0). \ldots$$

For studying bifurcation problems, we need to consider situations with parameters:

$$\dot{u}(t) = L(\alpha)(u_t) + F(u_t, \alpha), \tag{8.54}$$

where $\alpha \in V$, V a neighborhood of zero in $\mathbb{R}^m$, $L : V \to \mathcal{L}(C; \mathbb{R}^n), F : C \times V \to \mathbb{R}^n$ are, respectively, C^{k-1} and C^k functions, $k \geq 2, F(0, \alpha) = 0, D_1 F(0, \alpha) = 0$, for all $\alpha \in V$. Introducing the parameter α as a variable by adding $\dot{\alpha} = 0$, we write (8.54) as

$$\dot{u}(t) = L_0(u_t) + (L(\alpha) - L_0)(u_t) + F(u_t, \alpha) \quad (\dot{\alpha}(t) = 0), \qquad (8.55)$$

where $L_0 := L(0)$. In an obvious way, the above procedure can be repeated for (8.55), noting however that the term $(L(\alpha) - L_0)(u_t)$ is no longer of the first order, since α is taken as a variable. On the other hand, as for Equation (8.34), the infinitesimal generator of the C_0-semigroup associated with the flow of the linear equation $\dot{u}(t) = L_0(u_t), \dot{\alpha}(t) = 0$ has only point spectrum, given by $\sigma(A) \cup \{0\}$ (A being the infinitesimal generator for $\dot{u}(t) = L_0(u_t)$). Now, $\lambda = 0$ is always an eigenvalue, whose associated generalized eigenspace is $\mathcal{M}_0(A) \times \mathbb{R}^m$, with the notation $\mathcal{M}_0(A) = \{0\}$ if $0 \in \rho(A)$. In order to consider the entire generalized eigenspace associated with $\lambda = 0$, the assumption

$$0 \in \Lambda, \quad \text{whenever } 0 \in \sigma(A) \qquad (8.56)$$

is required; and the *non-resonance conditions relative to* Λ read now as

$$(q, \bar{\lambda}) \neq \mu, \quad \text{for all } \mu \in \sigma(A) \setminus \Lambda, q \in \mathbb{N}_0^p, |q| \geq 0. \qquad (8.57)$$

Writing the formal Taylor expansions $L(\alpha) = L_0 + L_1(\alpha) + \frac{1}{2}L_2(\alpha) + \cdots$, $F(v, \alpha) = \frac{1}{2}F_2(v, \alpha) + \frac{1}{3!}F_3(v, \alpha) + \cdots$, with L_j, F_j the jth Fréchet derivative of L and F in the variables α and (v, α), respectively, we note that $f_j = (f_j^1, f_j^2), j \geq 2$, are now defined by (see [50] for details)

$$\begin{aligned} f_j^1(x, y, \alpha) &= \Psi(0)[jL_{j-1}(\alpha)(\Phi x + y) + F_j(\Phi x + y, \alpha)] \\ f_j^2(x, y, \alpha) &= (I - \pi)X_0[jL_{j-1}(\alpha)(\Phi x + y) + F_j(\Phi x + y, \alpha)]. \end{aligned} \qquad (8.58)$$

Normal forms have also been developed by Weedermann [201] for *neutral functional differential equations* of the form

$$\frac{d}{dt}\Big[Du_t - G(u_t)\Big] = Lu_t + F(u_t), \qquad (8.59)$$

where $u_t \in C$, D, L are bounded linear operators from C to $\mathbb{R}^n$, with $D\phi = \phi(0) - \int_{-r}^0 d[\mu(\theta)]\phi(\theta)$ and μ is non-atomic at zero, $\mu(0) = 0$, and G, F are C^k functions , $k \geq 2$. Following the ideas above, in [201] the infinitesimal generator A for the solution operator of the linear equation

$$\frac{d}{dt}Du_t = Lu_t$$

was naturally extended to

$$\tilde{A} : C^1 \subset BC \to BC, \quad \tilde{A}\varphi = \dot{\varphi} + X_0[L(\varphi) - D\varphi'], \qquad (8.60)$$

where $\varphi' = \frac{d}{d\theta}\varphi$ and $D(\tilde{A}) = C^1$. Let $C = P \oplus Q$ and $BC = P \oplus \ker\pi$ be decomposed by a nonempty finite set $\Lambda \subset \sigma(A)$. Since the projection π in (8.40) still commutes in C^1 with the operator $\tilde{A}$ given by (8.60), in BC (8.59) is written as the abstract ODE

$$\dot{v} = \tilde{A}v + X_0[F(v) + G'(v)\dot{v}]$$

and decomposition (8.41) yields

$$\dot{x} = Bx + \Psi(0)[F(\Phi x + y) + G'(\Phi x + y)(\Phi \dot{x} + \dot{y})]$$

$$\dot{y} = A_{Q^1}y + (I - \pi)X_0[F(\Phi x + y) + G'(\Phi x + y)(\Phi \dot{x} + \dot{y})], \quad x \in \mathbb{R}^p, \quad y \in Q^1,$$

where A_{Q^1} is the restriction of $\tilde{A}$ to Q^1 interpreted as an operator acting in the Banach space $Ker(\pi)$. Suppose now that there is an invariant manifold tangent to the generalized space P associated with Λ. Define again the operators M_j^1, M_j^2 by expression (8.50), where now $\tilde{A}$ is as in (8.60). Lemmas 8.2.2 and 8.2.3 are true in this setting. The algorithm to compute normal forms on invariant manifolds for neutral equations (8.59) is developed along the lines of the method for FDE (8.34), and the conclusions of Theorems 8.2.4 and 8.2.6 follow. See [201] for details.

8.3 Applications to Hopf bifurcation

In this section, we show some applications of the theory of normal forms to the study of Hopf bifurcations. First, we consider the general case of scalar RFDE with a Hopf singularity, and obtain explicit formulas for the relevant coefficients of the equation giving the flow within the center manifold, following results in [50]. A second illustration concerns a Hopf bifurcation for a predator-prey system with one or two delays, where one of the delays is taken as the bifurcation parameter. This model will be generalized to a delayed predator-prey system with diffusion and Neumann conditions in Section 8.6. We mention that normal forms have been applied to the study of Hopf bifurcations for RFDE modeling situations that include neural networks and several equations appearing in population dynamics.

8.3.1 Hopf Bifurcation for Scalar RFDE: The General Case

Consider a scalar RFDE in $C([-r, 0]; \mathbb{R})$ in the form (8.54),

$$\dot{u}(t) = L(\alpha)(u_t) + F(u_t, \alpha), \tag{8.61}$$

with $\alpha \in V$, V a neighborhood of zero in $\mathbb{R}$, $L : V \to \mathcal{L}(C; \mathbb{R})$ a C^2 function and $F : C \times V \to \mathbb{R}$ a C^3 function such that $F(0, \alpha) = 0, D_1F(0, \alpha) = 0$, for all $\alpha \in V$. For $\alpha = 0$, let $L_0 = L(0)$ and A be the infinitesimal generator of the semigroup defined by the linear equation $\dot{u}(t) = L_0 u_t$. Since our purpose is to discuss the Hopf singularity for a general scalar FDE, for the characteristic equation

$$\Delta(\lambda, \alpha) := \lambda - L(\alpha)e^\lambda = 0 \tag{8.62}$$

we now assume the following hypotheses:

[i] there is a pair of simple characteristic roots $\gamma(\alpha) \pm i\omega(\alpha)$ of (8.61) crossing transversaly the imaginary axis at $\alpha = 0$:

$$\gamma(0) = 0, \quad \omega := \omega(0) > 0, \quad \gamma'(0) \neq 0 \quad (Hopf \ condition);$$

(8.63)

[ii] the characteristic equation $\Delta(\lambda, 0) = 0$ has no other roots with zero real parts.

(8.64)

Define Λ as the set of eigenvalues of A on the imaginary axis, $\Lambda = \{i\omega, -i\omega\}$. For computing normal forms for (8.61), it is useful to consider complex coordinates in order to obtain a diagonal matrix B in the normal form (8.49). Thus, we take C as the complex Banach space $C = C([-r, 0]; \mathbb{C})$. Decomposing C and the enlarged phase space BC by Λ as $C = P \oplus Q$, $BC = P \oplus Ker \, \pi$, we choose bases Φ, Ψ for P, P^*:

$$P = span \, \Phi, \quad \Phi(\theta) = (\varphi_1(\theta), \varphi_2(\theta)) = (e^{i\omega\theta}, e^{-i\omega\theta}), \quad -r \leq \theta \leq 0$$
$$P^* = span \, \Psi, \quad \Psi(s) = col \, (\psi_1(s), \psi_2(s)) = col \, (\psi_1(0)e^{-i\omega s}, \psi_2(0)e^{i\omega s}),$$
$$0 \leq s \leq r,$$

with $(\Psi, \Phi) = I$ if

$$\psi_1(0) = (1 - L_0(\theta e^{i\omega\theta}))^{-1}, \quad \psi_2(0) = \overline{\psi_1(0)}. \tag{8.65}$$

Note that $\psi_1(0)$ is well-defined since $\frac{d}{d\lambda}\Delta(i\omega, 0) = 1 - L_0(\theta e^{i\omega\theta}) \neq 0$, because $i\omega$ is a simple root of the characteristic equation $\Delta(\lambda, 0) = 0$. We have $\dot{\Phi} = \Phi B$, $-\dot{\Psi} = B\Psi$ for $B = diag \, (i\omega, -i\omega)$. Since B is diagonal, it follows that the operators M_j^1 defined in (1.17) have a diagonal matrix representation in the canonical basis of $V_j^3(\mathbb{C}^2)$, so that $V_j^3(\mathbb{C}^2) = Im \, (M_j^1) \oplus Ker \, (M_j^1)$, $j \geq 2$.

We write the Taylor formulas

$$F(\varphi, \alpha) = \frac{1}{2}F_2(\varphi, \alpha) + \frac{1}{3!}F_3(\varphi, \alpha) + h.o.t$$

$$L(\alpha) = L_0 + \alpha L_1 + \frac{\alpha^2}{2}L_2 + h.o.t.$$

and define $f_j^1(x, y, \alpha), f_j^2(x, y, \alpha), j = 2, 3$, as in (8.58). Remark that here, $f_j^1(x, y, \alpha)$ and $f_j^2(x, y, \alpha)$ are homogeneous polynomials in (x, y, α) of degree j, $j = 2, 3$, with coefficients in $\mathbb{C}^2, Ker \, \pi$, respectively. Since the Hopf bifurcation is generically determined at third order, for (8.61) we shall compute a normal form on the center manifold up to third order terms. This normal form is written as

$$\dot{x} = Bx + \frac{1}{2}g_2^1(x, 0, \alpha) + \frac{1}{3!}g_3^1(x, 0, \alpha) + h.o.t., \tag{8.66}$$

where $h.o.t.$ stands for higher order terms, and g_2^1, g_3^1 are the second and third order terms in (x, α), respectively, given by

$$\frac{1}{2}g_2^1(x,0,\alpha) = \frac{1}{2}Proj_{Ker(M_2^1)}f_2^1(x,0,\alpha)$$

$$\frac{1}{3!}g_3^1(x,0,\alpha) = \frac{1}{3!}Proj_{Ker(M_3^1)}\tilde{f}_3^1(x,0,\alpha).$$

Consider the operators M_j^1 defined in (8.50). For the present situation, we get

$$M_j^1(\alpha^l x^q e_k) = i\omega(q_1 - q_2 + (-1)^k)\alpha^l x^q e_k, \qquad (8.67)$$

for $l + q_1 + q_2 = j, k = 1,2, j = 2,3, q = (q_1,q_2) \in \mathbb{N}_0^2, l \in \mathbb{N}_0$ and $\{e_1,e_2\}$ the canonical basis for $\mathbb{C}^2$. Hence,

$$Ker\,(M_2^1) = span\left\{\begin{pmatrix}x_1\alpha\\0\end{pmatrix}, \begin{pmatrix}0\\x_2\alpha\end{pmatrix}\right\}$$

$$Ker\,(M_3^1) = span\left\{\begin{pmatrix}x_1^2 x_2\\0\end{pmatrix}, \begin{pmatrix}x_1\alpha^2\\0\end{pmatrix}, \begin{pmatrix}0\\x_1 x_2^2\end{pmatrix}, \begin{pmatrix}0\\x_2\alpha^2\end{pmatrix}\right\}. \qquad (8.68)$$

From (8.58), we have

$$\frac{1}{2}f_2^1(x,y,\alpha) = \Psi(0)[\alpha L_1(\Phi x + y) + \frac{1}{2}F_2(\Phi x + y,\alpha)]$$

$$\frac{1}{2}f_2^2(x,y,\alpha) = (I - \pi)X_0[\alpha L_1(\Phi x + y) + \frac{1}{2}F_2(\Phi x + y,\alpha)]$$

$$\frac{1}{3!}f_3^1(x,y,\alpha) = \Psi(0)[\frac{\alpha^2}{2}L_2(\Phi x + y) + \frac{1}{3!}F_3(\Phi x + y,\alpha)].$$

Since $F(0,\alpha) = 0, D_1F(0,\alpha) = 0$ for all $\alpha \in \mathbb{R}$, we write

$$F_2(\Phi x,\alpha) = F_2(\Phi x,0) = A_{(2,0,0)}x_1^2 + A_{(1,1,0)}x_1 x_2 + A_{(0,2,0)}x_2^2$$

$$F_3(\Phi x,\alpha) = \sum_{|(q_1,q_2,l)|=3} A_{(q_1,q_2,l)}x_1^{q_1} x_2^{q_2}\alpha^l,$$

where $A_{(q_2,q_1,l)} = \overline{A_{(q_1,q_2,l)}}$ for $q_1,q_2,l \in \mathbb{N}_0, |(q_1,q_2,l)| = q_1 + q_2 + l = j, j = 2,3$, and $A_{(1,0,2)} = A_{(0,1,2)} = A_{(0,0,3)} = 0$. Thus, the second order terms in (α,x) of the normal form (8.66) are given by

$$\frac{1}{2}g_2^1(x,0,\alpha) = \begin{pmatrix}B_1 x_1\alpha\\\bar{B}_1 x_2\alpha\end{pmatrix}, \quad \text{with} \quad B_1 = \psi_1(0)L_1(\varphi_1). \qquad (8.69)$$

The coefficient B_1 can be expressed in terms of the characteristic value $\lambda(\alpha) = \gamma(\alpha) + i\omega(\alpha)$ which crosses the positive imaginary axis at $\alpha = 0$. In fact, an implicit function argument applied to the equation $\Delta(\lambda,\alpha) = 0$ near the solution $(i\omega,0)$ shows that $\lambda'(0) = L_1(e^{i\omega\theta})/(1 - L_0(\theta e^{i\omega\theta})) = B_1$.

Now we compute the meaningful cubic terms for the ODE in normal form giving the flow on the center manifold. From (8.68) we first note that

$$g_3^1(x,0,\alpha) = Proj_{Ker(M_3^1)}\tilde{f}_3^1(x,0,\alpha) = Proj_S\tilde{f}_3^1(x,0,0) + O(|x|\alpha^2),$$

for $S := span\left\{ \begin{pmatrix} x_1^2 x_2 \\ 0 \end{pmatrix}, \begin{pmatrix} 0 \\ x_1 x_2^2 \end{pmatrix} \right\}$. The term $\tilde{f}_3^1(x,0,\alpha)$ is given in Remark 8.2.10. Since $g_2^1(x,0,0) = 0$, the expression of $g_3^1(x,0,0)$ is simplified and given by

$$g_3^1(x,0,0) = Proj_S f_3^1(x,0,0)+$$
$$+ \frac{3}{2} Proj_S \left[(D_x f_2^1)(x,0,0)U_2^1(x,0) + (D_y f_2^1)(x,0,0)U_2^2(x,0) \right], \quad (8.70)$$

where

$$U_2^1(x,0) = U_2^1(x,\alpha)|_{\alpha=0} = (M_2^1)^{-1} Proj_{Im(M_2^1)} f_2^1(x,0,0)$$

so,

$$U_2^1(x,0) = (M_2^1)^{-1} f_2^1(x,0,0)$$

and $U_2^2(x,0) = U_2^2(x,\alpha)|_{\alpha=0}$ is determined by the equation

$$(M_2^2 U_2^2)(x,0) = f_2^2(x,0,0).$$

The elements of the canonical basis of $V_2^3(\mathbb{C}^2)$ are

$$\begin{pmatrix} x_1^2 \\ 0 \end{pmatrix}, \begin{pmatrix} x_1 x_2 \\ 0 \end{pmatrix}, \begin{pmatrix} x_2^2 \\ 0 \end{pmatrix}, \begin{pmatrix} x_1 \alpha \\ 0 \end{pmatrix}, \begin{pmatrix} x_2 \alpha \\ 0 \end{pmatrix}, \begin{pmatrix} \alpha^2 \\ 0 \end{pmatrix}$$

$$\begin{pmatrix} 0 \\ x_1^2 \end{pmatrix}, \begin{pmatrix} 0 \\ x_1 x_2 \end{pmatrix}, \begin{pmatrix} 0 \\ x_2^2 \end{pmatrix}, \begin{pmatrix} 0 \\ x_1 \alpha \end{pmatrix}, \begin{pmatrix} 0 \\ x_2 \alpha \end{pmatrix}, \begin{pmatrix} 0 \\ \alpha^2 \end{pmatrix}.$$

From (8.67) we deduce that the images of these elements under $\frac{1}{i\omega} M_2^1$ are, respectively,

$$\begin{pmatrix} x_1^2 \\ 0 \end{pmatrix}, -\begin{pmatrix} x_1 x_2 \\ 0 \end{pmatrix}, -3\begin{pmatrix} x_2^2 \\ 0 \end{pmatrix}, \begin{pmatrix} 0 \\ 0 \end{pmatrix}, -2\begin{pmatrix} x_2 \alpha \\ 0 \end{pmatrix}, -\begin{pmatrix} \alpha^2 \\ 0 \end{pmatrix}$$

$$3\begin{pmatrix} 0 \\ x_1^2 \end{pmatrix}, \begin{pmatrix} 0 \\ x_1 x_2 \end{pmatrix}, -\begin{pmatrix} 0 \\ x_2^2 \end{pmatrix}, 2\begin{pmatrix} 0 \\ x_1 \alpha \end{pmatrix}, \begin{pmatrix} 0 \\ 0 \end{pmatrix}, \begin{pmatrix} 0 \\ \alpha^2 \end{pmatrix}.$$

Hence,

$$U_2^1(x,0) = \frac{1}{i\omega} \begin{pmatrix} \psi_1(0)(A_{(2,0,0)}x_1^2 - A_{(1,1,0)}x_1 x_2 - \frac{1}{3}A_{(0,2,0)}x_2^2) \\ \psi_2(0)(\frac{1}{3}A_{(2,0,0)}x_1^2 + A_{(1,1,0)}x_1 x_2 - A_{(0,2,0)}x_2^2) \end{pmatrix}.$$

A few computations show that

$$Proj_S f_3^1(x,0,0) = \begin{pmatrix} \psi_1(0)A_{(2,1,0)}x_1^2 x_2 \\ \psi_2(0)A_{(1,2,0)}x_1 x_2^2 \end{pmatrix}$$

and

$$Projs[(D_x f_2^1)(x,0,0)U_2^1(x,0)] = \begin{pmatrix} C_1 x_1^2 x_2 \\ \overline{C_1} x_1 x_2^2 \end{pmatrix},$$

with

$$C_1 = \frac{i}{\omega}\left(\psi_1^2(0)A_{(2,0,0)}A_{(1,1,0)} - |\psi_1(0)|^2(\frac{2}{3}|A_{(2,0,0)}|^2 + A_{(1,1,0)}^2)\right). \quad (8.71)$$

Now we determine $Projs[(D_y f_2^1)(x,0,0)U_2^2(x,0)]$. Let $H_2 : C \times C \to \mathbb{C}$ be the bilinear symmetric form such that $F_2(u,\alpha) = F_2(u,0) = H_2(u,u)$ for $u \in C$ (i.e., $F_2(u,0)$ is the quadratic form associated with $H_2(u,v)$). Then,

$$(D_y f_2^1)|_{y=0,\alpha=0}(h) = 2\Psi(0)H_2(\Phi x, h).$$

Define $h = h(x)(\theta)$ by $h(x) = U_2^2(x,0)$, and write

$$h(x) = h_{20}x_1^2 + h_{11}x_1 x_2 + h_{02}x_2^2,$$

where $h_{20}, h_{11}, h_{02} \in Q^1$. The coefficients of h are determined by $(M_2^2 h)(x) = f_2^2(x,0,0)$, which is equivalent to (see (8.50))

$$D_x h(x)Bx - A_{Q^1}(h(x)) = (I - \pi)X_0 F_2(\Phi x, 0).$$

Applying the definition of A_{Q^1} and π, we get

$$\dot{h}(x) - D_x h(x)Bx = \Phi\Psi(0)F_2(\Phi x, 0)$$

$$\dot{h}(x)(0) - L_0 h(x) = F_2(\Phi x, 0),$$

where $\dot{h}$ denotes the derivative of $h(x)(\theta)$ relative to θ. Matching the coefficients of $x_1^2, x_1 x_2$ and x_2^2 of these equations, one can verify that $\overline{h_{02}} = h_{20}$ and $\overline{h_{11}} = h_{11}$, and that h_{20}, h_{11} are respectively the solutions of the equations

$$\dot{h}_{20} - 2i\omega h_{20} = A_{(2,0,0)}\Phi\Psi(0)$$
$$\dot{h}_{20}(0) - L_0(h_{20}) = A_{(2,0,0)} \quad (8.72)$$

and

$$\dot{h}_{11} = A_{(1,1,0)}\Phi\Psi(0)$$
$$\dot{h}_{11}(0) - L_0(h_{11}) = A_{(1,1,0)}. \quad (8.73)$$

From this, we deduce that

$$Projs[(D_y f_2^1)h](x,0,0) = \begin{pmatrix} C_2 x_1^2 x_2 \\ \overline{C_2} x_1 x_2^2 \end{pmatrix},$$

with

$$C_2 = 2\psi_1(0)[H_2(\varphi_1, h_{11}) + H_2(\varphi_2, h_{20})]. \quad (8.74)$$

Hence

$$\frac{1}{3!}g_3^1(x,0,0) = \begin{pmatrix} B_2 x_1^1 x_2 \\ \overline{B_2} x_1 x_2^2 \end{pmatrix},$$

with

$$B_2 = \frac{1}{3!}\psi_1(0)A_{(2,1,0)} + \frac{1}{4}(C_1 + C_2).$$

Therefore, the normal form (8.66) reads as

$$\dot{x} = Bx + \begin{pmatrix} B_1 x_1 \alpha \\ \overline{B_1} x_2 \alpha \end{pmatrix} + \begin{pmatrix} B_2 x_1^2 x_2 \\ \overline{B_2} x_1 x_2^2 \end{pmatrix} + O(|x|\alpha^2 + |x|^4). \tag{8.75}$$

The change to real coordinates w, where $x_1 = w_1 - iw_2, x_2 = w_1 + iw_2$, followed by the use of polar coordinates (ρ, ξ), $w_1 = \rho \cos \xi, w_2 = \rho \sin \xi$, transforms this normal form into

$$\begin{cases} \dot{\rho} = K_1 \alpha \rho + K_2 \rho^3 + O(\alpha^2 \rho + |(\rho, \alpha)|^4) \\ \dot{\xi} = -\omega + O(|(\rho, \alpha)|), \end{cases} \tag{8.76}$$

with $K_1 = Re\, B_1$, $K_2 = Re\, B_2$.

If $K_2 \neq 0$, which is the case of the generic Hopf bifurcation, it is well-known that the direction of the bifurcation and the stability of the nontrivial periodic orbits are determined by the sign of $K_1 K_2$ and of K_2, respectively (e.g. [30], [32]). The computation of K_1, K_2 requires now the resolution of (8.72) and (8.73). By solving these systems, we get

$$h_{22}(\theta) = A_{(2,0,0)}\left(\frac{e^{2i\omega\theta}}{2i\omega - L_0(e^{2i\omega\theta})} - \frac{\psi_1(0)e^{i\omega\theta}}{i\omega} - \frac{\overline{\psi_1(0)}e^{-i\omega\theta}}{3i\omega} \right)$$

$$h_{11}(\theta) = A_{(1,1,0)}\left(-\frac{1}{L_0(1)} + \frac{1}{i\omega}\left(\psi_1(0)e^{i\omega\theta} - \overline{\psi_1(0)}e^{-i\omega\theta} \right) \right).$$

Plugging these functions in the formula for C_2 and using (8.69), (8.71), (8.74) and (8.75), the above considerations lead to the following result:

Theorem 8.3.1. *For Equation* (8.61), *assume hypotheses* (8.63) *and* (8.64). *Then a Hopf bifurcation occurs from* $\alpha = 0$ *on a 2-dimensional local center manifold of the origin. On this manifold, the flow is given in polar coordinates by equation* (8.76), *with*

$$K_1 = Re\,(\psi_1(0)L_1(e^{i\omega\theta}))$$

$$K_2 = \frac{1}{3!}Re\,[\psi_1(0)A_{(2,1,0)}] - \frac{A_{(1,1,0)}}{2L_0(1)}Re\,[\psi_1(0)H_2(e^{i\omega\theta},1)] \tag{8.77}$$

$$+ \frac{1}{2}Re\left[\frac{\psi_1(0)A_{(2,0,0)}}{2i\omega - L_0(e^{2i\omega\theta})} H_2(e^{-i\omega\theta}, e^{2i\omega\theta}) \right].$$

We observe that the constants appearing in the coefficients K_1, K_2 are directly evaluated in terms of the functions L, F in (8.61):

$$\psi_1(0) = (1 - L_0(\theta e^{i\omega\theta}))^{-1}$$

$$K_1 = Re\,(\psi_1(0)L_1(e^{i\omega\theta})) = \gamma'(0)$$

$$A_{(2,0,0)} = F_2(e^{i\omega\theta}, 0)$$

$$A_{(1,1,0)} = 2H_2(e^{i\omega\theta}, e^{-i\omega\theta}) = 2[F_2(\cos\omega\theta, 0) + F_2(\sin\omega\theta, 0)]$$

$$H_2(e^{i\omega\theta}, 1) = \tfrac{1}{4}[F_2(1 + e^{i\omega\theta}, 0) - F_2(1 - e^{i\omega\theta}, 0)]$$

$$H_2(e^{-i\omega\theta}, e^{2i\omega\theta}) = \tfrac{1}{4}[F_2(e^{-i\omega\theta} + e^{2i\omega\theta}, 0) - F_2(e^{-i\omega\theta} - e^{2i\omega\theta}, 0)]$$

$$A_{(2,1,0)} = \tfrac{1}{8}[F_3(2e^{i\omega\theta} + e^{-i\omega\theta}, 0) - F_3(2e^{i\omega\theta} - e^{-i\omega\theta}, 0) -$$

$$-2F_3(e^{-i\omega\theta}, 0)].$$

Example 8.3.2. Consider the Wright equation

$$\dot{u}(t) = -au(t-1)[1 + u(t)]. \tag{8.78}$$

This equation has been studied by many authors (e.g. [32]). Nevertheless, we use it here as an illustration of Theorem 8.3.1. The characteristic equation for the linear part $\dot{u}(t) = -au(t-1)$ is $\lambda + ae^{-\lambda}$, with simple imaginary roots $\omega = \omega_N$ if and only if $a = a_N$, where

$$a_N = (-1)^N \omega_N, \quad \omega_N = \frac{\pi}{2} + N\pi, \quad N \in \mathbb{N}_0.$$

Let $\alpha = a - a_N$. Thus, (8.78) has the form (8.61), with assumptions (8.63) and (8.64) fulfilled, where $L(\alpha)v = -(a_N + \alpha)v(-1)$ and $F(v, \alpha) = -(a_N + \alpha)v(0)v(-1)$. Thus, $F_2(\Phi x, 0) = 2\omega_N i(x_1^2 - x_2^2)$, $F_3(\Phi x, 0) = 0$. From Theorem 8.3.1, we derive that the equation on the center manifold of the origin near $a = a_N$ is given in polar coordinates by (8.76), with

$$K_1 = \frac{a_N}{1 + a_N^2}$$

$$K_2 = \frac{\omega_N}{5(1 + a_N^2)}[(-1)^N - 3\omega_n] < 0, \quad for \quad N \in \mathbb{N}_0. \tag{8.79}$$

Thus, a generic Hopf bifurcation for the Wright equation occurs from $u = 0, a = a_N$, with the bifurcating periodic orbits being stable on the center manifold for all $N \in \mathbb{N}_0$. Since $K_1 > 0$ if N even and $K_1 < 0$ if N odd, then the bifurcation is supercritical for N even and subcritical for N odd.

8.3.2 Hopf bifurcation for a delayed predator-prey system

Consider the following Lotka-Volterra predator-prey system:

$$\dot{u}(t) = u(t)[r_1 - au(t) - a_1 v(t - \sigma)]$$
$$\dot{v}(t) = v(t)[-r_2 + b_1 u(t - \tau) - bv(t)], \tag{8.80}$$

where τ, r_1, r_2, a_1, b_1 are positive constants and σ, a, b are non-negative constants.

In biological terms, $u(t)$ and $v(t)$ can be interpreted as the densities of prey and predator populations, respectively, and a, b self-limitation constants. In the absence of predators, the prey species follows the logistic equation $\dot{u}(t) = u(t)[r_1 - au(t)]$. In the presence of predators, there is a hunting term, $a_1 v(t - \sigma), a_1 > 0$, with a certain delay σ, called the hunting delay, and the growth rate of preys decreases linearly. In the absence of prey species, the predators decrease. The positive feedback $b_1 u(t - \tau)$ has a positive delay τ which is the delay in the predator maturation. Systems of type (8.80) or similar, and also predator-prey models with distributed delays, have been studied by many authors (e.g., [16], [47], [97], [114], [210] and references therein). However, most of the literature considers $\sigma = b = 0$ or some further constraints on the constants.

Related with the present model, we note that in [206] the following delayed predator-prey system with nonmonotonic functional response was considered:

$$\dot{u}(t) = ru(t)\left[1 - \frac{u(t)}{K} - \frac{v(t)}{a + u^2(t)}\right]$$

$$\dot{v}(t) = v(t)\left[\frac{\mu u(t - \tau)}{a + u^2(t - \tau)} - D\right],$$

where r, K, μ, D and τ are positive constants. For this model, Xiao and Ruan [206] also used normal forms techniques to obtain versal unfoldings for the Bogdanov–Takens and Hopf singularities that they proved to occur for some parameters values.

Through the change of variables $u \to b_1 u, v \to a_1 v$, we may assume that $a_1 = b_1 = 1$ in (8.80). Also normalizing the delay τ by the time-scaling $t \to t/\tau$, (8.80) is transformed into

$$\dot{u}(t) = \tau u(t)[r_1 - au(t) - v(t - r)]$$
$$\dot{v}(t) = \tau v(t)[-r_2 + u(t - 1) - bv(t)], \tag{8.81}$$

where $r = \sigma/\tau$. Without loss of generality, let $\max(1, r) = 1$. With the assumptions

$$r_1 > 0, r_2 > 0, r \geq 0, a \geq 0, \ b \geq 0, \ r_1 - ar_2 > 0, \tag{8.82}$$

there is a unique positive equilibrium E_* for (8.81) $E_* = (u_*, v_*)$, with

$$u_* = \frac{r_2 + br_1}{ab + 1}, \quad v_* = \frac{r_1 - ar_2}{ab + 1}. \tag{8.83}$$

By the translation $z(t) = (u(t), v(t)) - E_* \in \mathbb{R}^2$, (8.81) is written as an FDE in $C = C([-1, 0]; \mathbb{R}^2)$ as

$$\dot{z}(t) = N(\tau)(z_t) + f_0(z_t, \tau), \tag{8.84}$$

where $z_t \in C$ and $N(\tau) : C \to \mathbb{R}^2$, $f_0 : C \times \mathbb{R}^+ \to \mathbb{R}^2$ are given by

$$N(\tau)(\varphi) = \tau \begin{pmatrix} -u_*(a\varphi_1(0) + \varphi_2(-r)) \\ v_*(\varphi_1(-1) - b\varphi_2(0)) \end{pmatrix},$$

$$f_0(\varphi, \tau) = \tau \begin{pmatrix} -\varphi_1(0)(a\varphi_1(0) + \varphi_2(-r)) \\ \varphi_2(0)(\varphi_1(-1) - b\varphi_2(0)) \end{pmatrix},$$

for $\varphi = (\varphi_1, \varphi_2) \in C$. The characteristic equation for the linear equation $\dot{z}(t) = N(\tau)(z_t)$ is

$$\Delta(\lambda, \tau) := \lambda^2 + A_*\tau\lambda + B_*\tau^2 + C_*\tau^2 e^{-\lambda(1+r)} = 0, \tag{8.85}$$

where $A_* = au_* + bv_*$, $B_* = abu_*v_*$, $C_* = u_*v_*$.

The delay $\tau > 0$ will be taken as the bifurcating parameter. Using the material in [47] and [114] pp. 74-82, one can prove the following results:

Theorem 8.3.3. *Assume* (8.82). *If $ab \geq 1$, all the roots of the characteristic equation $\Delta(\lambda, \tau) = 0, \tau > 0$, have negative real parts. If $0 \leq ab < 1$, let $\rho_* = \rho_*(r_1, r_2, a, b)$ be the unique real positive solution of $\rho^4 + (a^2 u_*^2 + b^2 v_*^2)\rho^2 + (a^2 b^2 - 1)u_*^2 v_*^2 = 0$. Then, for $\sigma > 0, \tau > 0$, $\Delta(i\sigma, \tau) = 0$ if and only if there is an $n \in \mathbb{N}_0$ such that $\tau = \tau_n$ and $\sigma = \sigma_n$, where*

$$\cos(\sigma_n(1 + r)) = \frac{\rho_*^2 - B_*}{C_*}, \quad \tau_n = \frac{\sigma_n}{\rho_*}, \tag{8.86}$$

and

$$\sigma_n = \frac{2(n+1)\pi}{1+r} \text{ if } a = b = 0, \quad \sigma_n \in \left(\frac{2n\pi}{1+r}, \frac{(2n+1)\pi}{1+r}\right) \text{ if } a^2 + b^2 > 0. \tag{8.87}$$

Furthermore, $\pm i\sigma_n$ are simple roots, $\mathrm{Re}\, \lambda'(\tau_n) > 0$ and a Hopf bifurcation occurs for (8.84) *at $z = 0, \tau = \tau_n$.*

One can prove that $ab \geq 1$ implies that the equilibrium E_* is asymptotically stable for all $\tau > 0$. Also, if (i) $ab < 1$ and $a^2 + b^2 > 0$, the equilibrium E_* is asymptotically stable for $0 < \tau < \tau_0$ and unstable for $\tau > \tau_0$; for (ii) $a = b = 0$, E_* is unstable for all $\tau > 0$. Assuming $ab < 1$, $\pm i\sigma_0$ are the only eigenvalues on the imaginary axis for the linearized equation at $\tau = \tau_0$, $(u, v) = E_*$ (see [47] for more information and proofs). Therefore, the center manifold theory for RFDE [86] guarantees the existence of a local center manifold of dimension 2, where a Hopf bifurcation takes place. This manifold is stable in case (i) and unstable in case (ii).

In the sequel, we shall assume (8.82) and $ab < 1$. Moreover, the delay σ in (8.80) will no longer be interpreted as a free parameter, in the sense that the ratio $r = \sigma/\tau$ is assumed constant.

For the equilibrium E_* of (8.80), or equivalently, for the zero solution of (8.84), we want now to determine the Hopf singularity at the first bifurcation point $\tau = \tau_0$, by using the normal form theory. Introducing the new parameter $\alpha = \tau - \tau_0$, (8.84) is rewritten as

$$\dot{z}(t) = N(\tau_0)z_t + F_0(z_t, \alpha), \qquad (8.88)$$

where $F_0(\varphi, \alpha) = N(\alpha)(\varphi) + f_0(\varphi, \tau_0 + \alpha)$. Let $\Lambda_0 = \{i\sigma_0, -i\sigma_0\}$ and P be the center space for $\dot{z}(t) = N(\tau_0)(z_t)$. Considering complex coordinates, $P = \text{span}\{\phi_1, \phi_2\}$, with $\phi_1(\theta) = e^{i\sigma_0\theta}v_1, \phi_2(\theta) = \overline{\phi_1(\theta)}, -1 \le \theta \le 0$, where $v_1 \in \mathbb{C}^2$ is such that

$$N(\tau_0)(\phi_1) = i\sigma_0 v_1. \qquad (8.89)$$

For $\Phi = [\phi_1 \; \phi_2]$, note that $\dot{\Phi} = \Phi B$, where B is the 2×2 diagonal matrix $B = diag(i\sigma_0, -i\sigma_0)$. Choose a basis Ψ for the adjoint space P^*, such that $(\Psi, \Phi) = (\psi_i, \phi_j)_{i,j=1}^2 = I_2$. Thus, $\Psi(s) = col(\psi_1(s), \psi_2(s)) = col(u_1^T e^{-i\sigma_0 s}, \bar{u}_1^T e^{i\sigma_0 s})$, $s \in [0, 1]$, for $u_1 \in \mathbb{C}^2$ satisfying

$$(\psi_1, \phi_1) = 1, (\psi_1, \phi_2) = 0. \qquad (8.90)$$

In the following, we use some notation and material in Subsection 8.3.1. Consider the direct sums $V_j^3(\mathbb{C}^2) = Im(M_j^1) \oplus Ker(M_j^1)$. For Equation (8.88) we obtain a normal form for the ODE describing the flow on the center manifold of the origin near $\alpha = 0$, written as (8.75), where $x = (x_1, x_2) \in \mathbb{C}^2$, or in polar coordinates as (8.76), with the coefficients in these equations related by $K_1 = Re\, B_1$, $K_2 = Re\, B_2$. Describing the generic Hopf bifurcation ($K_2 \ne 0$) amounts to determining the sign of K_1 and K_2. We point out that normal form computations are particularly difficult here, because the original equation is two dimensional rather than scalar, and because there are two delays.

For equation (8.88), we have

$$f_2^1(x, y, \alpha) = 2\Psi(0)[N(\alpha)(\Phi x + y) + f_0(\Phi x + y, \tau_0)], \qquad (8.91)$$

and the second order terms in (α, x) of the normal form (8.75) are given by

$$\frac{1}{2}g_2^1(x, 0, \alpha) = \frac{1}{2}Proj_{Ker(M_2^1)}f_2^1(x, 0, \alpha) =$$

$$= Proj_{Ker(M_2^1)}\Psi(0)\left(N(\alpha)(\phi_1)x_1 + N(\alpha)(\phi_2)x_2\right)$$

Note that $N(\alpha) = \frac{\alpha}{\tau_0}N(\tau_0)$. Therefore, condition (8.89) gives

$$\frac{1}{2}g_2^1(x,0,\alpha) = \begin{pmatrix} B_1 x_1 \alpha \\ B_1 x_2 \alpha \end{pmatrix}, \tag{8.92}$$

with $B_1 = \frac{1}{\tau_0}u_1^T N(\tau_0)(\phi_1) = i\rho_* u_1^T v_1$, for ρ_* defined as in Theorem 8.3.3. The cubic terms $g_3^1(x,0,0)$ are given by (8.70). Here $f_3^1(x,0,0) = g_2^1(x,0,0) = 0$, $U_2^1 = (M_2^1)^{-1}f_2^1(x,0,0)$ and $U_2^2(x,0) = h(x)$ is evaluated by the system

$$\dot{h}(x) - D_x h(x)Bx = 2\Phi\Psi(0)F_0(\Phi x,0)$$
$$h(x)(0) - N(\tau_0)(h(x)) = 2F_0(\Phi x, 0), \tag{8.93}$$

where $\dot{h}$ denotes the derivative of $h(x)(\theta)$ relative to θ.

After computing f_2^1 and U_2^1, the expression of $g_3^1(x,0,0)$ is simplified and we get

$$g_3^1(x,0,0) = 4\sigma_0 i \begin{pmatrix} -|c_2|^2 x_1^2 x_2 \\ |c_2|^2 x_1 x_2^2 \end{pmatrix} + \frac{3}{2}Proj_S[(D_y f_2^1)h](x,0,0), \tag{8.94}$$

where $c_2 = u_1^T \overline{\Sigma}, \Sigma = \begin{pmatrix} v_{1,1}^2/u_* \\ v_{1,2}^2/v_* \end{pmatrix}$ and $v_1 = \begin{pmatrix} v_{1,1} \\ v_{1,2} \end{pmatrix}$. However, the first term in the right hand side of (8.94) does not interfere in the computation of the coefficient $K_2 = Re\,B_2$ in (8.76). To determine $Proj_S[(D_y f_2^1)h](x,0,0)$, we start by writing h as $h(x) = h_{20}x_1^2 + h_{11}x_1x_2 + h_{02}x_2^2$. With h written in this form, it is easy to see that $h_{11} = 0$ and $h_{02} = \overline{h}_{20}$. A few more computations lead to the following statement.

Theorem 8.3.4. *([47]) The flow on the center manifold of the origin for (8.88) near $\alpha = 0$ is given in polar coordinates by equation (8.75), with*

$$K_1 = Re\,(i\rho_* u_1^T v_1) \tag{8.95}$$

$$K_2 = \frac{\tau_0}{2} Re\,c_3, \tag{8.96}$$

where: $\rho_* = \sigma_0/\tau_0$; v_1, u_1 are vectors in $\mathbb{C}^2$ such that (8.89) and (8.90) hold; c_3 is given by

$$c_3 = u_1^T \begin{pmatrix} -[(a + i\frac{\rho_*}{u_*})\zeta_1(0) + \zeta_2(-r)]\bar{v}_{1,1} \\ [\zeta_1(-1) - (b + i\frac{\rho_*}{v_*})\zeta_2(0)]\bar{v}_{1,2} \end{pmatrix}, \tag{8.97}$$

where $v_1 = \begin{pmatrix} v_{1,1} \\ v_{1,2} \end{pmatrix}$; $\Sigma = \begin{pmatrix} v_{1,1}^2/u_* \\ v_{1,2}^2/v_* \end{pmatrix}$; $h_{20} = \begin{pmatrix} \zeta_1 \\ \zeta_2 \end{pmatrix}$ is the solution of

$$\begin{cases} \dot{h}_{20} - 2i\sigma_0 h_{20} = 2i\sigma_0(u_1^T \Sigma \phi_1 + \bar{u}_1^T \Sigma \phi_2) \\ h_{20}(0) - N(\tau_0)(h_{20}) = 2i\sigma_0 \Sigma. \end{cases} \tag{8.98}$$

Example 8.3.5. Take system (8.81) with $b = 0, r = 0, r_2 = 1, r_1 - a = 1$:

$$\dot{u}(t) = \tau u(t)[a + 1 - au(t) - v(t)]$$
$$\dot{v}(t) = \tau v(t)[-1 + u(t-1)], \tag{8.99}$$

also considered in [16], [128] for $a = 1$. Then, the positive equilibrium for (8.99) is $E_* = (1,1)$ and the first bifurcation point is $\tau_0 = \frac{\sigma_0}{\rho_*}$, with ρ_* given by $[(-a^2 + \sqrt{a^4+4})/2]^{1/2}$ and $\sigma_0 = \arcsin(a\rho_*)$; also $e^{i\sigma_0} = \rho_*(\rho_* + ia)$. From (8.89) and (8.90), we can choose

$$\Phi(\theta) = \begin{pmatrix} e^{i\sigma_0\theta} & e^{-i\sigma_0\theta} \\ -\frac{ie^{i\sigma_0(\theta-1)}}{\rho_*} & \frac{ie^{-i\sigma_0(\theta-1)}}{\rho_*} \end{pmatrix}, \quad \Psi(0) = \begin{pmatrix} -i\rho_* c & c \\ i\rho_* \bar{c} & \bar{c} \end{pmatrix},$$

where $c = [-a + \rho_*\sigma_0 - i(a\sigma_0 + 2\rho_*)]^{-1}$.

We now compute K_1, K_2 according to Theorem 8.3.4. From (8.95) we get

$$K_1 = Re\,[c(2\rho_*^2 - ia\rho_*)] = \frac{2\sigma_0^4/\tau_0 + a^2\sigma_0\tau_0}{(\sigma_0^2 - a\tau_0^2)^2 + \sigma_0^2(a\tau_0 + 2)^2} > 0.$$

On the other hand, $\Sigma = \begin{pmatrix} 1 \\ a^2 - \rho_*^2 + 2ia\rho_* \end{pmatrix}$, and

$$c_3 = c[(-\rho_*^2 + ia\rho_*)\zeta_1(0) + (-a + i\rho_*)\zeta_1(-1) + \rho_*(\rho_* + ia + i)\zeta_2(0)], \tag{8.100}$$

where ζ_1, ζ_2 satisfy the system

$$\begin{cases} \dot{\zeta}_1(\theta) - 2i\sigma_0\zeta_1(\theta) = 2\tau_0 C_1 e^{i\sigma_0\theta} + 2\tau_0 C_2 e^{-i\sigma_0\theta} \\ \dot{\zeta}_2(\theta) - 2i\sigma_0\zeta_2(\theta) = 2\tau_0 C_1(-a - i\rho_*)e^{i\sigma_0\theta} + 2\tau_0 C_2(-a + i\rho_*)e^{-i\sigma_0\theta}, \end{cases} \tag{8.101}$$

where $C_1 = c\rho_*[(1-2a)\rho_* + i(a^2 - \rho_*^2)]$, $C_2 = \bar{c}\rho_*[(-1-2a)\rho_* + i(a^2 - \rho_*^2)]$, with the constraints

$$\begin{cases} \dot{\zeta}_1(0) + \tau_0(a\zeta_1(0) + \zeta_2(0)) = 2i\sigma_0 \\ \dot{\zeta}_2(0) - \tau_0\zeta_1(-1) = 2\sigma_0[-2a\rho_* + i(a^2 - \rho_*^2)]. \end{cases} \tag{8.102}$$

From (8.101) and (8.102), we obtain the matricial system

$$\begin{pmatrix} a + 2i\rho_* & 1 \\ -e^{-2i\sigma_0} & 2i\rho_* \end{pmatrix}\begin{pmatrix} \zeta_1(0) \\ \zeta_2(0) \end{pmatrix} = 2\begin{pmatrix} i\rho_* - (C_1 + C_2) \\ D \end{pmatrix}, \quad \text{where}$$

$$D = \rho_*[-2a\rho_* + i(a^2 - \rho_*^2)] + [2a + 2i\rho_* - i\rho_*(\rho_* - ia)^2]C_1 + $$
$$+ [2a - 2i\rho_* - i\rho_*(\rho_* - ia)^2]\frac{C_2}{3}. \tag{8.103}$$

Although we have explicit formulas to compute $K_2 = \frac{\tau_0}{2}Re\,c_3$, it is complicated to find the sign of K_2 for a general a, as one can see from the calculus above. As an illustration, here we complete the calculus only for case $a = 1$ considered in [16], [128]. Simplifying the above formulas and using MAPLE V, for $a = 1$ we obtain $Re\,c_3 \approx -.5$, thus $K_2 < 0$.

For this rather particular situation, the Hopf bifurcation analysis is completed, since Theorem 8.3.4 implies the following statement:

Theorem 8.3.6. *Consider $a = 1, b = 0, r = 0, r_2 = 1, r_1 = 2$, and let τ_0 be defined as above. Then, for Equation (8.81) at $\tau = \tau_0$ there exists a generic supercritical Hopf bifurcation on a locally stable 2-dimensional center manifold of the positive equilibrium $E_* = (1,1)$; moreover, the associated non-trivial periodic solutions are stable.*

8.4 Applications to Bogdanov–Takens bifurcation

In this section, the theory of normal forms is first applied to the general situation of a scalar RFDE with a generic Bogdanov–Takens bifurcation (whose dynamics are then determined by the 2-jet). A second application is the study of square and pulse waves for a scalar FDE with two delays addressed in [93]. For this second example, the Bogdanov–Takens singularity is not determined at second order, implying more difficult calculus. To see how powerful the normal form technique is, the reader may compare the work of Hale and Tanaka [93], where the theory of normal forms for RFDE was used, with related studies of square and pulse waves for RFDE with only one delay [31], [81], [82] that did not use this approach.

8.4.1 Bogdanov–Takens bifurcation for scalar RFDE: The general case

Consider a scalar RFDE in $C([-r, 0]; \mathbb{R})$ in the form (8.54),

$$\dot{u}(t) = L(\alpha)(u_t) + F(u_t, \alpha), \qquad (8.104)$$

where $\alpha = (\alpha_1, \alpha_2)$ is in a neighborhood of zero $V \subset \mathbb{R}^2$, $L : V \to \mathcal{L}(C; \mathbb{R})$ is C^1-smooth and $F : C \times V \to \mathbb{R}$ is C^2-smooth such that $F(0, \alpha_1, \alpha_2) = 0, D_1 F(0, \alpha_1, \alpha_2) = 0$, for all $(\alpha_1, \alpha_2) \in V$. Let $L_0 = L(0)$. For the linear equation $\dot{u}(t) = L_0 u_t$ we suppose that:

[i] $\lambda = 0$ is a double characteristic value of $\dot{u}(t) = L_0 u_t$;   (8.105)

[ii] all other characteristic values of $\dot{u}(t) = L_0 u_t$ have non-zero real parts.
                                                                            (8.106)

Note that (8.106) means that $L_0(1) = 0, L_0(\theta) = 1, L_0(\theta^2) \neq 0$.

Let $\Lambda = \{0\}$ and consider the enlarged phase space BC decomposed by Λ as $BC = P \oplus Ker(\pi)$, where P is the center space for $\dot{u}(t) = L_0 u_t$. For P and its dual P^*, we have

$$P = \text{span}\, \Phi, \quad \Phi(\theta) = (1, \theta), \quad \theta \in [-r, 0],$$

$$P^* = \text{span}\, \Psi, \quad \Psi(s) = \begin{pmatrix} \psi_1(0) - s\psi_2(0) \\ \psi_2(0) \end{pmatrix}, \quad s \in [0, r],$$

with $(\Psi, \Phi) = I$ if

$$\psi_1(0) = L_0(\theta^2/2)^{-2} L_0(\theta^3/3!), \quad \psi_2(0) = -L_0(\theta^2/2)^{-1}.$$

The matrix B satisfying $\dot{\Phi} = \Phi B, -\dot{\Psi} = B\Psi$ is $B = \begin{pmatrix} 0 & 1 \\ 0 & 0 \end{pmatrix}$. So for $\alpha = 0$ the 2-dimensional ODE in the center manifold has a Bogdanov–Takens singularity, and, consequently, its dynamics in a neighborhood of $u = 0$ are generically determined by the quadratic terms of this equation.

Consider the Taylor formulas $L(\alpha) = L_0 + L_1(\alpha) + O(|\alpha|^2)$, $F(u, \alpha) = \frac{1}{2} F_2(u, \alpha) + O(|u|^3 + |\alpha||u|^2)$. The normal form for (8.104) on the center manifold of the origin near $\alpha = 0$ has the form

$$\dot{x} = Bx + \frac{1}{2} g_2^1(x, 0, \alpha) + h.o.t.,$$

where $x = (x_1, x_2) \in \mathbb{R}^2$, $h.o.t.$ stands for higher order terms and (see (8.51) and Remark 8.2.10)

$$g_2^1(x, 0, \alpha) = (I - P_2^1) f_2^1(x, 0, \alpha) \qquad (8.107)$$

where $f_2^1(x, 0, \alpha) = \Psi(0)[2L_1(\alpha)(\Phi x) + F_2(\Phi x, \alpha)]$. For B as above, the operator M_2^1 in (8.50) is given by

$$M_2^1 \begin{pmatrix} p_1 \\ p_2 \end{pmatrix} = \begin{pmatrix} \frac{\partial p_1}{\partial x_1} x_2 - p_2 \\ \frac{\partial p_2}{\partial x_1} x_2 \end{pmatrix}.$$

Consider the canonical basis of $V_2^4(\mathbb{R}^2)$, with 20 elements. Some of them are irrelevant for our computations. For the elements

$$\begin{pmatrix} x_1^2 \\ 0 \end{pmatrix}, \begin{pmatrix} x_1 x_2 \\ 0 \end{pmatrix}, \begin{pmatrix} x_2^2 \\ 0 \end{pmatrix}, \begin{pmatrix} x_1 \alpha_i \\ 0 \end{pmatrix}, \begin{pmatrix} x_2 \alpha_i \\ 0 \end{pmatrix}, \quad i = 1, 2$$

$$\begin{pmatrix} 0 \\ x_1^2 \end{pmatrix}, \begin{pmatrix} 0 \\ x_1 x_2 \end{pmatrix}, \begin{pmatrix} 0 \\ x_2^2 \end{pmatrix}, \begin{pmatrix} 0 \\ x_1 \alpha_i \end{pmatrix}, \begin{pmatrix} 0 \\ x_2 \alpha_i \end{pmatrix}, \quad i = 1, 2, \quad (8.108)$$

their images under M_2^1 are, respectively

$$\begin{pmatrix} 2x_1 x_2 \\ 0 \end{pmatrix}, \begin{pmatrix} x_2^2 \\ 0 \end{pmatrix}, \begin{pmatrix} 0 \\ 0 \end{pmatrix}, \begin{pmatrix} x_2 \alpha_i \\ 0 \end{pmatrix}, \begin{pmatrix} 0 \\ 0 \end{pmatrix}, \quad i = 1, 2$$

$$\begin{pmatrix} -x_1^2 \\ 2x_1 x_2 \end{pmatrix}, \begin{pmatrix} -x_1 x_2 \\ x_2^2 \end{pmatrix}, \begin{pmatrix} -x_2^2 \\ 0 \end{pmatrix}, \begin{pmatrix} -x_1 \alpha_i \\ x_2 \alpha_i \end{pmatrix}, \begin{pmatrix} -x_2 \alpha_i \\ 0 \end{pmatrix}, \quad i = 1, 2 \quad (8.109)$$

It is easy to check that one can choose the decomposition $V_2^4(\mathbb{R}^2) = Im(M_2^1) \oplus (Im(M_2^1))^c$, with complementary space $(Im(M_2^1))^c$ defined by

$$(Im\,(M_2^1))^c = span\left\{ \begin{pmatrix} 0 \\ x_1^2 \end{pmatrix}, \begin{pmatrix} 0 \\ x_1x_2 \end{pmatrix}, \begin{pmatrix} 0 \\ x_1\alpha_1 \end{pmatrix}, \begin{pmatrix} 0 \\ x_1\alpha_2 \end{pmatrix}, \begin{pmatrix} 0 \\ x_2\alpha_1 \end{pmatrix}, \right.$$

$$\left. \begin{pmatrix} 0 \\ x_2\alpha_2 \end{pmatrix}, \begin{pmatrix} 0 \\ \alpha_1^2 \end{pmatrix}, \begin{pmatrix} 0 \\ \alpha_1\alpha_2 \end{pmatrix}, \begin{pmatrix} 0 \\ \alpha_2^2 \end{pmatrix} \right\}. \quad (8.110)$$

On the other hand, we write $F_2(\Phi x, \alpha) = F_2(\Phi x, 0) = A_{(2,0)}x_1^2 + A_{(1,1)}x_1x_2 + A_{(0,2)}x_2^2$. The decomposition above yields

$$\frac{1}{2}g_2^1(x, 0, \alpha) = Proj_{(Im\,(M_2^1))^c}\, f_2^1(x, 0, \alpha) =$$

$$= \begin{pmatrix} 0 \\ \lambda_1 x_1 + \lambda_2 x_2 \end{pmatrix} + \begin{pmatrix} 0 \\ B_1 x_1^2 + B_2 x_1 x_2 \end{pmatrix}, \quad (8.111)$$

where

$$B_1 = \frac{1}{2}\psi_2(0)A_{(2,0)} \quad \text{and}$$

$$B_2 = \psi_1(0)A_{(2,0)} + \frac{1}{2}\psi_2(0)A_{(1,1)}$$

$$(8.112)$$

and the bifurcating parameters are given by

$$\lambda_1 = \psi_2(0)L_1(\alpha)(1)$$
$$\lambda_2 = \psi_1(0)L_1(\alpha)(1) + \psi_2(0)L_1(\alpha)(\theta)$$

$$(8.113)$$

These results lead to the following statement (see e.g. [23], [30]):

Theorem 8.4.1. *Consider Equation* (8.104) *and assume that* (8.105) *and* (8.106) *hold. Then, for α small there is 2-dimensional local center manifold of the origin, on which the flow is given by*

$$\begin{cases} \dot{x}_1 = x_2 + h.o.t \\ \dot{x}_2 = \lambda_1 x_1 + \lambda_2 x_2 + B_1 x_1^2 + B_2 x_1 x_2 + h.o.t, \end{cases} \quad (8.114)$$

where the coefficients $\lambda_1, \lambda_2, B_1, B_2$ are given by (8.112) *and* (8.113). *If $B_1 B_2 \neq 0$ and λ_1, λ_2 are linearly independent, then* (8.104) *exhibits a generic Bogdanov–Takens bifurcation from $u = 0, \alpha = 0$.*

Example 8.4.2. Consider the scalar delay-differential equation

$$\dot{u}(t) = (1+\alpha_1)u(t)-(1+\alpha_2)u(t-1)+au^2(t)+bu(t)u(t-1)+cu^2(t-1), \quad (8.115)$$

with $\alpha_1, \alpha_2, a, b, c \in \mathbb{R}$. For $\alpha_1 = \alpha_2 = 0$, the linear equation $\dot{u}(t) = u(t) - u(t-1)$ satisfies (8.105) and (8.106). In this example, $\psi_1(0) =$

2/3, $\psi_2(0) = 2$, $L_1(\alpha_1, \alpha_2)(\varphi) = \alpha_1\varphi(0) - \alpha_2\varphi(-1)$ for $\varphi \in C = C([-1, 0]; \mathbb{R})$ and $F_2(\Phi x, 0)/2 = (a + b + c)x_1^2 - (b + 2c)x_1x_2 + cx_2^2$. Thus, the flow on the center manifold of the origin near $\alpha = 0$ is given by (8.114), with the following coefficients and bifurcating parameters:

$$B_1 = 2(a + b + c), \quad B_2 = \frac{2}{3}(2a - b - 4c)$$

$$\lambda_1 = 2(\alpha_1 - \alpha_2), \quad \lambda_2 = \frac{2}{3}(\alpha_1 + 2\alpha_2).$$

For $B_1 B_2 \neq 0$, (8.114) undergoes a generic Bogdanov-Takens bifurcation on the center manifold of the origin. For instance, suppose that $B_1 < 0$ and $B_2 > 0$. Then in the (λ_1, λ_2)-bifurcation diagram, the Hopf bifurcation curve H and the homoclinic bifurcation curve HL lie in the region $\lambda_1 > 0, \lambda_2 < 0$, with H to the left of HL, and both the homoclinic loop and the periodic orbit are asymptotically stable. The analysis of the other cases is similar (see e.g. [23], [30]).

8.4.2 Square and pulse waves

In [93] the following RFDE with two delays was considered:

$$\epsilon\dot{x}(t) + x(t) = \lambda - bx(t - 2) - (1 + b)x(t - 1) - x^2(t - 1), \qquad (8.116)$$

where $b \in (-1, 1)$ and $\epsilon > 0, \lambda$ are small parameters. This equation arises as a perturbation of the Hénon map on the plane. In a more general setting, for scalar equations that have the form

$$\dot{x}(t) + x(t) = f(x(t - r_1), x(t - r_2)), \qquad (8.117)$$

suppose that the two delays $r_1, r_2 > 0$ become unbounded in a certain direction, that is, $(r_1, r_2) = (1/\epsilon, d/\epsilon)$, with d a fixed constant and $\epsilon \to 0$. Rescaling time, (8.117) is transformed into

$$\epsilon\dot{x}(t) + x(t) = f(x(t - 1), x(t - d)). \qquad (8.118)$$

For the case of one delay ($d = 1$), the behavior of solutions of (8.118), namely the existence of square and pulse waves, has been extensively studied (see [31], [81], [82] and references therein). For the above equation (8.116), a situation with two delays, Hale and Tanaka [93] addressed similar questions, using ideas and results in [81], [82], and also the normal form theory described in Section 8.2.

For $\lambda > -(1 + b)^2$, (8.116) has an equilibrium $\delta(\lambda) = -(1 + b) + \sqrt{(1 + b)^2 + \lambda}$. Translating $\delta(\lambda)$ to the origin, we get

$$\epsilon\dot{y}(t) + y(t) = -by(t - 2) - (1 + b + 2\delta(\lambda))y(t - 1) - y^2(t - 1). \qquad (8.119)$$

By studying the characteristic equation for its linearized equation at zero, it was shown in [93] that, in the neighborhood of the origin and for ϵ small, there is a curve in the (λ, ϵ)–plane, along which there exists a generic Hopf bifurcation with respect to ϵ that is supercritical for $b \in (-1, 1/3)$ and subcritical for $b \in (1/3, 1)$.

To establish the existence of periodic solutions and their limiting profile, Hale and Tanaka used several techniques. First, by a rescaling, the problem of existence of periodic solutions with period close to 2 was transformed into the problem of finding periodic solutions for a 2–dimensional system with two new parameters. Next, for this system and if $b \neq 1/3$, the ODE giving the flow on the 2–dimensional center manifold was *directly* obtained (up to third order terms), by using the normal form theory for retarded FDE described here. In this aspect, the approach in [93] is completely different from the one in [31], [81], where the normal form computation was accomplished in two steps: first, the approximate vector field on the center manifold was determined, and second, the normal form for the ODE describing the flow on the center manifold was computed. The work in [93] is summarized below, without proofs and omitting computation details.

If $y(t)$ is a periodic solutions of (8.119) with period near two, consider $2 + 2(r_0 + h)\epsilon$ as the period of $y(t)$, with the new parameters r_0, h to be determined, and effect the transformation

$$
\begin{cases}
u(t) = y(-\epsilon(r_0 + h)t) \\
v(t) = y(-\epsilon(r_0 + h)t + 1 + \epsilon(r_0 + h)) \\
l = 2\delta(\lambda)
\end{cases}
$$

For $z = (u, v)$, equation (8.119) becomes

$$\dot{z}(t) = L_0(z_t) + [L_\alpha(z_t) - L_0(z_t)] + F(z_t, \alpha), \tag{8.120}$$

where

$$[L_\alpha(\varphi) = (r_0 + h)[\varphi(0) + b\varphi(-2) + (1 + b + l)A\varphi(-1)$$

$$A = \begin{pmatrix} 0 & 1 \\ 1 & 0 \end{pmatrix}, \quad \alpha = (h, l)$$

$$F(\varphi, \alpha) = (r_0 + h)A\varphi^2(-1), \quad \text{for } \varphi^2 := (\varphi_1^2, \varphi_2^2).$$

So we look now for periodic solutions of the 2-dimensional system (8.120) with the new parameters l, h. The characteristic equation for the linear equation $\dot{z}(t) = L_0(z_t)$ is $det\, \Delta(\mu, r_0) = 0$ where

$$det\, \Delta(\mu, r_0) = (\mu - r_0)^2 - 2(\mu - r_0)r_0 b e^{-2\mu} + r_0^2 b^2 e^{-4\mu} - r_0^2(1 + b)^2 e^{-2\mu}. \tag{8.121}$$

Theorem 8.4.3. *[93] For $b \in (-1, 1)$ let $r_0 = 1/(1-b)$. Then, $\mu = 0$ is the only root of (3.14) on the imaginary axis. Furthermore,*

(i) if $b \in (-1, 1), b \neq 1/3$, then $\mu = 0$ is a root of multiplicity 2.

(ii) $b = 1/3$, then $\mu = 0$ is a root of multiplicity 3.

From [86], there exists a local center manifold of the origin for (8.120) near $\alpha = 0$, and small periodic solutions of (8.120) must lie on it. For $b \neq 1/3$, the next step is to obtain the ODE giving the flow on the 2–dimensional center manifold by using the normal form theory for retarded FDE. With the previous notations, in the phase space $C = C([-2, 0]; \mathbb{R}^2)$, Φ and Ψ are normalized dual bases for P and P^*, with

$$\Phi = [\varphi_1 \; \varphi_2], \quad \varphi_1(\theta) = \begin{pmatrix} 1 \\ -1 \end{pmatrix}, \quad \varphi_2(\theta) = \theta \begin{pmatrix} 1 \\ -1 \end{pmatrix}, \quad -2 \leq \theta \leq 0$$

$$\Psi = [\psi_1 \; \psi_2]^T, \quad \psi_1(s) = \frac{1-b}{(1-3b)^2} \left(\frac{1-7b}{3} + (3b-1)s \right) [1 \; -1],$$

$$\psi_2(s) = \frac{1-b}{(1-3b)} [1 \; -1], \quad 0 \leq s \leq 2,$$

and the matrix B such $\dot{\Phi} = \Phi B$ is $B = \begin{pmatrix} 0 & 1 \\ 0 & 0 \end{pmatrix}$. Choosing the complementary space $(Im\,(M_2^1))^c$ of $Im\,(M_2^1))$ in $V_2^4(\mathbb{R}^2)$ as in (8.110), the normal form for the ODE in the center manifold up to the quadratic terms has the form (8.114). But, in this situation the coefficients B_1, B_2 are both zero, thus (8.114) is not a versal unfolding of the Bogdanov–Takens singularity, and higher order terms have to be computed. One can easily prove that

$$g_3^1(x, 0, \alpha) = Proj_{(Im\,(M_2^1))^c}\, \tilde{f}_3^1(x, 0, \alpha) =$$

$$= Proj_S\, \tilde{f}_3^1(x, 0, 0) + O(|\alpha|^2 |x|), \quad (8.122)$$

for $S = span\left\{ \begin{pmatrix} 0 \\ x_1^3 \end{pmatrix}, \begin{pmatrix} 0 \\ x_1^2 x_2 \end{pmatrix} \right\}$. To compute the cubic term $g_3^1(x, 0, 0)$ ones requires the resolution of several ODE, as for the situation of the Hopf bifurcation treated in Section 8.3. However, in this case the application of the normal form algorithm is significantly more difficult, because the operator M_3^1 is not diagonal and also because (8.120) is a FDE on the plane, instead of a scalar equation. In [93] the normal form was computed up to cubic terms:

$$\begin{cases} \dot{x}_1 = x_2 + h.o.t \\ \\ \dot{x}_2 = \epsilon_1 x_1 + \epsilon_2 x_2 + \frac{\delta_1}{6} x_1^3 + \delta_2 x_1^2 x_2 + h.o.t, \end{cases} \quad (8.123)$$

where

$$\epsilon_1 = \frac{2\alpha_2}{3b-1}, \quad \epsilon_2 = -\frac{2(1-b)^2\alpha_1}{3b-1} + \frac{4(1-b)\alpha_2}{3(3b-1)^2}$$

$$\delta_1 = -\frac{12}{(1+b)(3b-1)}, \quad \delta_2 = \frac{8(b-1)}{(1+b)(3b-1)^2}. \quad (8.124)$$

For the 2-dimensional ODE (8.123), we look now for the regions in the new parameters plane in which exist bifurcating periodic solutions encircling the origin. The idea is to then relate these periodic solutions with the periodic solutions with period ≈ 2 of the original equation, and hence to derive the sectorial regions of the plane for which they exist, as well as their limiting profiles. For this final analysis, the arguments in [79] can be applied. See [93] for more results and proofs.

Theorem 8.4.4. *[93] For $b \in (-1,1)$, $b \neq 1/3$, there exists a sector S_b in the (λ, ϵ)-plane, which closure contains a segment of the axis $\epsilon = 0$, such that for $(\lambda, \epsilon) \in S_b$ there is a unique periodic orbit of (8.119) with period close to 2. Furthermore, this orbit is stable (resp. unstable) and approaches a square (resp. pulse) wave as $\epsilon \to 0$ if $b \in (-1, 1/3)$ (resp. $b \in (1/3, 1)$).*

8.5 Singularity with a pure imaginary pair and a zero as simple eigenvalues

The case of this singularity is of particular interest as it is shown by Faria and Magalhães [52] that there occur restrictions on the flows defined by RFDE with nonlinearities involving just one delayed value of the solutions, when compared with the possible flows for ODE with the same singularity. The restrictions found have an important geometrical significance, since they imply the impossibility of observing for those RFDE the homoclinic orbits which occur for ODE in arbitrary small neighborhoods of the considered singularity.

The example becomes even more interesting when we notice this is the simplest nondegenerate singularity of local flows determined by normal forms up to finite order for which we observe restrictions on the local flows of scalar RFDE with the same singularity and nonlinearities involving just one delay.

In fact, the simplest singularities with the stated property are, in this order, the Hopf singularity, the Bogdanov–Takens singularity and the singularity considered in this section. It was shown in [52], using normal forms for RFDE, that there are no restrictions for the Hopf and the Bogdanov–Takens singularities (the restriction noted at the end of the discussion of the Bogdanov–Takens singularity in Section 8.1 occurs because there the linear parts of the considered RFDE were just of the form $\dot{x} = x(t) - x(t-1)$).

Consider a scalar RFDE in $C([-r, 0], \mathbb{R})$ in the form

$$\dot{u} = L(u_t) + F(u(t-r_0)) \quad (8.125)$$

where $L \in \mathcal{L}(C, \mathbb{R})$, $0 < r_0 \leq r$, $F \in C^2(\mathbb{R}, \mathbb{R})$ is such that $F(0) = 0$ and $F'(0) = 0$. We suppose that $\lambda = 0$ and $\lambda = \pm i\omega$, with $\omega \neq 0$, are simple characteristic values of the linear equation $\dot{u} = L(u_t)$, and that there are no other characteristic values of this equation in the imaginary axis.

We take $\Lambda = \{i\omega, -i\omega, 0\}$ and apply the normal form theory for RFDE presented in Section 8.2. As the singularity involves a conjugate pair of pure imaginary eigenvalues, it is convenient to consider (8.125) in $C([-r, 0]; \mathbb{C})$, still denoted here by C, extending L and F to complex functions in the natural way.

Let us consider the decomposition $BC = P \oplus \ker \pi$, where P is the center space of $\dot{u} = L(u_t)$. For $P = span\Phi$ and its dual $P^* = span\Psi$ we have

$$\Phi(\theta) = (\varphi_1(\theta), \varphi_2(\theta), \varphi_3(\theta)) = (e^{i\omega\theta}, e^{-i\omega\theta}, 1), \quad -r \leq \theta \leq 0$$

$$\Psi(s) = col\ (\psi_1(s),\ \psi_2(s),\ \psi_3(s))$$
$$= col\ (\psi_1(0)e^{-i\omega s}, \psi_2(0)e^{i\omega s}, \psi_3(0)), \qquad 0 \leq s \leq r \quad ,$$

with $(\Psi, \Phi) = I$. This last identity is verified if and only if

$$\psi_1(0) = (1 - L(\theta\ e^{i\omega\theta}))^{-1}, \quad \psi_2(0) = \overline{\psi_1(0)}, \quad \psi_3(0) = (1 - L(\theta))^{-1}, \tag{8.126}$$

where we use $L(\varphi) = L(\varphi(\theta))$. We have of course $\dot{\Phi} = \Phi B$ and $\dot{\Psi} = -B\Psi$, where $B = diag\ (i\omega, -i\omega, 0)$.

Since B is a diagonal matrix, the operators M_j^1, $j \geq 2$, defined in (2.10) have a diagonal representation relative to the canonical basis $\{x^q e_k : k = 1, 2, 3, q \in \mathbb{N}_0^3, |q| = j\}$ of $V_j^3(\mathbb{C}^3) = \{\sum_{|q|=j} c_q x^q : q \in \mathbb{N}_0^3, c_q \in \mathbb{C}^3\}$, where (e_1, e_2, e_3) is the canonical basis of $\mathbb{C}^3$.

We write the Taylor formula for F as

$$F(v) = \frac{1}{2} A_2 v^2 + h.o.t$$

with $A_2 \in \mathbb{R}$. We also write $u = z_t$ as $u = \Phi x + y$, with $x \in \mathbb{C}^3$ and $y \in Q^1 = Q \cap C^1$, according to the decomposition $C = P \oplus Q$. Defining $f_2 = (f_2^1, f_2^2) \in V_2^3(\mathbb{C}^3) \times V_2^3(\ker \pi)$, we have

$$f_2(x, 0) = \begin{bmatrix} \Psi(0) \\ (I - \pi)X_0 \end{bmatrix} [A_2(\Phi(-r_0)x)^2] \quad.$$

We also have

$$\ker(M_j^1) = span\ \{x^q e_k : (q, \overline{\lambda}) = \lambda_k,\ k = 1, 2, 3,\ q \in \mathbb{N}_0^3,\ |q| = j\},$$

with $\overline{\lambda} = (\lambda_1, \lambda_2, \lambda_3) = (i\omega, -i\omega, 0)$. So

$$\ker(M_2^1) = span\ \left\{ \begin{pmatrix} x_1 x_3 \\ 0 \\ 0 \end{pmatrix}, \begin{pmatrix} 0 \\ x_2 x_3 \\ 0 \end{pmatrix}, \begin{pmatrix} 0 \\ 0 \\ x_1 x_2 \end{pmatrix}, \begin{pmatrix} 0 \\ 0 \\ x_3^2 \end{pmatrix} \right\} \quad.$$

The normal form for (8.125) on the center manifold of the origin has the form

$$\dot{x} = Bx + \frac{1}{2}g_2^1(x,0) + h.o.t, \tag{8.127}$$

where $g_2^1(x,0)$ is given by the projection of $f_2^1(x,0)$ on $\mathrm{Ker}(M_2^1)$ along $\mathrm{Im}(M_2^1)$. In order to simplify the notation, we introduce the operators $\mathcal{S}: V_j^3(\mathbb{C}) \to V_j^3(\mathbb{C}^2)$ such that $\mathcal{S}(h+k) = \mathcal{S}(h) + \mathcal{S}(k)$ and

$$\mathcal{S}(c\, x_1^{q_1}\, x_2^{q_2}\, x_3^{q_3}) = \begin{bmatrix} c\, x_1^{q_1}\, x_2^{q_2}\, x_3^{q_3} \\ \bar{c}\, x_1^{q_2}\, x_2^{q_1}\, x_3^{q_3} \end{bmatrix},$$

$c \in \mathbb{C}$, $(q_1, q_2, q_3) \in \mathbb{N}_0^3$, $|(q_1, q_2, q_3)| = j$. We obtain

$$g_2^1(x,0) =$$

$$= \begin{bmatrix} \mathcal{S}(\psi_1(0)2A_2 e^{-i\omega r_0}) \\ \psi_3(0)[2A_2 x_1 x_2 + A_2 x_3^2] \end{bmatrix}.$$

The normal form (8.127) can now be written in real coordinates w, through the change of variables $x_1 = w_1 - iw_2$, $x_2 = w_1 + iw_2$, $x_3 = w_3$, and changing to cylindrical coordinates according to $w_1 = \rho\cos\xi$, $w_2 = \rho\sin\xi$, $w_3 = \zeta$, we obtain

$$\begin{aligned} \dot{\rho} &= a_1\rho\zeta + O(\rho|(\rho,\zeta)|^2) \\ \dot{\zeta} &= b_1\rho^2 + b_2\zeta^2 + O(|(\rho,\zeta)|^3) \\ \dot{\xi} &= -\omega + O(|(\rho,\zeta)|) \quad , \end{aligned} \tag{8.128}$$

with

$$a_1 = A_2 Re[\psi_1(0)e^{-i\omega r_0}]$$

$$b_1 = A_2\psi_3(0) \tag{8.129}$$

$$b_2 = \tfrac{1}{2}\psi_3(0).$$

Writing equation (8.128) up to second order terms and eliminating the equation in ξ, since the right hand side of (8.128) is independent of this variable, we get the equation in the plane (ρ, ζ)

$$\begin{aligned} \dot{\rho} &= a_1\, \rho\, \zeta \\ \dot{\zeta} &= b_1\, \rho^2 + b_2\, \zeta^2 \quad . \end{aligned} \tag{8.130}$$

As it is shown in Takens [197], the singularity considered is determined to second order, provided $a_1, b_1, b_2 \neq 0$ and $a_1 \neq b_2$. In fact, it is known (see Takens [197]) that the possible phase portraits in a neighborhood of zero are completely determined by the signs of a_1, b_1, b_2 and $(b_2 - a_1)$. From equations (8.129) we have always

$$b_1 = 2b_2$$

$$a_1 = 2[1 - L(\theta)]Re\left(\frac{e^{-i\omega r_0}}{[1-L(\theta e^{i\omega\theta})]}\right)b_2 \quad,$$

(8.131)

where b_2 can assume arbitrary values. These restriction impose drastic limitations on the flows that can be observed.

In fact, the possible topological types of the phase portraits around the origin for equations (8.130) with $b_1 > 0$ are sketched in Takens [197] as appears in Figure 8.1, and for $b_1 < 0$ they can be obtained by just reversing the arrows.

From (8.131) it follows that the cases corresponding to $b_1 b_2 < 0$ cannot occur for RFDE whose nonlinearities involve just one delay. So, of the five phase portraits sketched on Figure 8.1, only the first two can occur for RFDE whose nonlinearities have just one delay. In particular, for ODE with the singularity considered, we can observe homoclinic orbits on arbitrary small neighborhoods of the origin, while this cannot occur for RFDE with nonlinearities involving just one delay. In fact, these do not have global orbits in sufficiently small neighborhoods of the origin except the origin itself.

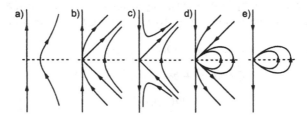

Fig. 8.1. a) $b_1 b_2 > 0, b_1(b_2 - a_1) > 0$, b) $b_1 b_2 > 0, b_1(b_2 - a_1) < 0, b_1 a_1 > 0$, c) $b_1 b_2 < 0, b_1(b_2 - a_1) < 0, b_1 a_1 > 0$, d) $b_1 b_2 < 0, b_1(b_2 - a_1) < 0, b_1 a_1 < 0$, e) $b_1 b_2 < 0, b_1(b_2 - a_1) > 0$.

8.6 Normal forms for RFDE in infinite dimensional spaces

In the following, X is a real or complex Hilbert space with inner product $\langle \cdot, \cdot \rangle$ and $C = C([-r, 0]; X)$ $(r > 0)$ is the Banach space of continuous maps from $[-r, 0]$ to X with the sup norm. As in Section 1, in order to simplify the notation we fix $\mathbb{R}$ as the scalar field, but all the theory can be applied with no changes to the case where X is a Hilbert space over $\mathbb{C}$. We write $u_t \in C$ for $u_t(\theta) = u(t + \theta), -r \leq \theta \leq 0$ and consider RFDE in X with an equilibrium point at the origin, given in abstract form as

$$\dot{u}(t) = A_T u(t) + L(u_t) + F(u_t) \qquad (t \geq 0), \qquad (8.132)$$

where $A_T : D(A_T) \subset X \to X$ is a linear operator, $L \in \mathcal{L}(\mathcal{C}; X)$, i.e., $L : \mathcal{C} \to X$ is a bounded linear operator, and $F : \mathcal{C} \to X$ is a C^k function $(k \geq 2)$ such that $F(0) = 0, DF(0) = 0$.

When considering equations in the form (8.132), we have in mind models involving both time delays and spatial diffusion, that increasingly arise from a variety of situations in population dynamics and other fields. In particular, reaction-diffusion equations with delays appearing in the reaction terms have been extensively used as models. Typically, delayed reaction-diffusion equations in $\mathbb{R}^n$ take the form (8.132), where a diffusion term $d\Delta v(t,x), d = (d_1, \ldots, d_n) \in \mathbb{R}^n$ constant, is given by $A_T u(t) = d\Delta v(t,x)$, for $u(t)(x) := v(t,x), x \in \mathbb{R}^n$.

For the linearized equation about the zero equilibrium

$$\dot{u}(t) = A_T u(t) + L(u_t), \qquad (8.133)$$

we assume the following hypotheses (see [125], [142], [205], [208]):

(H1) A_T generates a C_0 semigroup of linear operators $\{T(t)\}_{t \geq 0}$ on X with $|T(t)| \leq Me^{\omega t}$ $(t \geq 0)$ for some $M \geq 1, \omega \in \mathbb{R}$, and $T(t)$ is a compact operator for $t > 0$;

(H2) the eigenfunctions $\{\beta_k\}_{k=1}^{\infty}$ of A_T, with eigenvalues $\{\mu_k\}_{k=1}^{\infty}$, form an orthonormal basis for X;

(H3) the subspaces $\mathcal{B}_k$ of $\mathcal{C}$, $\mathcal{B}_k := \{\varphi \beta_k \,|\, \varphi \in C([-r,0]; \mathbb{R})\}$ satisfy $L(\mathcal{B}_k) \subset \text{span}\{\beta_k\}$, $k \in \mathbb{N}$;

(H4) L can be extended to a bounded linear operator from BC to X, where $BC = \{\psi : [-r,0] \longrightarrow X \,|\, \psi$ is continuous on $[-r,0), \exists \lim_{\theta \to 0^-} \psi(\theta) \in X\}$, with the sup norm.

For a straightforward generalization of these assumptions to equations (8.132) and (8.133) in Banach spaces see [205]. Hypothesis (H3) roughly states that the operator L only recognizes those spatial variations described by eigenvectors β_k of A_T and does not mix them. This condition restricts strongly the application of the normal form theory presented here. In fact, in many problems arising from population dynamics (H3) fails. In [46] normal forms are developed for RFDE in Banach spaces without imposing (H2)-(H3) (see Section 8.7. However, it is important to consider the situation where (H1)-(H4) hold, since it allows us to define an *associated RFDE* in $\mathbb{R}^n$, whose normal forms on invariant manifolds coincide with normal forms on invariant manifolds for Equation (8.132), up to some order, and hence to take advantage of the method for RFDE in $\mathbb{R}^n$ described in Section 8.2.

Let A be the infinitesimal generator associated with the semiflow of the linearized equation (8.133). It is known (e.g. [200]) that A is given by

$$(A\phi)(\theta) = \dot{\phi}(\theta),$$

$$D(A) = \{\phi \in \mathcal{C} : \dot{\phi} \in \mathcal{C}, \phi(0) \in D(A_T), \dot{\phi}(0) = A_T\phi(0) + L\phi\},$$

that the spectrum $\sigma(A)$ of A coincides with its point spectrum $\sigma_P(A)$ and that $\lambda \in \mathbb{C}$ is in $\sigma_P(A)$ if and only if λ satisfies the characteristic equation

$$\Delta(\lambda)y := \lambda y - A_T y - L(e^{\lambda \cdot} y) = 0, \quad y \in D(A_T) \setminus \{0\}. \qquad (8.134)$$

For any $a \in \mathbb{R}$, the number of solutions of (8.134) such that $\mathrm{Re}\,\lambda \geq a$ is finite. Since (H3) states that L does not mix the modes of eigenvalues of A_T, and (H2) allows us to decompose X by $\{\beta_k\}_{k=1}^{\infty}$, we deduce that equation $\Delta(\lambda)y = 0$ is equivalent to the sequence of "characteristic equations"

$$\lambda - \mu_k - L_k(e^{\lambda \cdot}) = 0 \quad (k \in \mathbb{N}), \qquad (8.135)$$

where $L_k : C \longrightarrow \mathbb{R}$, $C := C([-r, 0]; \mathbb{R})$, are defined by $L_k(\psi)\beta_k = L(\psi\beta_k)$, for $k \in \mathbb{N}$.

Let $\Lambda \neq \emptyset$ be a finite subset of $\sigma(A)$. If $\Lambda = \{\lambda \in \sigma(A) : \mathrm{Re}\,\lambda = 0\}$, the center space $\mathcal{P}$ for (8.133) is the generalized eigenspace associated with Λ. In [125], the existence of a local center manifold for (8.132) tangent to $\mathcal{P}$ at zero was proven under (H1)-(H4), following the approach in [142]. Its dimension is equal to the dimension of $\mathcal{P}$, that is, it is equal to the number of $\lambda \in \sigma(A)$ with real part zero, counting multiplicities. Instead of center manifolds, we can consider different invariant manifolds associated with other finite subsets Λ of $\sigma(A)$. E.g., Memory [142] established the existence of unstable manifolds, corresponding to the choice $\Lambda = \{\lambda \in \sigma(A) : \mathrm{Re}\,\lambda > 0\}$. Throughout this section, we assume that the set $\Lambda = \{\lambda \in \sigma(A) : \mathrm{Re}\,\lambda = 0\}$ contains exactly the solutions on the imaginary axis of the first N equations (8.135), i.e., $\Lambda = \{\lambda \in \mathbb{C} : \lambda$ is a solution of (8.135) with $\mathrm{Re}\,\lambda = 0$, for some $k \in \{1, \ldots, N\}\}$.

Now, we describe briefly an adjoint theory for RFDE in X as in [125], [142], [208], assuming the former hypotheses. The main idea is to relate the eigenvalues of the infinitesimal generator A (in this case, only the elements of Λ) with the eigenvalues of certain scalar FDE. On $\mathcal{B}_k$, the linear equation $\dot{u}(t) = A_T u(t) + L(u_t)$ is equivalent to the FDE on $\mathbb{R}$

$$\dot{z}(t) = \mu_k z(t) + L_k z_t, \qquad (8.136)$$

with characteristic equation given by (8.135), $k \in \mathbb{N}$. By this identification, the standard adjoint theory for RFDE is used to decompose $\mathcal{C}$, in the following way. For $1 \leq k \leq N$, define $(\cdot, \cdot)_k$ as the adjoint bilinear form on $C^* \times C$, $C^* := C([0, r]; \mathbb{R})$, and decompose C by $\Lambda_k := \{\lambda \in \mathbb{C} : \lambda$ satisfies (8.135) and $\mathrm{Re}\,\lambda = 0\}$ as in Section 8.2:

$$C = P_k \oplus Q_k, \quad P_k = \mathrm{span}\Phi_k, \quad P_k^* = \mathrm{span}\Psi_k,$$
$$(\Psi_k, \Phi_k)_k = I, \quad \dim P_k = \dim P_k^* := m_k, \dot{\Phi}_k = \Phi_k B_k,$$

where P_k is the generalized eigenspace for (8.136) associated with Λ_k and B_k is an $m_k \times m_k$ constant matrix. We use the above decompositions to decompose C by Λ:

$$\mathcal{C} = \mathcal{P} \oplus \mathcal{Q}, \quad \mathcal{P} = \operatorname{Im} \pi, \quad \mathcal{Q} = \operatorname{Ker} \pi,$$

where $\dim \mathcal{P} = \sum_{k=1}^{N} m_k := M$ and $\pi : \mathcal{C} \longrightarrow \mathcal{P}$ is the canonical projection defined by

$$\pi\phi = \sum_{k=1}^{N} \Phi_k (\Psi_k, \langle \phi(\cdot), \beta_k \rangle)_k \beta_k . \qquad (8.137)$$

We now develop a normal form theory for RFDE in X based on the theory in Section 8.2, so we omit some details. First, the phase space $\mathcal{C}$ is enlarged in such a way that (8.132) can be written as an abstract ODE in a Banach space. Rewriting (8.132) with $v = u_t \in \mathcal{C}$, we obtain the system

$$\frac{dv}{dt}(0) = A_T v(0) + L(v) + F(v), \quad \frac{dv}{dt}(\theta) = \dot{v}(\theta), \ \theta \in [-r,0).$$

It is therefore natural to consider for phase space the Banach space BC introduced in (H4). In terms of the function X_0 defined by $X_0(0) = I, X_0(\theta) = 0, \ -r \le \theta < 0$, the elements of BC have the form $\psi = \phi + X_0\alpha$, with $\phi \in \mathcal{C}, \alpha \in X$, so that $BC \equiv \mathcal{C} \times X$. The abstract ODE in BC equivalent to (8.132) is

$$\frac{d}{dt}v = \tilde{A}v + X_0 F(v), \qquad (8.138)$$

where $\tilde{A}$ is an extension of the infinitesimal generator A,

$$\tilde{A} : \mathcal{C}_0^1 \subset BC \longrightarrow BC \quad , \quad \tilde{A}\phi = \dot{\phi} + X_0[L(\phi) + A_T\phi(0) - \dot{\phi}(0)],$$

defined on $\mathcal{C}_0^1 := \{\phi \in \mathcal{C} : \dot{\phi} \in \mathcal{C}, \phi(0) \in D(A_T)\}$. On the other hand, it is easy to see that π, as defined in (4.6), is extended to a continuous projection (which we still denote by π), $\pi : BC \longrightarrow \mathcal{P}$, by defining

$$\pi(X_0\alpha) = \sum_{k=1}^{N} \Phi_k \Psi_k(0) \langle \alpha, \beta_k \rangle \beta_k, \quad \alpha \in X. \qquad (8.139)$$

The projection π leads to the topological decomposition

$$BC = \mathcal{P} \oplus \operatorname{Ker} \pi, \qquad (8.140)$$

with $\mathcal{Q} \subset_{\ne} \operatorname{Ker} \pi$, $\mathcal{P} \subset \mathcal{C}_0^1$ and π commuting with $\tilde{A}$ in $\mathcal{C}_0^1$. This allows us to decompose (8.138) as a system of abstract ODE on $\mathbb{R}^M \times \operatorname{Ker} \pi$, with linear and nonlinear parts separated and with finite and infinite dimensional variables also separated in the linear term, as in (8.42). More precisely, decomposing $v \in \mathcal{C}_0^1$ as $v(t) = \sum_{k=1}^{N} \Phi_k x_k(t)\beta_k + y(t)$, where $x_k(t) = (\Psi_k, \langle v(t)(\cdot), \beta_k \rangle)_k \in \mathbb{R}^{m_k}, 1 \le k \le N, y(t) \in \mathcal{C}_0^1 \cap \operatorname{Ker} \pi = \mathcal{C}_0^1 \cap \mathcal{Q} := \mathcal{Q}^1$, in $BC \equiv \mathbb{R}^M \times \operatorname{Ker} \pi$, (8.138) is equivalent to the system

$$\dot{x} = Bx + \Psi(0)\left(\langle F(\sum_{k=1}^{N} \Phi_k x_k \beta_k + y), \beta_p \rangle\right)_{p=1}^{N}$$

$$\dot{y} = A_{\mathcal{Q}^1} y + (I - \pi) X_0 F(\sum_{k=1}^{N} \Phi_k x_k \beta_k + y),$$

(8.141)

for $x = (x_1, \ldots, x_N) \in \mathbb{R}^M$, $y \in \mathcal{Q}^1$, where B is the $M \times M$ constant matrix $B = \mathrm{diag}\,(B_1, \ldots, B_N)$, Φ is the $N \times M$ matrix $\Phi = \mathrm{diag}(\Phi_1, \ldots, \Phi_N)$, Ψ is the $M \times N$ matrix $\Psi = \mathrm{diag}\,(\Psi_1, \ldots, \Psi_N)$ and $A_{\mathcal{Q}^1} : \mathcal{Q}^1 \subset \mathrm{Ker}\,\pi \longrightarrow \mathrm{Ker}\,\pi$, is defined by $A_{\mathcal{Q}^1} \phi = \tilde{A}\phi$, for $\phi \in \mathcal{Q}^1$.

We consider now the formal Taylor expansion

$$F(v) = \sum_{j \geq 2} \frac{1}{j!} F_j(v) \quad, v \in \mathcal{C},$$

where F_j is the jth Fréchet derivative of F. Then, (8.141) is written as

$$\dot{x} = Bx + \sum_{j \geq 2} \frac{1}{j!} f_j^1(x, y)$$

$$\dot{y} = A_{\mathcal{Q}^1} y + \sum_{j \geq 2} \frac{1}{j!} f_j^2(x, y),$$

(8.142)

where $x = (x_1, \ldots, x_N) \in \mathbb{R}^M$, $y \in \mathcal{Q}^1$ and $f_j = (f_j^1, f_j^2), j \geq 2$, are defined by

$$f_j^1(x, y) = \Psi(0)\left(\langle F_j(\sum_{k=1}^{N} \Phi_k x_k \beta_k + y), \beta_p \rangle\right)_{p=1}^{N}$$

$$f_j^2(x, y) = (I - \pi) X_0 F_j\left(\sum_{k=1}^{N} \Phi_k x_k \beta_k + y\right).$$

Normal forms for (8.141) are again obtained by a recursive process of change of variables. Formally, the operators M_j^1, M_j^2 appearing in the normal form process, and that are associated with the sequence of changes of variables, are still given by (8.50), with $A_{\mathcal{Q}^1}$ and π replaced by $A_{\mathcal{Q}^1}$ and π as defined above in this section. One can prove that the results stated in Lemmas 8.2.2 and 8.2.3 are still valid in this setting. This procedure is summarized in the next theorem and we refer the reader to [45] for its complete description and proofs.

Theorem 8.6.1. *Suppose that (H1)-(H4) hold and $\Lambda = \{\lambda \in \sigma(A) : \mathrm{Re}\,\lambda = 0\} \neq \emptyset$. Then, there exists a formal change of variables $(x, y) = (\bar{x}, \bar{y}) + O(|\bar{x}|^2)$ such that:*
(i) Equation (8.142) is transformed into

$$\dot{\bar{x}} = B\bar{x} + \sum_{j \geq 2} \frac{1}{j!} g_j^1(\bar{x}, \bar{y})$$

$$\dot{\bar{y}} = A_{Q^1}\bar{y} + \sum_{j \geq 2} \frac{1}{j!} g_j^2(\bar{x}, \bar{y}) , \tag{8.143}$$

and the new terms of order j, $g_j = (g_j^1, g_j^2), j \geq 2$, satisfy $g_j^2(\bar{x}, 0) \equiv 0, j \geq 2$
;

(ii) *a local center manifold for* (8.132) *at zero satisfies $\bar{y} = 0$ and the flow on it is given by the M-dimensional ODE*

$$\dot{\bar{x}} = B\bar{x} + \sum_{j \geq 2} \frac{1}{j!} g_j^1(\bar{x}, 0), \tag{8.144}$$

which is in normal form (in the usual sense of ODE).

Equation (8.143) is called a **normal form relative to Λ** for Equation (8.132).

We note that in particular the operators $f_j^1 \longmapsto g_j^1$ coincide with those operators appearing in the computation of normal forms for FDE of the form (8.34) and therefore also with the operators involved in the computation of ODE in $\mathbb{R}^M$.

For the sake of simplicity and because of the applications, we have considered $\Lambda = \{\lambda \in \sigma(A) : \text{Re }\lambda = 0\}$. Now, let Λ be another nonempty finite subset of $\sigma(A)$ and suppose, as before, that $\mathcal{P}$ is the invariant space of the linearized equation $\dot{u}(t) = A_T u(t) + L(u_t)$ associated with Λ. For this case, we also obtain the above decomposition $C = \mathcal{P} \oplus \mathcal{Q}$, as well as the procedure leading to (8.143). If there exists a local invariant manifold $\mathcal{M}_{A,F}$ for (8.132) tangent to $\mathcal{P}$ at zero, for that manifold we achieve a similar result to the one stated in Theorem 8.6.1 for the center manifold, if we guarantee that the following **nonresonance conditions relative to Λ** are satisfied:

$$(q, \bar{\lambda}) \neq \mu \quad , \text{ for all } \mu \in \sigma(A) \setminus \Lambda, \ q \in \mathbb{N}_0^M, |q| \geq 2, \tag{8.145}$$

where $\bar{\lambda} = (\lambda_1, \ldots, \lambda_M), \lambda_1, \ldots, \lambda_M$ are the elements of Λ, each of them appearing as many times as its multiplicity as a root of the associated characteristic equation, $(q, \bar{\lambda}) = q_1\lambda_1 + \cdots + q_M\lambda_M, |q| = q_1 + \cdots + q_M$, for $q = (q_1, \ldots, q_M) \in \mathbb{N}_0^M$. For instance, if $\Lambda = \{\lambda \in \sigma(A) : \text{Re }\lambda > 0\} \neq \emptyset$, then Theorem 8.6.1 is valid with "unstable manifold" instead of "center manifold".

Equations with parameters are more interesting from the point of view of normal forms, because of their applications to the study of bifurcations. Consider

$$\dot{u}(t) = A_T u(t) + L(\alpha)(u_t) + F(u_t, \alpha), \tag{8.146}$$

where $\alpha \in \mathbb{R}^p, L : \mathbb{R}^p \longrightarrow \mathcal{L}(C; X), F : C \times \mathbb{R}^p \longrightarrow X$ are C^k functions, $k \geq 2$, with $F(0, \alpha) = 0, D_1 F(0, \alpha) = 0$ for all $\alpha \in \mathbb{R}^p$. As for the case of FDE in finite-dimensional spaces, normal forms for these equations are computed in a similar way by introducing the parameter α as a variable with $\dot{\alpha}(t) = 0$.

We shall now take advantage of the similarity between (8.42) and (8.141), by defining an RFDE in a finite-dimensional space with an equilibrium at zero, in such a way that, on one hand, the eigenvalues on the imaginary axis for its linearized equation at zero are precisely the elements of Λ; and, on the other hand, and eventually under appropriate additional hypotheses, the normal form on the center manifold for that RFDE can be chosen so that it coincides with the normal form (8.144) on the center manifold of the original equation (8.132), at least up to same finite order. Here, we shall only state what happens up to second and third orders, because of applications to equations for which a Bogdanov–Takens or a Hopf bifurcation occur — so, the singularities are generically determined to second or to third orders, respectively.

It is natural to associate an RFDE in $\mathbb{R}^N$ with the original RFDE (8.132). Formally, if one dropped the eigenfunctions β_k and the inner product in (8.141), one sees that the system would become system (8.42) for the following RFDE in $C_N := C([-r, 0]; \mathbb{R}^N)$:

$$\dot{x}(t) = R(x_t) + \left(\langle F(\sum_{k=1}^{N} x_{t,k} \beta_k), \beta_p \rangle \right)_{p=1}^{N}, \qquad (8.147)$$

where $x(t) = \left(x_k(t)\right)_{k=1}^{N}, x_t = \left(x_{t,k}\right)_{k=1}^{N}$, and $R \in \mathcal{L}(C_N; \mathbb{R}^N)$ is the linear operator defined by

$$R(\phi) = \left(\mu_k \phi_k(0) + L_k(\phi_k)\right)_{k=1}^{N},$$

for $\phi = (\phi_1, \dots, \phi_N) \in C_N$.

Definition 8.6.2. *Equation (8.147) is called the FDE in $\mathbb{R}^N$ associated with equation (8.132) by Λ at zero.*

Note that the characteristic equation for the linearized equation $\dot{x}(t) = R(x_t)$ is $\det [\lambda I - R(e^{\lambda \cdot} I)] = 0$, thus λ is a root of the characteristic equation if and only if $\lambda - \mu_k - L_k(e^{\lambda \cdot}) = 0$, for some $1 \leq k \leq N$. Consequently, $\Lambda = \{\lambda \in \sigma(A) : \operatorname{Re} \lambda = 0\}$ is also the set of characteristic eigenvalues of R with real part zero. The next result establishes a relationship between (8.132) and (8.147).

Theorem 8.6.3. *[45] Suppose that (H1)-(H4) hold and let $\Lambda = \{\lambda \in \sigma(A) : \operatorname{Re} \lambda = 0\} \neq \emptyset$. Consider (8.132) with F of class C^2. Then, for a suitable change of variables, the normal forms on the center manifold for both (8.132) and (8.147) are the same, up to second order terms. If F is of class C^3, $F(v) = \frac{1}{2!} F_2(v) + \frac{1}{3!} F_3(v) + o(|v|^3)$, and assuming also hypothesis (H5):*

$$\langle DF_2(u)(\phi \beta_j), \beta_p \rangle = 0 , \text{ for } 1 \leq p \leq N, j > N \text{ and all } u \in \mathcal{P}, \ \phi \in C,$$

*then, for a suitable change of variables, the normal forms on the center man-
ifold for both* (8.132) *and* (8.147) *are the same, up to third order terms.*

Proof: The proof of this theorem is based on the possibility of identification of
the operators M_j^1, M_j^2 of the normal form algorithm for (8.132) with the corresponding operators appearing in the computation of normal forms relative
to Λ for (8.147). To explain the main idea as to why this result is true, write
(8.132) as $\dot{u}(t) = A_T u(t) + \mathcal{F}(u_t)$. We observe that (H3) imposes a condition
that can be interpreted as a nonresonance condition between the first order
terms of $\mathcal{F}$ and the eigenspaces for A_T, while (H5) imposes a nonresonance
condition between the second order terms of $\mathcal{F}$ and the eigenspaces for A_T.
Of course, this approach could be pursued in order to derive nonresonance
conditions between the j-order terms, $j \leq k - 1$, of $\mathcal{F}$ and the eigenspaces
for A_T that assure that the normal forms on the center manifold for both
(8.132) and (8.147) are the same up to k order terms. Moreover, we point
out that even when (H5) fails, the associated RFDE by Λ still carries much
information that can be transferred to the computation of normal forms for
the original RFDE (8.132) ([45]).   ■

Example 8.6.4. Consider the Hutchinson equation with diffusion (see [45],
[125], [142], [208]):

$$\frac{\partial u(t,x)}{\partial t} = d\frac{\partial^2 u(t,x)}{\partial x^2} - au(t-1,x)[1 + u(t,x)], \quad t > 0, x \in (0,\pi)$$

$$\frac{\partial u(t,x)}{\partial x} = 0, \quad x = 0, \pi, \quad (8.148)$$

where $d > 0, a > 0$. This equation can be written in abstract form in $\mathcal{C} = C([-1,0]; X)$ as

$$\frac{d}{dt}u(t) = d\Delta u(t) + L(a)(u_t) + f(u_t, a) \quad (8.149)$$

where

$$X = \{v \in W^{2,2}(0,\pi) : \frac{dv}{dx} = 0 \text{ at } x = 0, \pi\},$$

$$L(a)(v) = -av(-1), f(v,a) = -av(0)v(-1).$$

However, in the following, we shall begin by considering the general case of
any $f : \mathcal{C} \times \mathbb{R} \to X$ such that $f \in C^3, f(0,a) = 0, D_1 f(0,a) = 0$, for $a > 0$.
The functions $\beta_k(x) = \frac{\cos(kx)}{\|\cos(kx)\|_{2,2}}$ are normalized eigenfunctions of $d\Delta = d\frac{\partial^2}{\partial x^2}$ on X, with corresponding eigenvalues $\mu_k = -dk^2, k \geq 0$, and (H1)-(H4)
hold for $a > 0$ ([125], [142]). By linearizing (8.149) about the equilibrium
$u = 0$, for this case the characteristic equations (8.135) (where here is $k \in \mathbb{N}_0$)
are

$$\lambda + ae^{-\lambda} + dk^2 = 0 , \quad (k = 0, 1, \dots). \quad (8.150)$$

Yoshida [208] showed that for $a < \pi/2$, all roots of all equations (8.150) have negative real parts, so the zero solution is stable; when $a = \pi/2$ and $k = 0$, (8.150) has the unique pair $\pm i\pi/2$ of (simple) solutions on the imaginary axis, and all other solutions of (8.150), $k \geq 0$, have negative real parts. To study the qualitative behavior near the critical point $a = \pi/2$, let $a = \pi/2+\alpha$. Again from [208], for $k = 0$ there is a pair of solutions $\lambda(\alpha), \overline{\lambda(\alpha)}$ of (8.150), with $\lambda(0) = i\pi/2$ and Re $\lambda'(0) > 0$ (Hopf condition) and then a Hopf bifurcation occurs at $\alpha = 0$.

Let $\Lambda = \{i\pi/2, -i\pi/2\}$. Defining $L := L(\pi/2)$ and $F(v, \alpha) = -\alpha v(-1) + f(v, \pi/2 + \alpha)$, (4.18) becomes

$$\frac{d}{dt}u(t) = d\Delta u(t) + L(u_t) + F(u_t, \alpha). \qquad (8.151)$$

Since $L(\psi\beta_0) = -\frac{\pi}{2}\psi(-1)\beta_0$, the operator $L_0 : C([-1, 0]; \mathbb{R}) \to \mathbb{R}$ in (8.135) corresponding to the eigenvalue $\mu_0 = 0$ is $L_0(\psi) = -\frac{\pi}{2}\psi(-1)$. The FDE associated with (8.151) by Λ at the equilibrium point $u = 0, \alpha = 0$ is $\dot{x}(t) = L_0(x_t)+ < F(x_t\beta_0, \alpha), \beta_0 >$, i.e., the RFDE in $C := C([-1, 0]; \mathbb{R})$

$$\dot{x}(t) = -\frac{\pi}{2}x(t - 1) + \langle F(x_t\beta_0, \alpha), \beta_0\rangle. \qquad (8.152)$$

With the former notations, except that now $k \in \mathbb{N}_0$ instead of $k \in \mathbb{N}$, we have $N = 0$ and $P = P_0$, dim $\mathcal{P} = $ dim $P_0 = 2$, $P_0 = $ span Φ, where in complex coordinates (see 8.3.1) we have

$$\Phi(\theta) = (\phi_1(\theta), \phi_2(\theta)) = (e^{i\frac{\pi}{2}\theta}, e^{-i\frac{\pi}{2}\theta}),$$

$$B = diag(i\pi/2, -i\pi/2)$$

$$\Psi(0) = \begin{pmatrix} \psi_1(0) \\ \psi_2(0) \end{pmatrix}$$

with

$$\psi_1(0) = \overline{\psi_2(0)} = \frac{1 - i\frac{\pi}{2}}{1 + \frac{\pi^2}{4}}.$$

Suppose that (H5) holds; since $\beta_0 = 1/\sqrt{\pi}$, that means

$$\langle D_1F_2(\Phi c, \alpha)(\psi\beta_k), 1\rangle = 0, \text{ for all } k \in \mathbb{N}, c \in \mathbb{C}^2, \psi \in C.$$

On the other hand, for (8.151) there exists a two–dimensional local center manifold tangent to $\mathcal{P}$ at $u = 0, \alpha = 0$, which is stable [125]. Theorem 8.6.3 allows us to conclude that the equation on the center manifold for (8.151) and (8.152) coincide up to third order terms — which are sufficient to determine a generic Hopf bifurcation. For (8.152), we are able to write that equation without additional calculus, using Theorem 8.3.1.

For instance, the Hutchinson equation with diffusion (8.148) is written as

$$\frac{d}{dt}u(t) = d\Delta u(t) - au(t-1)[1 + u(t)] ,$$

$a, d > 0$. Then $F(v, \alpha) = -\alpha v(-1) - (\frac{\pi}{2} + \alpha)v(-1)v(0)$, and

$$F_2(v, \alpha) = -2\alpha v(-1) - \pi v(-1)v(0),$$
$$D_1 F_2(v, \alpha)(u) = -2\alpha u(-1) - \pi\big(u(-1)v(0) + v(-1)u(0)\big),$$

and clearly (H5) is fulfilled. In this case, the associated FDE (8.152) in C is

$$\dot{x}(t) = -ax(t-1)[1 + \frac{1}{\sqrt{\pi}}x(t)],$$

which is the well–known Wright equation under the change $x \longmapsto \frac{1}{\sqrt{\pi}}x$, so we can apply the results in Example 8.3.2. Its equation on the center manifold is given in polar coordinates (ρ, ξ) by

$$\dot{\rho} = K_1 \alpha \rho + \frac{1}{\pi}K_2 \rho^3 + O(\alpha^2 \rho + |(\rho, \alpha)|^4)$$

$$\dot{\xi} = -\frac{\pi}{2} + O(|(\rho, \alpha)|) , \quad (8.153)$$

where $K_1 = \mathrm{Re}\,\lambda'(0)$ and K_2 are as in (8.79) with N=0:

$$K_1 = \frac{2\pi}{4 + \pi^2} > 0 , \quad K_2 = \frac{\pi(2 - 3\pi)}{5(4 + \pi^2)} < 0 . \quad (8.154)$$

Theorem 8.6.3 tells us that the flow on the center manifold for (8.148) at $u = 0, a = \pi/2$ is also given by (8.153) where (8.154) holds. Therefore, the periodic solutions associated with the generic supercritical Hopf bifurcation for (8.148) are stable, because $K_2 < 0$.

A more general framework in which to develop a normal form theory on invariant manifolds of equations (8.132) was set in [44]. Hypothesis (H3) was slightly weakened, by imposing that the eigenvalues of A_T can be organized in blocks, in such a way that the modes of the generalized eigenspaces for A_T generated by the eigenvalues in each block are not mixed by L. For the sake of simplicity of exposition, we shall also restate our hypothesis (H2) in another form. For (8.132), assume (H1), (H4) and the following two conditions:

(H2') Let $\{\mu_k^{i_k} : k \in \mathbb{N}, i_k = 1, \ldots, p_k\}$ be the eigenvalues of A_T (distinct or not) and $\beta_k^{i_k}$ be eigenfunctions corresponding to $\mu_k^{i_k}$, such that $\{\beta_k^{i_k} : k \in \mathbb{N}, i_k = 1, \ldots, p_k\}$ form an orthonormal basis for X;

(H3') the subspaces $\mathcal{B}_k$ of $\mathcal{C}$, $\mathcal{B}_k := \mathrm{span}\{\varphi\beta_k^{i_k} \mid \varphi \in C([-r, 0]; \mathbb{R}), i_k = 1, \ldots, p_k\}$ satisfy $L(\mathcal{B}_k) \subset \mathrm{span}\{\beta_k^1, \ldots, \beta_k^{p_k}\}$.

With these assumptions, for each $k \in \mathbb{N}$ define the linear operators L_k : $C_{p_k} \longrightarrow \mathbb{R}^{p_k}$, $C_{p_k} = C([-r, 0]; \mathbb{R}^{p_k})$, by

$$L_k(\psi) = (L_k^1(\psi), \ldots, L_k^{p_k}(\psi))$$

$$L(\psi_1 \beta_k^1 + \cdots + \psi_{p_k} \beta_k^{p_k}) = \sum_{i_k=1}^{p_k} L_k^{i_k}(\psi)\beta_k^{i_k}, \quad (8.155)$$

where $\psi = (\psi_1, \ldots, \psi_{p_k}) \in C_{p_k}$. Note that L_k are well defined because of (H3′). Then, (H2′) implies that the characteristic equation (8.134) is equivalent to the sequence of characteristic equations $\det \Delta_k(\lambda) = 0$, where

$$\Delta_k(\lambda) := \lambda I - M_k - L_k(e^{\lambda \cdot} I) \quad (k \in \mathbb{N}), \quad (8.156)$$

and $M_k = \mathrm{diag}\,(\mu_k^1, \ldots, \mu_k^{p_k})$. Since (8.156) is the characteristic equation for the retarded FDE in C_{p_k}

$$\dot{z}(t) = M_k z(t) + L_k z_t \quad (k \in \mathbb{N}), \quad (8.157)$$

the above procedure can be followed with a few changes, described below.

Suppose once more that the set Λ of the eigenvalues of A on the imaginary axis is given by $\Lambda = \{\lambda \in \mathbb{C} : \lambda \text{ is a solution of (8.156) with } \mathrm{Re}\,\lambda = 0, \text{ for some } k \in \{1, \ldots, N\}\}$, and that $\Lambda_k := \{\lambda \in \mathbb{C} : \lambda \text{ satisfies (8.156) and } \mathrm{Re}\,\lambda = 0\}$. We adopt the same notations as before with some necessary modifications. Namely, we have now C_{p_k} decomposed by Λ_k, $C_{p_k} = P_k \oplus Q_k$, the adjoint bilinear form is $(\cdot, \cdot)_k : C_{p_k}^* \times C_{p_k} \to \mathbb{R}$, Φ_k is a $p_k \times m_k$ matrix and Ψ_k is an $m_k \times p_k$ matrix (instead of $1 \times m_k$ and $m_k \times 1$, respectively). The decomposition $C = P \oplus Q$, $P = \mathrm{Im}\,\pi$, $Q = \mathrm{Ker}\,\pi$, is now achieved by a projection $\pi : C \to P$ with a more complicated expression to take into account the coefficients of $\beta_k^1, \ldots, \beta_k^{p_k}$ all together. The projection $\pi : BC \to P$ is written as (cf. (8.137) and (8.139))

$$\pi(\phi) = \sum_{k=1}^{N} \left[\Phi_k \left(\Psi_k, \left(\langle \phi(\cdot), \beta_k^{i_k} \rangle \right)_{i_k=1}^{p_k} \right)_k \right]^T \begin{pmatrix} \beta_k^1 \\ \vdots \\ \beta_k^{p_k} \end{pmatrix}, \quad \phi \in C,$$

$$\pi(X_0 \alpha) = \sum_{k=1}^{N} \left[\Phi_k \Psi_k(0) \left(\langle \alpha, \beta_k^{i_k} \rangle \right)_{i_k=1}^{p_k} \right]^T \begin{pmatrix} \beta_k^1 \\ \vdots \\ \beta_k^{p_k} \end{pmatrix}, \quad \alpha \in X. \quad (8.158)$$

It is easy to check that decomposing $v \in C_0^1$ according to (8.140) as

$$v(t) = \sum_{k=1}^{N} \left[\Phi_k x_k(t) \right]^T \begin{pmatrix} \beta_k^1 \\ \vdots \\ \beta_k^{p_k} \end{pmatrix} + y(t), \quad (8.159)$$

where $x_k(t) = \left(\Psi_k, \left(\langle v(t)(\cdot), \beta_k^{i_k} \rangle \right)_{i_k=1}^{p_k} \right)_k \in \mathbb{R}^{m_k}, 1 \leq k \leq N, y(t) \in Q^1$, (8.138) is equivalent to the system on $BC \equiv \mathbb{R}^M \times \mathrm{Ker}\,\pi$

$$\dot{x} = Bx + \Psi(0)\left(\left(\left\langle F\left(\sum_{k=1}^{N}\left[\Phi_k x_k\right]^T\begin{pmatrix}\beta_k^1\\ \vdots\\ \beta_k^{p_k}\end{pmatrix}+y\right),\beta_n^{i_n}\right\rangle\right)_{i_n=1}^{p_n}\right)_{n=1}^{N}$$

$$\dot{y} = A_{Q^1}y + (I-\pi)X_0 F\left(\sum_{k=1}^{N}\left[\Phi_k x_k\right]^T\begin{pmatrix}\beta_k^1\\ \vdots\\ \beta_k^{p_k}\end{pmatrix}+y\right)$$

$$\tag{8.160}$$

where $x = (x_1,\ldots,x_N) \in \mathbb{R}^M, y \in \mathcal{Q}^1$ (cf. (8.141)). Normal form computations follow now as previously described. In this case, the finite-dimensional FDE associated by Λ is

$$\dot{x}(t) = R(x_t) + G(x_t), \tag{8.161}$$

where: $x(t) = \left(x_k(t)\right)_{k=1}^{N}, x_k \in \mathbb{R}^{p_k}$, and $R : C_J \to \mathbb{R}^J, G : C_J \to \mathbb{R}^J$, for $J := \sum_{k=1}^{N} p_k$, are defined by

$$R(\phi) = \left(M_k\phi_k(0) + L_k(\phi_k)\right)_{k=1}^{N},$$

$$G(\phi) = \left(\left(\left\langle F\left(\sum_{k=1}^{N}[\phi_k]^T\begin{pmatrix}\beta_k^1\\ \vdots\\ \beta_k^{p_k}\end{pmatrix}\right),\beta_n^{i_n}\right\rangle\right)_{i_n=1}^{p_n}\right)_{n=1}^{N},$$

for $\phi = (\phi_1,\ldots,\phi_N) \in C_J, \phi_k \in C_{p_k}, k = 1,\ldots,N$. Clearly, hypothesis (H5) is the same, but reformulated according to (H2) rewritten as (H2'):

$$\langle DF_2(u)(\phi\beta_j^{ij}),\beta_n^{i_n}\rangle = 0, \quad for \quad 1 \le n \le N, \quad 1 \le i_n \le p_n,$$
$$j > N, 1 \le i_j \le p_j \quad and \quad all \quad u \in \mathcal{P}, \phi \in C. \ \textbf{(H5')} \quad (8.162)$$

Within this context, we present the new version of Theorem 8.6.3.

Theorem 8.6.5. . *Suppose (H1), (H2'), (H3'), (H4) hold and let $\Lambda = \{\lambda \in \sigma(A) : \text{Re}\,\lambda = 0\} \ne \emptyset$. Consider equation (8.132) with F of class C^2. Then, for a suitable change of variables, the normal forms on the center manifold for both (8.132) and (8.161) are the same, up to second order terms. Furthermore, if F is of class C^3, $F(v) = \dfrac{1}{2!}F_2(v) + \dfrac{1}{3!}F_3(v) + o(|v|^3)$ and (H5') holds, also the third order terms of those equations on the center manifold for both (8.132) and (8.161) are the same.*

Example 8.6.6. Recall the predator-prey system (8.151). Models involving delays and also spatial diffusion are increasingly applied to the study of a variety of situations. For this reason, along with equation (8.151) we consider a second model, the delayed reaction-diffusion system with Neumann conditions resulting from considering one spatial variable and adding diffusion terms $d_1 \Delta u, d_2 \Delta v$, $d_1, d_2 > 0$, respectively, to the first and second equations of (8.151):

$$\frac{\partial u(t, x)}{\partial t} = d_1 \frac{\partial^2 u(t, x)}{\partial x^2} + u(t, x)[r_1 - au(t, x) - a_1 v(t - \sigma, x)]$$

$$\frac{\partial v(t, x)}{\partial t} = d_2 \frac{\partial^2 v(t, x)}{\partial x^2} + v(t, x)[-r_2 + b_1 u(t - \tau, x) - b(t, x)],$$

$$t > 0, x \in (0, \pi);$$

$$\frac{\partial u(t, x)}{\partial x} = \frac{\partial v(t, x)}{\partial x} = 0 \ , \ x = 0, \pi,$$

(8.163)

where $d_1, d_2, \tau, r_1, r_2, a_1, b_1$ are positive constants and σ, a, b are non-negative constants. Similarly to what was done for (8.80), we may assume $a_1 = b_1 = 1$. If (8.82) holds, the positive equilibrium $E_* = (u_*, v_*)$ of (8.80) is now the unique positive stationary solution of (8.163).

Our purpose, in the sequel, is to relate the dynamics of the two systems (without and with diffusion) in the neighborhood of E_*, and determine the effect of the diffusion terms, regarding the stability and the Hopf bifurcation, near the first bifurcation point $\tau = \tau_0$. After the time-scaling $t \to t/\tau$, (8.163) is given in abstract form as

$$\frac{d}{dt} u(t) = d_1 \tau \Delta u(t) + \tau u(t)[r_1 - au(t) - v(t - r)]$$

$$\frac{d}{dt} v(t) = d_2 \tau \Delta v(t) + \tau v(t)[-r_2 + u(t - 1) - bv(t)],$$

(8.164)

where $r = \sigma/\tau$ and, for simplification of notation, we use $u(t)$ for $u(t, \cdot)$, $v(t)$ for $v(t, \cdot)$, and $(u(t), v(t)) = (u(t, \cdot), v(t, \cdot))$ is in the Hilbert space $X = \{(u, v) : u, v \in W^{2,2}(0, \pi), \frac{du}{dx} = \frac{dv}{dx} = 0 \text{ at } x = 0, \pi\}$, with the inner product $\langle \cdot, \cdot \rangle$ induced by the inner product of the Sobolev space $W^{2,2}(0, \pi)$. Translating E_* to the origin by setting $U(t) = (u(t), v(t)) - E_* \in X$, (8.164) is transformed into the equation in $\mathcal{C} := C([-1, 0]; X)$

$$\frac{d}{dt} U(t) = \tau d\Delta U(t) + L(\tau)(U_t) + f(U_t, \tau),$$

(8.165)

where $d\Delta = (d_1 \Delta, d_2 \Delta)$ and $L(\tau) : \mathcal{C} \to \mathbb{R}^2, f : \mathcal{C} \times \mathbb{R}^+ \to \mathbb{R}^2$ are given by

$$L(\tau)(\varphi) = \tau \begin{pmatrix} -u_*(a\varphi_1(0) + \varphi_2(-r)) \\ v_*(\varphi_1(-1) - b\varphi_2(0)) \end{pmatrix},$$

$$f(\varphi, \tau) = \tau \begin{pmatrix} -\varphi_1(0)(a\varphi_1(0) + \varphi_2(-r)) \\ \varphi_2(0)(\varphi_1(-1) - b\varphi_2(0)) \end{pmatrix},$$

for $\varphi = (\varphi_1, \varphi_2) \in C$ and u_*, v_* as in (8.83). The characteristic equation for the linearized equation $\frac{d}{dt}U(t) = \tau d\Delta U(t) + L(\tau)(U_t)$ is

$$\lambda y - \tau d\Delta y - L(\tau)(e^{\lambda \cdot} y) = 0, \quad y \in D(\Delta), y \neq 0. \tag{8.166}$$

The eigenvalues of $\tau d\Delta$ on X are $\mu_k^i = -d_i \tau k^2, i = 1, 2, \ k = 0, 1, 2, \ldots,$ with corresponding normalized eigenfunctions β_k^i, where

$$\beta_k^1 = \begin{pmatrix} \gamma_k \\ 0 \end{pmatrix}, \ \beta_k^2 = \begin{pmatrix} 0 \\ \gamma_k \end{pmatrix}, \ \gamma_k(x) = \frac{\cos(kx)}{\|\cos(kx)\|_{2,2}}, \ k \in \mathbb{N}_0.$$

In this case (H3) fails. However, we have

$$L(\tau)(\varphi_1 \beta_k^1 + \varphi_2 \beta_k^2) = -u_* \tau (a\varphi_1(0) + \varphi_2(-r))\beta_k^1 + v_* \tau (\varphi_1(-1) - b\varphi_2(0))\beta_k^2,$$

or, equivalently,

$$L(\tau)(\varphi^T \begin{pmatrix} \beta_k^1 \\ \beta_k^2 \end{pmatrix}) = \left[N(\tau)(\varphi) \right]^T \begin{pmatrix} \beta_k^1 \\ \beta_k^2 \end{pmatrix},$$

where $N(\tau)$ is as in (8.84), implying that $L(\tau)$ does not mix the modes of the generalized eigenspaces $span\{\beta_k^1, \beta_k^2\}$. Thus (H3') holds with $\mathcal{B}_k = \{\varphi_1 \beta_k^1 + \varphi_2 \beta_k^2 : (\varphi_1, \varphi_2) \in C([-1, 0]; \mathbb{R}^2)\}$. For any $y \in X$, consider now its Fourier series relative to the basis $\{\beta_k^i : i = 1, 2; k = 0, 1, \ldots\}$, written in such a way that the Fourier coefficients relative to β_k^1, β_k^2 are kept together:

$$y = \sum_{K=0}^{\infty} Y_k^T \begin{pmatrix} \beta_k^1 \\ \beta_k^2 \end{pmatrix}, \quad Y_k = \begin{pmatrix} angley, \beta_k^1 \rangle \\ \langle y, \beta_k^2 \rangle \end{pmatrix}.$$

Using this decomposition, we note that for $y \in D(\Delta), y \neq 0$, the characteristic equation (8.94) is equivalent to

$$\sum_{K=0}^{\infty} Y_k^T \left[\lambda I - \tau \begin{pmatrix} -k^2 d_1 & 0 \\ 0 & -k^2 d_2 \end{pmatrix} - \tau \begin{pmatrix} -au_* & -u_* e^{-\lambda r} \\ v_* e^{-\lambda} & -bv_* \end{pmatrix} \right] \begin{pmatrix} \beta_k^1 \\ \beta_k^2 \end{pmatrix} = 0.$$

Hence, we conclude that (8.94) is equivalent to the sequence of characteristic equations

$$\Delta_k(\lambda, \tau) := \lambda^2 + (d_1 k^2 + d_2 k^2 + au_* + bv_*)\tau\lambda +$$
$$+ (d_1 k^2 + au_*)(d_2 k^2 + bv_*)\tau^2 + u_* v_* \tau^2 e^{-\lambda(1+r)} = 0, \quad k = 0, 1, 2, \ldots. \tag{8.167}$$

It is important to remark that for $k = 0$ the above equation (8.167) is the characteristic equation (8.85)obtained for the system without diffusion. The analysis of the characteristic equations (8.167), for $k \geq 1$, shows that all their roots have Re $\lambda < 0$, provided some additional conditions are made. See [47] for proofs.

Theorem 8.6.7. *Assume* (8.92) *and define* τ_0, σ_0 *as in Theorem 8.3.3. Suppose also that*

$$a^2 + b^2 > 0 \quad \text{and} \quad ab(au_* + bv_*)^2 \le u_* v_*, \tag{8.168}$$

or

$$a = b = 0, \quad \text{and} \quad d_1 + d_2 \ge 2\pi\sqrt{r_1 r_2} \quad \text{or} \quad d_1 d_2 \ge r_1 r_2. \tag{8.169}$$

Then, for $0 < \tau \le \tau_0$ *and* $k \ge 1$, *all the roots of the characteristic equations* (8.167) *have negative real parts.*

Remark 8.6.8. If $a = b = 0$ and the coefficients of the diffusion terms d_1, d_2 are small, the instability of the stationary solution E_* of the reaction-diffusion equation (8.164) at τ_0 might increase. Actually, one can prove that for $a = b = 0$, $d_1 = d_2$ and $\tau = \tau_0 = 2\pi/((1+r)\sqrt{r_1 r_2})$, if $k \ge 1$ is such that $d_1 k^2 \le \sqrt{3r_1 r_2}/2$, then $\Delta_k(\lambda, \tau_0) = 0$ has at least a pair of complex conjugated roots with positive real parts [47]. This shows the effect of small diffusion terms creating more instability of E_* at the first bifurcation point τ_0.

Reasoning as in [125], the existence of a local center manifold of the stationary point E_* follows for the model presented here. In the sequel, we always assume (8.82), and either (8.168) or (8.169). Now, we show how the Hopf bifurcation analysis for (8.164) near E_* and τ_0 (if it is generic) can be deduced from the case without diffusion. Again let $\tau = \tau_0 + \alpha$ and $\Lambda_0 = \{-i\sigma_0, i\sigma_0\}$. Theorem 8.6.7 implies that Λ_0 is the set of eigenvalues on the imaginary axis of the infinitesimal generator associated with the flow of

$$\frac{d}{dt}U(t) = \tau_0 d\Delta U(t) + L(\tau_0)(U_t). \tag{8.170}$$

Defining $F(v, \alpha) := \alpha d\Delta v(0) + L(\alpha)(v) + f(v, \tau_0 + \alpha)$, (8.165) is written as

$$\frac{d}{dt}U(t) = \tau_0 d\Delta U(t) + L(\tau_0)(U_t) + F(U_t, \alpha). \tag{8.171}$$

Let the phase space $\mathcal{C}$ be decomposed by Λ_0, $\mathcal{C} = \mathcal{P} \oplus \mathcal{Q}$, where $\mathcal{P}$ is the center space for (8.170), given by $\mathcal{P} = span\{\Phi^T \begin{pmatrix} \beta_0^1 \\ \beta_0^2 \end{pmatrix}\}$, and $\Phi = [\phi_1 \ \phi_2]$ is the basis for the space $P \subset \mathcal{C} = C([-1,0]; \mathbb{C}^2)$ defined as in subsection 8.3.2. Note that $u \in \mathcal{P}$ necessarily has the form $u = \frac{1}{\sqrt{\pi}}(\varphi_1, \varphi_2)$, with $(\varphi_1, \varphi_2) \in C$. For (8.171) we define its associated FDE by Λ_0 at $U = 0, \alpha = 0$, as the retarded FDE in C:

$$\dot{x}(t) = R(x_t) + G(x_t, \alpha), \tag{8.172}$$

where R, G are given by

$$L(\tau_0)\left(\varphi^T \begin{pmatrix} \beta_0^1 \\ \beta_0^2 \end{pmatrix}\right) = \left[R(\varphi)\right]^T \begin{pmatrix} \beta_0^1 \\ \beta_0^2 \end{pmatrix},$$

$$G(\varphi, \alpha) = \begin{pmatrix} \langle F\left(\varphi^T \begin{pmatrix} \beta_0^1 \\ \beta_0^2 \end{pmatrix}\right), \alpha\rangle, \beta_0^1 \rangle \\ \langle F\left(\varphi^T \begin{pmatrix} \beta_0^1 \\ \beta_0^2 \end{pmatrix}\right), \alpha\rangle, \beta_0^2 \rangle \end{pmatrix}, \quad \varphi \in C.$$

We observe that $R = N(\tau_0)$ as in (8.88). Also, we have $\Delta\beta_0^i = 0$, $i = 1, 2$, $\beta_0^1 = (1/\sqrt{\pi}, 0), \beta_0^2 = (0, 1/\sqrt{\pi}), \langle\beta_0^i, \beta_0^i\rangle = 1, \langle\beta_0^i, \beta_0^j\rangle = 0$, if $i \neq j, i, j = 1, 2$. Therefore, $G(\varphi, \alpha) = N(\alpha)(\varphi) + \frac{1}{\sqrt{\pi}} f_0(\varphi, \tau_0 + \alpha)$ and (8.172) becomes

$$\dot{x}(t) = N(\tau_0 + \alpha)(x_t) + \frac{1}{\sqrt{\pi}} f_0(x_t, \tau_0 + \alpha). \tag{8.173}$$

The scaling $x = \sqrt{\pi} z$ transforms (8.173) into (8.88). This proves that the dynamics for (8.88) and for (8.173) are the same near the origin and $\alpha = 0$.

On the other hand, the nonresonance condition on the second order terms of F (H5') is translated here as

$$\langle D_1 F_2(u, \alpha)(\delta\beta_k^i), \beta_0^j\rangle = 0,$$
$$\text{for } i, j = 1, 2, u \in P, \delta \in C([-1, 0]; \mathbb{R}) \text{ and } k \geq 1, \tag{8.174}$$

where $\frac{1}{2}F_2(v, \alpha)$ are the quadratic terms of F in (v, α). The above definition of F yields

$$\frac{1}{2}F_2(v, \alpha) = \alpha d\Delta v(0) + L(\alpha)(v) + \tau_0 \begin{pmatrix} -v_1(0)(av_1(0) + v_2(-r)) \\ v_2(0)(v_1(-1) - bv_2(0)) \end{pmatrix},$$

for $v = (v_1, v_2) \in C$. Using $\langle\beta_k^i, \beta_0^j\rangle = 0$, $k \geq 1, i, j = 1, 2$, one can see that (8.174) is fulfilled.

Since the terms up to third order are sufficient to determine the dynamics of a generic Hopf bifurcation, the above arguments and Theorem 8.6.5 lead to the following conclusion:

Theorem 8.6.9. *Assume (8.92), and either (8.168) or (8.169); let τ_0 be as in Theorem 8.3.3. Then, for a suitable change of variables, the equations on the center manifold of E_* at $\tau = \tau_0$ for both (8.81) and (8.164) are the same, up to third order terms. In particular, if the Hopf bifurcation on the 2-dimensional locally center manifold for (8.81) is generic, the same is true for (8.164) and the bifurcation direction and the stability of the periodic orbits on the center manifold for (8.81) and (8.164) are the same.*

Under the above conditions, these results show that the local stability of E_* for $0 < \tau \leq \tau_0$, as well as the Hopf singularity at τ_0, if generic, are reduced to the case without diffusion. In this sense, the diffusion terms are irrelevant in our model.

As a particular situation, recall Example 8.3.5 and consider (8.81) and (8.164) with $a = 1, b = 0, r = 0, r_1 = 2, r_2 = 1$. The conclusions of Theorem 8.3.6 apply to the model with diffusion, and we deduce that the Hopf bifurcation occurring on the stable center manifold of the equilibrium $E_* = (1, 1)$ at $\tau = \tau_0$ is supercritical and that the bifurcating periodic orbits are stable.

8.7 Normal forms for periodic RFDE on $\mathbb{R}^n$ and autonomous RFDE on Banach spaces

This section is a short overview of further developments on normal forms for periodic RFDE in $\mathbb{R}^n$ and for autonomous RFDE in Banach spaces.

The normal form theories presented here, either for finite or infinite dimensional RFDE, were developed near an *equilibrium point* of an *autonomous* equation. An important generalization needed is the construction of normal forms around periodic orbits.

For an autonomous equation

$$\dot{u}(t) = f(u_t) \tag{8.175}$$

in $C = C([-r,0]; \mathbb{R}^n)$, where $f : C \to \mathbb{R}^n$ is $C^k, k \geq 1$, suppose that $p(t)$ is a ω-periodic solution of this equation. In order to study the properties of the flow near $p(t)$, consider the change of variables $u(t) = p(t) + x(t)$, leading to

$$\dot{x}(t) = L(t)x_t + F(t, x_t), \tag{8.176}$$

where $L(t) = Df(p_t)$, $F(t, \cdot) = f(p_t + \cdot) - f(p_t) - Df(p_t)$ are ω-periodic functions. In [43], the normal form theory was extended to periodic RFDE of the form $\dot{x}(t) = Lx_t + F(t, x_t)$, with $L \in \mathcal{L}(C; \mathbb{R}^n), F \in C^k(\mathbb{R} \times C; \mathbb{R}^n), k \geq 2, F(0) = 0, DF(0) = 0$. This only covers equations in the form (8.175) with an *autonomous* linear part, i.e., equations that satisfy

$$L(t) = L \qquad \text{for all} \quad t \in \mathbb{R}, \tag{8.177}$$

clearly a very restrictive assumption. The application of the normal form technique requires a suitable system of coordinates. The reason for imposing such a restriction is that in [43] the author was not able to provide a suitable system of coordinates in the neighborhood of periodic solutions for (8.175). We recall that this cannot be accomplished by the Floquet theory, as for the case of ODE in $\mathbb{R}^n$, since there is no complete Floquet theory for periodic RFDE but only a Floquet representation on generalized eigenspaces of characteristic multipliers (see [86]). Although the hypothesis (8.177) is not convenient from the point of view of applications, it provides a strong result: it was proven in [43] that if certain nonresonance conditions are satisfied, the normal form on center manifolds (or other invariant manifolds) of the origin for (8.176) coincides with the normal form for the *averaged equation*

$$\dot{x}(t) = Lx_t + F_0(x_t),$$

where $F_0(\phi) = \frac{1}{\omega} \int_0^\omega F(t, \phi)dt$ for $\phi \in C$.

Recently, Hale and Weedermann [94] have established a local system of coordinates near a periodic orbit of (8.175). In fact, they considered the more general case of neutral FDE. The key idea to set this system is summarize below. We refer the reader to [86] Chapter 8, for notation and general theory for periodic FDE.

Suppose $\Gamma = \{p_t : 0 \le t \le \omega\}$ is a nontrivial periodic orbit of (8.175), and consider the linear variational equation

$$\dot{x}(t) = L(t)x_t. \tag{8.178}$$

Since 1 is always a Floquet multiplier of (8.178), consider a decomposition of the phase space C with respect to $\mu = 1$. One possible choice for the basis of $E_1(s)$ is $\Phi_1(s) = \{\dot{p}_s, \phi_2, \ldots, \phi_{d_1}\}$, because $\dot{p}_t$ is a corresponding eigenvector for $\mu = 1$. Let $T(t,s), t \ge s$, be the solution operators for (8.178), defined by $T(t,s)\phi = x_t(s,\phi)$ for all $t \ge s$ and $\phi \in C$, where $x = x(s,\phi)$ is the solution of (8.178) such that $x_s = \phi$. Assume that Γ is non-degenerate, that is, $\mu = 1$ is a simple Floquet multiplier of (8.176). In this case a decomposition of C is given by

$$C = [\dot{p}_s] \oplus Q_1(s),$$

where $Q_1(s)$ is such that $T(t,s)Q_1(s) = Q_1(t)$. So the above decomposition is a natural *moving coordinate system* in C. Every element in C has a unique representation of the form $\phi = \langle q^s, \phi \rangle_s \dot{p}_s + \phi^{Q_1(s)}$, where q^s is the dual solution corresponding to $\dot{p}_s$. Let M_o be a co-dimension 1 subspace of C, for instance $Q_1(0)$. For an element w of $M_o \subset C$ we obtain $w = \langle q^s, w \rangle_s \dot{p}_s + L_s w$, where L_s is explicitly defined as $L_s w = w - \langle q^s, w \rangle_s \dot{p}_s$.

Lemma 8.7.1. *[94] Suppose* $\Gamma = \{p_s \, | \, 0 \le s \le \omega\}$ *is a nontrivial periodic orbit of (8.175) of minimal period* ω. *Then there exists a neighborhood* V *of* Γ *and an* $\epsilon > 0$ *such that, for every* $\phi \in V$, *there exists a unique pair* $(s,w) \in [0,\omega) \times M_{o,\epsilon}$, $M_{o,\epsilon} = \{w \in M_o : |w|_C < \epsilon\}$, *such that* $\phi = p_s + L_s w$.

This proves the existence of a local system of coordinates around Γ, which is the starting point to deduce a natural decomposition of the enlarged phase space BC defined in Section 8.2. After writing (8.175) as an abstract ODE in BC, it can be decomposed into a system of two equations. In fact, the introduction of the new coordinates (s,w) transforms the abstract ODE in BC into

$$\dot{s} = 1 + g_1(s,w)$$

$$\dot{w} = A(s)w + X_0 g_2(s,w)$$

where $A(s) = \tilde{A} + X_0 L(s)$, $L(s) = Df(p_s)$ and $\tilde{A}$ is the extended infinitesimal generator for $\dot{x} = 0$, that is (cf.(8.38)), $A(s)\phi = \dot{\phi} + X_0[L(s) - \dot{\phi}(0)]$. The second equation of this system can now be decomposed as usual.

Now we give some considerations on a general normal form theory for abstract FDE of the form

$$\dot{u}(t) = A_T u(t) + L u_t + F(u_t), \tag{8.179}$$

where now X is a Banach space and A_T, L and F are as in (8.132). In many situations, the operator L does mix the modes of generalized eigenspaces for

A_T, and the decomposition (8.140) based on the decompositions of the characteristic equation (8.134) into the sequence of equations (8.135) or (8.156) ($k \in \mathbb{N}$) as described in Section 8.6 is not possible. For instance, consider the very simple case of the scalar partial FDE

$$\frac{\partial u(t,x)}{\partial t} = \frac{\partial^2 u(t,x)}{\partial x^2} + b(x)u(t-1,x) + O(x) \quad, t \geq 0, x \in (a_1, a_2),$$

with Neumann or Dirichlet conditions, where b is a continuous function in $[a_1, a_2] \subset \mathbb{R}$ and $O(x)$ stands for nonlinear reaction terms. Suppose that we choose X as a Hilbert space that takes into account the Neumann or Dirichlet conditions, and write the above equation in the form (8.179), where $A_T = \Delta, Lv = bv(-1), v \in \mathcal{C}$. The characteristic equation for the linearized equation is $\Delta(\lambda)y = 0$, $y \in D(\Delta) \setminus \{0\}$, where

$$\Delta(\lambda)y = \lambda y - \Delta y - be^{-\lambda}y.$$

For a generic function b, it is not possible to relate the eigenvalues of the infinitesimal generator A associated with the flow of the linearized equation (i.e., the solutions λ of the above characteristic equation) with the eigenvalues of Δ, because there is no general method to solve linear ODE with variable coefficients. So the equation does not fit into the framework of Section 8.6. It is therefore convenient to develop a normal form theory without imposing the very unrealistic restriction (H3) or (H3'). Also, assumption (H2) is too restrictive, since in applications one might have operators A_T whose eigenvectors do not form a basis of X.

A normal form theory for RFDE in infinite dimensional spaces X should be solely based on the idea of finding an adequate system of coordinates, in order to obtain a decomposition of the phase space $\mathcal{C} = C([-r, 0]; X)$ by a nonempty finite set Λ of characteristic values. On the other hand, it is necessary to have invariant manifolds results, in particular a setting that guarantees the existence of center manifolds. Otherwise normal forms are not useful in terms of applications.

In [46], a normal form theory for RFDE in Banach spaces was addressed, under weaker hypotheses than those that usually appear in the literature dealing with semigroups for abstract FDE. For equations of the form

$$\dot{u}(t) = A_T u(t) + Lu_t, \tag{8.180}$$

in [46] it is only assumed (H1) and that L is given by $L(\varphi) = \int_{-r}^{0} d\eta(\theta)\varphi(\theta)$, $\varphi \in \mathcal{C} = C([-r, 0]; X)$, for a mapping $\eta : [-r, 0] \rightarrow \mathcal{L}(X, X)$ of bounded variation. The approach there on one hand relies on a theorem of existence of center manifolds of (8.179) and, on the other hand, a complete formal adjoint theory for linear RFDE in Banach spaces of the form (8.180), two key points that were established in [48] under the above assumptions only. These two intermediate technical tools set up the necessary framework to construct a

normal form theory that extends to equations (8.179) the main results stated
in Sections 8.2 and 8.6. See [46] for the complete theory and applications.
The bases of this *adjoint theory* for linear FDE in Banach spaces (8.180) go
back as far as the work of Travis and Webb [200], where the *formal duality*
as well as the *formal adjoint equation* had already been defined. However, to
our knowledge, the adjoint theory was not completed until the work reported
in [48], where it was used to derive the decomposition of the phase space
$\mathcal{C}$ by a nonempty finite set Λ of eigenvalues of the infinitesimal generator
corresponding to (8.180), $\mathcal{C} = \mathcal{P} \oplus \mathcal{Q}$, where $\mathcal{P}$ is the generalized eigenspace
associated with Λ and $\mathcal{Q}$ is the *orthogonal* space to the *dual* space P^* relative
to the formal duality.

8.8 A Viscoelastic model

In the discussion of the Levin-Nohel equation in Chapter 5, we observed that
there was no generic Hopf bifurcation. We also remarked that this could be
true if the kernel was permitted to belong to a larger class of functions. This
section, due to Teresa Faria, contains the details of the computations.

Let $a_0(x) = 1 - x$, $g_0(x) = 4\pi^2 x$, $x \in [0,1]$, and consider the class of
perturbations $(a_\alpha, g_{\beta,\gamma})$ of (a_0, g_0) in $C^2([0,1]; \mathbb{R})$,

$$
\begin{aligned}
a_\alpha(x) &= (1-x)(\alpha x^2 + 1) \\
g_{\beta,\gamma} &= 4\pi^2 x + \beta x^2 + \gamma x^3,
\end{aligned}
\tag{8.181}
$$

where $\alpha, \beta \in \mathbb{R}, \gamma > 0$. We note that $(a_\alpha, g_{\beta,\gamma}) \to (a_0, g_0)$ in the C^2-topology
when $(\alpha, \beta, \gamma) \to 0$ and that the functions $(a_\alpha, g_{\beta,\gamma})$ satisfy the following
conditions:

$$
a_\alpha(0) = 1, \ a_\alpha(1) = 0, \ a'_\alpha(x) \le 0, \ x \in [0,1], \quad \text{for } \alpha \in [-1,3]
$$

$$
G_{\beta,\gamma}(x) = \int_0^x g_{\beta,\gamma}(s)ds \to \infty, \quad \text{as } |x| \to \infty. \tag{8.182}
$$

We point out that $a''_\alpha(x)$ changes sign at $x = 1/3$ for all α.

Consider the FDE in $C := C([-1,0]; \mathbb{R})$

$$
\dot{x}(t) = -\int_{-1}^0 a_\alpha(-\theta)g_{\beta,\gamma}(x(t+\theta))d\theta, \tag{8.183}
$$

with $\alpha, \beta, \gamma \in V$, V a neighborhood of zero in $\mathbb{R}$, $\gamma > 0$.

For each $\alpha \in V$, the linearized equation around the equilibrium zero is

$$
\dot{y}(t) = -4\pi^2 \int_{-1}^0 a_\alpha(-\theta)y(t+\theta)d\theta, \tag{8.184}
$$

with characteristic equation

$$\lambda + 4\pi^2 \int_{-1}^{0} (1+\theta)(\alpha\theta^2 + 1)e^{\lambda\theta}d\theta = 0. \qquad (8.185)$$

For $\alpha = 0$, $\lambda = \pm 2\pi i$ are the only (simple) eigenvalues on the imaginary axis and all the other roots of (8.185) have negative real parts. From the computations below we shall see that the Hopf condition holds, i.e., the roots of (8.185) are crossing transversaly the imaginary axis at $\alpha = 0$.

To describe the Hopf bifurcation, we now apply the method of normal forms for FDE. We start by writing FDE (8.183) in abstract form in C as

$$\dot{x}(t) = L(\alpha)x_t + f_{\alpha,\beta,\gamma}(x_t)$$

where $L(\alpha), f_{\alpha,\beta,\gamma} : C \to \mathbb{R}$ are defined by

$$L(\alpha)\varphi = -4\pi^2 \int_{-1}^{0} a_\alpha(-\theta)\varphi(\theta)d\theta,$$

$$f_{\alpha,\beta,\gamma}(\varphi) = -\int_{-1}^{0} a_\alpha(-\theta)(\gamma\varphi^3(\theta) + \beta\varphi^2(\theta))d\theta.$$

Since we want to use $\alpha \in V$ as the bifurcation parameter, we introduce α as a variable and rewrite the above equation as

$$\dot{x}(t) = L_0(x_t) + F_{\beta,\gamma}(x_t, \alpha) \qquad (8.186)$$

with $L_0 = L(0)$ and

$$F_{\beta,\gamma}(\varphi, \alpha) = (L(\alpha) - L_0)\varphi + f_{\alpha,\beta,\gamma}(\varphi)$$

$$= -4\pi^2\alpha \int_{-1}^{0} \theta^2(1+\theta)\varphi(\theta)d\theta + f_{\alpha,\beta,\gamma}(\varphi).$$

Consider $\Lambda = \{-2\pi i, 2\pi i\}$ and let $C = P \oplus Q$ be decomposed by Λ. With the usual notations,

$$P = span\Phi, \quad \Phi(\theta) = (e^{2\pi i\theta} \quad e^{-2\pi i\theta}), \quad \theta \in [-1, 0]$$

$$P^* = span\Psi, \quad \psi(s) = \begin{pmatrix} \psi_1(0)e^{-2\pi i s} \\ \overline{\psi_1(0)}e^{2\pi i s} \end{pmatrix}, \quad s \in [0, 1],$$

with

$$\psi_1(0) = [1 - L_0(\theta e^{2\pi i\theta})]^{-1} = \frac{1}{3}.$$

In the enlarged phase space $BC \equiv C \times \mathbb{R}$ decomposed by Λ as $BC = P \oplus Ker \pi$, in complex coordinates equation (8.186) becomes

$$\begin{cases} \dot{x} &= Bx + \Psi(0)F_{\beta,\gamma}(\Phi x + y, \alpha) \\ \dot{y} &= A_{Q^1}y + (I - \pi)X_0 F_{\beta,\gamma}(\Phi x + y, \alpha), \end{cases}$$

where $x \in \mathbb{C}^2, y \in Q^1 := Q \cap C^1([-1,0]; \mathbb{C})$ and $B = \begin{pmatrix} 2\pi i & 0 \\ 0 & -2\pi i \end{pmatrix}$ (see chapter on normal forms for further notation and results). Write

$$F_{\beta,\gamma}(\varphi, \alpha) = \frac{1}{2} F_{\beta,\gamma,2}(\varphi, \alpha) + \frac{1}{3!} F_{\beta,\gamma,3}(\varphi, \alpha) + O(|\alpha||\varphi|^3),$$

where $F_{\beta,\gamma,2}, F_{\beta,\gamma,3}$ are the Fréchet derivatives of F of second and third orders in (φ, α), respectively, given by

$$\frac{1}{2} F_{\beta,\gamma,2}(\varphi, \alpha) = -4\pi^2 \alpha \int_{-1}^{0} \theta^2 (1+\theta)\varphi(\theta)d\theta - \beta \int_{-1}^{0} (1+\theta)\varphi^2(\theta)d\theta,$$

$$\frac{1}{3!} F_{\beta,\gamma,3}(\varphi, \alpha) = -\gamma \int_{-1}^{0} (1+\theta)\varphi^3(\theta)d\theta - \alpha\beta \int_{-1}^{0} \theta^2 (1+\theta)\varphi^2(\theta)d\theta.$$

In the following, in order to simplify notation, we often drop the dependence on β, γ. For each $\beta, \gamma \in V, \gamma > 0$, define

$$f_j^1(x, y, \alpha) = \Psi(0) F_{\beta,\gamma,j}(\Phi x + y, \alpha)$$
$$f_j^2(x, y, \alpha) = (I - \pi) X_0 F_{\beta,\gamma,j}(\Phi x + y, \alpha), \quad j = 2, 3, \quad (8.187)$$

for $x \in \mathbb{C}^2, y \in Q^1, \alpha \in V$. From the normal form theory, the ODE giving the flow on the 2-dimensional center manifold of the origin for (8.186) is written in normal form (up to third order terms) as

$$\dot{x} = Bx + \frac{1}{2} g_2^1(x, 0, \alpha) + \frac{1}{3!} g_3^1(x, 0, 0) + O(\alpha^2|x| + \alpha|x|^3) \qquad (8.188)$$

where $g_j^1 = g_{\beta,\gamma,j}^1, j = 2, 3$ are given by

$$g_2^1(x, 0, \alpha) = Proj_{Ker(M_2^1)} f_2^1(x, 0, \alpha)$$

$$g_3^1(x, 0, \alpha) = Proj_S \tilde{f}_3^1(x, 0, 0),$$

for

$$Ker(M_2^1) = span\left\{ \begin{pmatrix} x_1\alpha \\ 0 \end{pmatrix}, \begin{pmatrix} 0 \\ x_2\alpha \end{pmatrix} \right\}, \quad S = span\left\{ \begin{pmatrix} x_1^2 x_2 \\ 0 \end{pmatrix}, \begin{pmatrix} 0 \\ x_1 x_2^2 \end{pmatrix} \right\}$$

and

$$f_3^1(\tilde{x}, 0, 0) = f_3^1(x, 0, 0) + \frac{3}{2} \left[(D_x f_2^1)(x, 0, 0)U_2^1(x, 0) \right.$$
$$\left. + (D_y f_2^1)(x, 0, 0)U_2^2(x, 0) - (D_x U_2^1)(x, 0)g_2^1(x, 0, 0) \right]. \qquad (8.189)$$

From (8.187), we get

$$\frac{1}{2}f_2^1(x,y,\alpha) = -\Psi(0)\left[4\pi^2\alpha\int_{-1}^0(1+\theta)\theta^2(e^{2\pi i\theta}x_1 + e^{-2\pi i\theta} + y)d\theta\right.$$

$$\left. + \beta\int_{-1}^0(1+\theta)(e^{2\pi i\theta}x_1 + e^{-2\pi i\theta}x_2 + y)^2d\theta\right]. \quad (8.190)$$

Hence

$$\frac{1}{2}g_2^1(x,0,\alpha) = \begin{pmatrix} A_1x_1\alpha \\ \overline{A}_1x_2\alpha \end{pmatrix},$$

with $A_1 = -(4\pi^2/3)\int_{-1}^0(1+\theta)\theta^2e^{2\pi i\theta}d\theta$, thus

$$A_1 = \frac{1}{3} - \frac{i}{\pi}. \quad (8.191)$$

We remark that the Hopf condition is clearly satisfied, since it is equivalent to $Re\, A_1 \neq 0$.

Now we compute the third order term $g_3^1(x,0,0)$. Note that $g_2^1(x,0,0) = 0$. On the other hand, since

$$f_2^1(x,0,0) = -2\beta\Psi(0)\left(-\frac{i}{4\pi}x_1^2 + x_1x_2 + \frac{i}{4\pi}x_2^2\right),$$

from the definition of U_2^1 and M_2^1 in the chapter of normal forms, we have

$$D_xf_2^1(x,0,0) = -\frac{2\beta}{3}\begin{pmatrix} -\frac{i}{2\pi}x_1 + x_2x_1 + \frac{i}{2\pi}x_2 \\ -\frac{i}{2\pi}x_1 + x_2x_1 + \frac{i}{2\pi}x_2 \end{pmatrix}$$

$$U_2^1(x,0) = (M_2^1)^{-1}f_2^1(x,0,0) = -\frac{\beta}{3\pi i}\begin{pmatrix} -\frac{i}{4\pi}x_1^2 - x_1x_2 - \frac{i}{12\pi}x_2^2 \\ -\frac{i}{12\pi}x_1^2 + x_1x_2 - \frac{i}{4\pi}x_2^2 \end{pmatrix}$$

and thus

$$Projs[(D_xf_2^1)(x,0,0)U_2^1(x,0)] = \begin{pmatrix} C_1x_1^2x_2 \\ \overline{C}_1x_1x_2^2 \end{pmatrix}$$

with

$$C_1 = \frac{2\beta^2}{9\pi}\left(\frac{1}{\pi} - i - \frac{i}{24\pi^2}\right). \quad (8.192)$$

Let

$$U_2^2(x,0) := h(x) = h_{20}x_1^2 + h_{11}x_1x_2 + h_{02}x_2^2,$$

with $h_{20}, h_{11}, h_{02} \in Q^1$. From (8.190), we obtain

$$D_yf_2^1(x,0,0)h(x) = -4\beta\Psi(0)\int_{-1}^0(1+\theta)(e^{2\pi i\theta}x_1 + e^{-2\pi i\theta}x_2)h(\theta)d\theta,$$

hence

$$Projs[D_yf_2^1(x,0,0)U_2^2(x,0)] =$$

$$= -\frac{4\beta}{3}\begin{pmatrix} \left(\int_{-1}^0(1+\theta)(e^{2\pi i\theta}h_{11}(\theta) + e^{-2\pi i\theta}h_{20}(\theta))d\theta\right)x_1^2x_2 \\ \left(\int_{-1}^0(1+\theta)(e^{-2\pi i\theta}h_{11}(\theta) + e^{2\pi i\theta}h_{02}(\theta))d\theta\right)x_1x_2^2 \end{pmatrix}. \quad (8.193)$$

Recall that $h(x)$ is defined as the solution of

$$\dot{h}(x) - D_x h(x)Bx = \Phi\Psi(0)F_{\beta,\gamma,2}(\Phi x, 0)$$

$$\dot{h}(x)(0) - L_0(h(x)) = F_{\beta,\gamma,2}(\Phi x, 0).$$

It follows that $h_{02} = \overline{h_{20}}$, h_{11} is real, and h_{20}, h_{11} are the solutions of the initial value problems

$$\begin{cases} \dot{h}_{20} - 4\pi i h_{20} = \frac{\beta i}{6\pi}(e^{2\pi i\theta} + e^{-2\pi i\theta}) \\ h_{20}(0) - L_0(h_{20}) = \frac{\beta i}{2\pi} \end{cases}$$

and

$$\begin{cases} \dot{h}_{11} = -\frac{\beta}{3}(e^{2\pi i\theta} + e^{-2\pi i\theta}) \\ h_{11}(0) - L_0(h_{11}) = -\beta \end{cases}$$

respectively. The resolution of these systems gives

$$h_{20}(\theta) = \frac{\beta}{18\pi^2}\left(3e^{4\pi i\theta} - \frac{1}{2}(3e^{2\pi i\theta} + e^{-2\pi i\theta})\right),$$

$$h_{11}(\theta) = \frac{\beta}{6\pi}\left(\frac{1}{\pi} + i(e^{2\pi i\theta} - e^{-2\pi i\theta})\right).$$

Thus, from (8.193) we derive

$$Projs[(D_y f_2^1)(x,0,0)U_2^1(x,0)] = \begin{pmatrix} C_2 x_1^2 x_2 \\ \overline{C}_2 x_1 x_2^2 \end{pmatrix}$$

with

$$C_2 = \frac{\beta^2 i}{9\pi}\left(\frac{25}{12\pi^2} + 1\right). \tag{8.194}$$

Also, (8.187) gives

$$\frac{1}{3!}f_3^1(x,0,0) = -\gamma\Psi(0)\int_{-1}^{0}(1+\theta)(e^{2\pi i\theta}x_1 + e^{-2\pi i\theta}x_2)^3 d\theta$$

and we deduce that

$$\frac{1}{3!}Projs\, f_3^1(x,0,0) = \begin{pmatrix} C_3 x_1^2 x_2 \\ \overline{C}_3 x_1 x_2^2, \end{pmatrix}$$

with

$$C_3 = \frac{\gamma i}{2\pi}. \tag{8.195}$$

From the definition of $g_3^1(x,0,0)$ in (8.188) and from formulas (8.192), (8.194) and (8.195), we finally get

$$\frac{1}{3!}g_3^1(x,0,0) = \begin{pmatrix} A_2 x_1^2 x_2 \\ A_2 x_1 x_2^2 \end{pmatrix},$$

with $A_2 = (C_1 + C_2)/4 + C_3$.

The change to real coordinates w, where $x_1 = w_1 - iw_2, x_2 = w_1 + iw_2$, followed by the use of polar coordinates (ρ, ξ), $w_1 = \rho \cos \xi, w_2 = \rho \sin \xi$, transforms the normal form (8.188) into

$$\begin{cases} \dot{\rho} = K_1 \alpha \rho + K_2 \rho^3 + O(\alpha^2 \rho + \alpha |\rho|^3) \\ \dot{\xi} \qquad = -2\pi + O(|(\rho, \alpha)|), \end{cases} \qquad (8.196)$$

with $K_1 = \operatorname{Re} A_1$, $K_2 = \operatorname{Re} A_2 = \operatorname{Re}(C_1/4)$, for A_1, C_1 given by (8.187) and (8.192). Thus

$$K_1 = \frac{1}{3} > 0$$

$$K_2 = \frac{\beta^2}{18\pi^2} > 0, \quad \text{for} \quad \beta \neq 0.$$

These computations lead us to the following conclusions: For the perturbed family $(a_\alpha, g_{\beta,\gamma})$, a Hopf bifurcation occurs at $\alpha = 0$ on a 2-dimensional local center manifold of the origin. On this manifold, the flow is given in polar coordinates by equation (8.196), with $K_1, K_2 > 0$, if $\beta \neq 0$, implying that the Hopf bifurcation is subcritical and that the non-trivial periodic orbits bifurcating from the stable equilibrium $x = 0$ are unstable. We also remark that if $\beta = 0$, then $K_2 = 0$, which means that the Hopf bifurcation is not generic and higher order terms for the normal form giving the flow on the center manifold must be computed, in order to have a complete picture of the bifurcation.

9 Attractor Sets as C^1-Manifolds

It is of some interest to determine when the set $A(F)$ of an RFDE F is a C^1-manifold, since it will then have a particularly simple geometric structure which will facilitate the study of qualitative properties of the flow. Results in this direction can be established through the use of C^k-*retractions* which are defined as C^k maps γ from a Banach manifold into itself such that $\gamma^2 = \gamma$, $k \geq 1$.

Lemma 9.0.1. *If B is a Banach manifold (without boundary) and $\gamma : B \to B$ is a C^1-retraction, then $\gamma(B)$ is a Banach C^1-submanifold of B (without boundary).*

Proof: Since $\gamma \cdot \gamma = \gamma$, the derivative T of γ at a point $p \in \gamma(B)$, $T = \gamma'_p$, satisfies $\gamma'_{\gamma(p)} \cdot \gamma'_p = \gamma'_p$, that is, $T^2 = T$. This implies $T = \gamma'_p$ is double splitting; in fact, taking $E = T_p B$, $T : E \to E$, then the image and kernel of T are, respectively, the kernel and image of $(I - T)$. The local representative theorem shows that with suitable local charts called there α and β, the map γ can be represented by

$$\bar{\gamma} : (u, v) \to \big(u, \eta(u, v)\big),$$

for $(u, v) \in B_1 \times B_2$ where B_1 and B_2 are the open unit balls in $E_1 =$ image of T and $E_2 =$ kernel of T and, the map $\eta(u, v)$ satisfies $D\eta(0,0) = 0$. Consider the points $(u, v) \in B_1 \times B_2$ such that $\eta(u, v) = v$. By the implicit function theorem, there exist open balls $\bar{B}_1 \subseteq B_1$ and $\bar{B}_2 \subseteq B_2$ such that the set of points in $\bar{B}_1 \times \bar{B}_2$ satisfying $\eta(u, v) = v$ is the graph of a function $v = \bar{v}(u)$, $u \in \bar{B}_1$. The Banach manifold $\bar{M}$ defined by that graph in B is locally contained in the image of $\bar{\gamma}$ since $\gamma\big(u, \bar{v}(u)\big) = \big(u, \eta\big(u, \bar{v}(u)\big)\big) = \big(u, \bar{v}(u)\big)$. On the other hand, the map γ can be also represented by

$$\bar{\bar{\gamma}} = \alpha \cdot \gamma \cdot \alpha^{-1} : (u, v) \to \big(f(u, v), g(u, v)\big)$$

and the partial derivative $D_v g(0,0)$ is zero since $v \in E_2 =$ kernel of T. The equation $g(u, v) = v$ has a local solution $v = \bar{v}(u)$ in an open neighborhood denoted again by $\bar{B}_1 \times \bar{B}_2$ and the corresponding graph contains the fixed points of $\bar{\bar{\gamma}}$ which are given by $g(u, v) = v$ and $f(u, v) = u$. The set of fixed points of $\bar{\bar{\gamma}}$ restricted to $\bar{B}_1 \times \bar{B}_2$ is the image of $\bar{\bar{\gamma}}$, since $\bar{\bar{\gamma}}$ is a retraction.

Therefore, the graph of $v = \bar{v}(u)$ defines in B another Banach manifold $\bar{\bar{M}}$ locally containing the image of γ. Since, locally, $\bar{M} \subseteq$ Image of $\gamma \subseteq \bar{\bar{M}}$ and both are E_1-Banach manifolds containing the point $p \in \gamma(B)$, we get, locally, $\bar{M} =$ Image of $\gamma = \bar{\bar{M}}$. ∎

In Example 3.2.2 a C^1 vector field X defined on a manifold M was used to define an RFDE on M by $F = X \circ \rho$. The map $\Sigma_X : M \to C^0(I, M)$ such that $\Sigma_X(p)$ is the restriction of the solution of X, through p at $t = 0$, to the interval $I = [-r, 0]$, is a cross-section with respect to ρ and $A(F) = \Sigma_X(M)$. The map $\gamma = \Sigma_X \cdot \rho$ is a C^1-retraction and commutes with the flow of F, in agreement with $A(F)$ being a C^1-manifold diffeomorphic to M and invariant under F.

Theorem 9.0.2. *(Oliva [151]) Let $F \in \mathcal{X}^1$ be an RFDE on a compact and connected manifold M and assume there exists a C^1-retraction $\gamma : C^0 \to C^0$ such that $A(F) = \gamma(C^0)$. Then, the attractor set $A(F)$ is a connected compact C^1-manifold. Besides, if γ is homotopic to the identity, $A(F)$ is diffeomorphic to M.*

Proof: We know that $A(F)$ is a connected and compact set; by Lemma 9.0.1 $A(F)$ is a C^1-manifold without boundary. Arguments like the ones used in Lemma 4.0.5 and Theorem 4.0.6 show that $A(f)$ is diffeomorphic to M. ∎

Theorem 9.0.3. *(Oliva [151]) Let $F \in \mathcal{X}^1$ be an RFDE on a compact and connected manifold M without boundary, and assume there is a constant $k > 0$ such that $\|d\Phi_t(\varphi)\| \leq k$ and $d\Phi_t$ has Lipschitz constant k for all $t \geq 0$ and $\varphi \in C^0$. Then, there exists a unique C^1-retraction γ of C^0 onto $A(F)$ which commutes with the flow, i.e., $\gamma \Phi_t = \Phi_t \gamma$, $t \geq 0$. The attractor set $A(F)$ is a connected compact C^1-manifold without boundary and the restriction of Φ_t, $t \geq 0$, to $A(F)$ is a one-parameter group of diffeomorphisms.*

Proof: Let t_n be a sequence of real numbers such that $t_n \to \infty$ and $s_n = (t_n - t_{n-1}) \to \infty$. Since M is bounded and $F \in \mathcal{X}^1$, the set $K = \Phi_r(C^0)$ is precompact. For n large enough, the restrictions of $\Phi_{t_n - r}$ to K belong to a set of equicontinuous functions and, for each $\varphi \in K$, the set of all $\Phi_{t_n - r}(\varphi)$ is relatively compact. Then by Ascoli's theorem, for a subsequence, denoted again by t_n, the $\Phi_{t_n - r}$ converge to a continuous map $\bar{\beta} : K \to C$, uniformly on K. It follows easily that Φ_{t_n} converges to the map $\beta = \bar{\beta} \cdot \Phi_r$, uniformly on the Banach manifold $C^0(I, M)$. Using the same argument, there is a subsequence of Φ_{s_n} which converges to a map γ, uniformly on $C^0(I, M)$. Then γ is continuous and Lipschitz with constant k since $\|d\Phi_t\|$ is bounded by k for all $t \geq 0$. For $\varphi \in C^0$ we have $\beta(\varphi) \in \omega(\varphi)$ and $\omega(\varphi) \subseteq A(F)$. On the other hand, given $\bar{\varphi} \in A(F)$ and $t_n \in \mathbb{R}$, there exist $\Psi_n \in A(F)$ such that $\bar{\varphi} = \Phi_{t_n}(\Psi_n)$. When $n \to \infty$, there is a subsequence (denoted with the same indices) such that $\Psi_n \to \Psi$ and $\Psi \in A(F)$ because $A(F)$ is invariant and closed. Therefore, locally and for n large enough,

$$\|\Phi_{t_n}(\Psi_n) - \beta(\Psi)\| \leq \|\Phi_{t_n}(\Psi_n) - \Phi_{t_n}(\Psi)\| + \|\Phi_{t_n}(\Psi) - \beta(\Psi)\|$$
$$\leq k\|\Psi_n - \Psi\| + \|\Phi_{t_n}(\Psi) - \beta(\Psi)\| < \varepsilon,$$

which implies $\beta(\Psi) = \bar{\varphi}$ and, consequently, $\beta : C^0 \to A(F)$ is onto.

Now, the relations $\Phi_{s_n} \cdot \Phi_{t_{n-1}} = \Phi_{t_n} = \Phi_{t_{n-1}} \cdot \Phi_{s_n}$ show that $\gamma \cdot \beta = \beta = \beta \cdot \gamma$. Also $\gamma(C^0) = A(F)$ and, then, the map $\gamma : C^0 \to A(F)$ is a retraction since, for any $\bar{\varphi} \in A(F)$, there exists Ψ such that $\beta(\Psi) = \bar{\varphi}$ which implies $\gamma(\bar{\varphi}) = \gamma(\beta(\Psi)) = \beta(\Psi) = \bar{\varphi}$. Finally γ commutes with Φ_t, $t \geq 0$ since

$$\Phi_t \cdot [\gamma(\varphi)] = \Phi_t (\lim \Phi_{s_n}(\varphi)) = \lim [\Phi_t \cdot \Phi_{s_n}(\varphi)]$$
$$= \lim [\Phi_{s_n} \Phi_t(\varphi)] = \gamma [\Phi_{t_n}(\varphi)].$$

If $\bar{\gamma}$ is another retraction onto $A(F)$ and $\bar{\gamma} \cdot \Phi_t = \Phi_t \cdot \bar{\gamma}$, $t \geq 0$, then $\bar{\gamma} \cdot \gamma = \gamma \cdot \bar{\gamma}$. Now $\gamma(\varphi) = \bar{\gamma}(\gamma(\varphi)) = \bar{\gamma}(\varphi)$, i.e. $\gamma = \bar{\gamma}$. This proves uniqueness.

We need to show that γ is C^1. Denote $M_1 = \{v \in TM : \|v\| \leq 1\}$ and $C^0(I, M_1) = \bar{M}_1$, the set of all $\psi \in C^0(I, TM)$ such that $\|\psi\| \leq 1$. Let $\Psi_t = T\Phi_t$ be the flow on $TC^0(I, M)$ of the first variational equation. The set $\Psi_r \bar{M}_1$ is relatively compact and its closure is a compact set K_1; this follows from the boundedness of $\|d\Phi_t(\varphi)\|$ and $F \in \mathcal{X}^1$. Consider now the sequence of functions $\Psi_{s_n - r} : K_1 \to TC^0$ which are equicontinuous since $d\Phi_t$ has Lipschitz constant k. Then for each $(\varphi, \psi) \in K_1$, the set of all $\Psi_{s_n - r}(\varphi, \psi)$ is relatively compact. Thus, there is a subsequence of $(\Psi_{s_n - r})$ which converges uniformly on K_1 and, therefore, (Ψ_{s_n}) converges uniformly on $\bar{M}_1$ to a map $\bar{\gamma}$ which must be the derivative of γ and, consequently, γ is C^1.

Now, Theorem 9.0.2 implies all the other statements in the theorem, except that Φ_t is a group of diffeomorphisms on $A(F)$, which is, therefore, the only thing that is left to prove.

The solution map $\Phi_t : A(F) \to A(F)$ is C^1 differentiable. Let φ and ψ be two elements of $A(F)$. If, for $t = \bar{t}$, $\Phi_{\bar{t}}(\varphi) = \Phi_{\bar{t}}(\psi)$ with $\varphi \neq \psi$ we get $\Phi_t(\varphi) = \Phi_t(\psi)$ for all $t \geq \bar{t}$. Using the sequence Φ_{s_n} which has defined the retraction γ, we get $\Phi_{s_n}(\varphi) = \Phi_{s_n}(\psi)$, and, therefore, $\gamma(\varphi) = \gamma(\psi)$ and $\varphi = \psi$. Since for each t, Φ_t is one-to-one on $A(F)$ and $A(F)$ is compact, Φ_t is a homeomorphism. Also, in the manifold $C^0(I, M_1)$, one has an attractor, defined in a similar way by a retraction, the derivative of γ, obtained by the uniform convergence of the flow Ψ_{s_n} on the manifold $\bar{M}_1 = C^0(I, M_1)$. This shows that the map ψ_t is also one-to-one, and, consequently, Φ_t is a diffeomorphism. Denoting by Φ_{-t} the inverse of Φ_t, $t \geq 0$, one obtains a one-parameter group of C^1-diffeomorphisms acting on the compact manifold $A(F)$. ∎

Remark 9.0.4. The hypothesis $\|d\Phi_t(\varphi)\| \leq k$ for all $t \geq 0$ and all $\varphi \in C^0(I, M)$, in Theorem 9.0.3 is assured by the following geometric condition: the first variational equation restricted to the manifold $\{\psi \in C^0(I, TM) : |\psi| \leq k\}$ is such that its values are vectors tangent to the manifold $\{v \in TM : |v| \leq k\}$, and at points ψ such that $|\psi(0)| = k$, the value of the first variational equation is an "inward" vector.

Example 9.0.5. (Oliva [150]) Consider the RFDE on the circle S^1 given by the scalar equation (see Section 3.2.10)

$$\dot{x} = b\big[\sin\big(x(t) - x(t-1)\big)\big], \qquad (9.1)$$

where $b : R \to R$ is a C^1 function with Lipschitz first derivative satisfying also $b(0) = 0$ and $\left|\frac{db}{dx}\right| \le \sigma < 1$. The global solutions of this equation are the constant functions. To see this we consider the map

$$T : z(t) \to \int_{t-1}^{t} b\big[\sin\big(z(u)\big)\big]\,du$$

acting in the Banach space of all bounded continuous functions with the sup norm. It is easy to see that T is a contraction and $z(t) \equiv 0$ is its fixed point. On the other hand, if $x(t)$ is a global solution, $[x(t) - x(t-1)]$ is bounded and

$$x(t) - x(t-1) = \int_{t-1}^{t} b\big[\sin\big(x(u) - x(u-1)\big)\big]\,du$$

which shows that $x(t) - x(t-1) = 0$ and, using the equation, $\dot{x}(t) \equiv 0$ and $x(t) = $ constant. $A(F)$ is in this case a circle in $C^0(I, S^1)$.

Let $x(t)$ be the solution defined by the initial condition φ at $t = 0$. For $n \ge 2$ one has

$$\max_{u\in[n-1,n]} |x(u) - x(u-1)| \le \sigma \cdot \max_{u\in[n-2,n-1]} |x(u) - x(u-1)| \le$$

$$\le \sigma^2 \cdot \max_{u\in[n-3,n-2]} |x(u) - x(u-1)| \le \cdots \le \sigma^{(n-2)} \cdot \max_{u\in[1,2]} |x(u) - x(u-1)|.$$

Then $\lim_{t\to+\infty} |x(t) - x(t-1)| = 0$ and $\lim_{t\to+\infty} \dot{x}(t) = 0$. Since

$$x(t) = \varphi(0) + \int_{0}^{t} b\big[\sin\big(x(u) - x(u-1)\big)\big]\,du,$$

one has

$$|x(t) - \varphi(0)| \le \int_{0}^{t} \sigma|x(u) - x(u-1)|\,du \le K\frac{1}{1-\sigma}$$

for a suitable K. Thus, $x(t)$ is bounded as $t \to +\infty$. Moreover, given $\varepsilon > 0$, there exists $T(\varepsilon)$ such that

$$|x(t) - x(t')| = |\dot{x}(\xi)| < \varepsilon \quad \text{for } t, t' > T(\varepsilon),$$

and the limit of $x(t)$ as $t \to +\infty$ exists.

The flow Φ_t has a limit:

$$\gamma(\varphi) = \lim_{t\to+\infty} \Phi_t(\varphi) = c \quad \text{(constant solution)}.$$

γ is a C^1-retraction, $\gamma^2 = \gamma$, and $\gamma \cdot \Phi_t = \Phi_t \cdot \gamma$. To prove that γ is C^1 with uniform Lipschitz constant, we need to consider the derivative $d\Phi_t$ which is the flow of the first variational equation:

$$\begin{cases} \dot{x} = b\big[\sin\big(x(t) - x(t-1)\big)\big] \\ \dot{y} = (b \circ \sin)'\big(x(t) - x(t-1)\big) \cdot \big[y(t) - y(t-1)\big]. \end{cases}$$

The critical points in this case are the elements of TS^1. It can be proved that

$$d\gamma(\varphi)\Psi = \lim_{t \to +\infty} d\Phi_t(\varphi)\Psi, \quad \Psi \in TC^0(I, S^1).$$

The retraction γ has $A(F)$ as image and the example satisfies the hypothesis of Theorem 9.0.3.

The hypothesis of Theorem 9.0.3 is very restrictive. However, using infinite dimensional analogues on the continuity properties of a certain class of attractors, it is possible to show that the attractor set of small perturbations of equations satisfying the above hypotheses are also C^1-manifolds. For this, we need some more notation.

Let $F \in \mathcal{X}^1$ be an RFDE on a compact manifold M, such that its attractor set satisfies $A(F) = \gamma(C^0)$ for some C^1-retraction $\gamma : C^0 \to C^0$. This implies that $A(F)$ is a compact C^1-manifold and there is a tubular neighborhood U of $A(F)$ in $C^0 = C^0(I, M) \subset C^0(I, \mathbb{R}^N)$ (see Lemma 9.0.1 and Theorem 3.1.1). For each $R, L > 0$, let $\mathcal{F}^{0,1}(R, L)$ be defined by

$$\mathcal{F}^{0,1}(R, L) = \Big\{ s \in C^0\big(A(F), U\big) : \gamma s = \mathrm{id}\big(A(F)\big),\ s' \overset{\mathrm{def}}{=} s - \mathrm{id}\big(A(F)\big)$$
$$\text{satisfies}\ \ \big|s'(u)\big| \le R \ \ \text{and} \ \ \big|s'(u) - s'(v)\big| \le L|u - v|$$
$$\text{for all}\ u, v \in A(F) \Big\}.$$

It is not difficult to show that $\mathcal{F}^{0,1}(R, L)$ with distance

$$d(s, s_1) = \sup\big\{ |s'(u) - s_1'(u)| : u \in A(F) \big\}$$

is a complete metric space. For $w_0 \in C^1(U, C^0)$ such that:

(i) $w_0\gamma = \gamma w_0$ and $w_0|A(F)$ is a diffeomorphism onto $A(F)$,
(ii) $\big\| d\big(w_0|\gamma^{-1}(p)\big)(p)\big\| \le \xi < 1$ for all $p \in A(F)$,
(iii) $\big\| d\big(w_0|\gamma^{-1}(p)\big)(p)\big\| \cdot \big\| d\big(w_0|A(F)\big)(p)^{-1}\big\| \le \xi < 1$ for all $p \in A(F)$,

define $\mathcal{L}_{w_0}^1(\bar{R}, \bar{L})$, for $\bar{R}, \bar{L} > 0$, to be the set

$$\mathcal{L}_{w_0}^1(\bar{R}, \bar{L}) = \Big\{ w \in C^1(U, C^0) : \big|w(u) - w_0(u)\big| \le \bar{R}$$
$$\text{and}\ \big\| Dw(u) - Dw_0(u)\big\| \le \bar{L}$$
$$\text{for all}\ u \in U \Big\}.$$

Lemma 9.0.6. *Let $F \in \mathcal{X}^1$ be an RFDE on a compact manifold M, such that the attractor $A(F) = \gamma(C^0)$ for some C^1-retraction γ and let $w_0 \in C^1(U, C^0)$ satisfy the above conditions (i), (ii) and (iii). If $w \in \mathcal{L}^1_{w_0}(\bar{R}, \bar{L})$ for $\bar{R}, \bar{L}$ sufficiently small, then there exists a C^1-manifold B_w diffeomorphic to $A(F)$ which is invariant under w, $B_w \to A(F)$ as $w \to w_0$ in the Hausdorff metric, the restriction of w to B_w is a diffeomorphism, and B_w is uniformly asymptotically stable for the discrete flow defined by w^n, $n = 1, 2, 3, \ldots$.*

Proof: The proof (see Oliva [151]) follows closely the computation in Kurzweil [117] and Lewowicz [124]. Let $w \in \mathcal{L}^1_{w_0}(\bar{R}, \bar{L})$. For each $s \in \mathcal{F}^{0,1}(R, L)$, define $H_s : A(F) \to A(F)$ by $H_s = \gamma w s$. For $\bar{R}, \bar{L}$ sufficiently small, H_s is close to $H_s^0 \overset{\text{def}}{=} \gamma w_0 s$, and, since w_0 commutes with γ and $\gamma_s = \text{id}(A(F))$, we have $H_s^0 = w_0 \cdot \text{id}(A(F))$, and, consequently, H_s is a $C^{0,1}$-homeomorphism. On the other hand, for $\bar{R}, \bar{L}$ sufficiently small, it can be shown after some computations that the map $\mathcal{K} : \mathcal{F}^{0,1} \to \mathcal{F}^{0,1}$ given by $\mathcal{K}(s) = w s H_s^{-1}$ is a contraction, and, therefore has a unique fixed point $\bar{s} = w \bar{s} H_{\bar{s}}^{-1}$. If we define $B_w = \bar{s}(A(F))$, then B_w is a C^1-manifold diffeomorphic to $A(F)$, and invariant under w because $w\bar{s} = \bar{s}H_{\bar{s}}$ implies $wB_w \subset B_w$. Letting $w \to w_0$, we have $\bar{s} \to \text{id}(A(F))$, implying that $B_w \to A(F)$. For $\bar{R}, \bar{L}$ sufficiently small, w is close to w_0 in the C^1-uniform norm, and, since w_0 is a diffeomorphism on $A(F)$, it follows that w is a diffeomorphism on B_w. It remains to prove the stability of B_w under $\{w^n\}$.

It is easy to see that, for $\bar{R}, \bar{L}$ sufficiently small and for each $y \in U$ sufficiently close to B_w, there exists $s \in \mathcal{F}^{0,1}$ such that $y = s\gamma(y)$. By the definition of the map $\mathcal{K}$, we have $\mathcal{K}^n(s)\gamma w^n s = w^n s$, which implies $w^n y = \mathcal{K}^n(s)\gamma w^n y$. Due to the properties of $\mathcal{F}^{0,1}$ and since $\mathcal{K}$ is a contraction, we have $\mathcal{K}^n(s) \to \bar{s}$ as $n \to \infty$, uniformly in $s \in \mathcal{F}^{0,1}$. As $B_w = \bar{s}(A(F))$, it follows that, for every $\varepsilon > 0$, there exists $N > 0$ such that $n \geq N$ implies $\text{dist}(\mathcal{K}^n(s)\psi, \bar{s}\psi) < \varepsilon$ for all $\psi \in A(F)$, and consequently, also $\text{dist}(w^n y, B^w) < \varepsilon$. Since the first inequality is uniform in s, it follows that the second is uniform in y. This proves that B_w is uniformly asymptotically stable under the flow $\{w^n\}$. ∎

Theorem 9.0.7. *(Oliva [151]) Let $F \in \mathcal{X}^1$ be an RFDE on a compact manifold M. Suppose there is a constant k such that $\|d\Phi_t(\varphi)\| \leq k$ and $d\Phi_t$ has Lipschitz constant k, for all $t \geq 0$ and $\varphi \in C^0$. Then, there is a neighborhood V of F in $\mathcal{X}^1$ such that $A(G)$ is diffeomorphic to $A(F)$ for $G \in V$ and $A(G) \to A(F)$ as $G \to F$. In particular, $A(G)$ is a C^1-manifold, and, if M is connected and without boundary, then $A(G)$ is a connected compact C^1-manifold without boundary and the restriction of Φ_t^G, $t \geq 0$, to $A(G)$ is a one-parameter group of diffeomorphisms.*

Proof: Let γ be the retraction onto $A(F)$ constructed in the proof of Theorem 9.0.3 and U a tubular neighborhood of $A(F)$ in C^0. By continuity of the semiflow map $\Phi(t, \varphi, F)$ and the continuity of the semiflow map for the first variational equation, given $\bar{R}, \bar{L} > 0$, there is a neighborhood V of F in $\mathcal{X}^1$ and some $T > 0$ such that $G \in V$ implies $\Phi_T^G \in \mathcal{L}_\gamma^1(\bar{R}, \bar{L})$. Taking $\bar{R}, \bar{L}$ sufficiently small, Lemma 9.0.6 can be applied with $w_0 = \gamma$ and $w = \Phi_T^G$ to give $B(G) = B_w$ diffeomorphic to $A(F)$, invariant under Φ_T^G, uniformly asymptotically stable for the flow $(\Phi_T^G)^n$, $n = 1, 2, \ldots$, with Φ_T^G being a diffeomorphism on $B(G)$ and $B(G) \to A(F)$ as $G \to F$.

For $t \geq 0$ small and G sufficiently close to F, Φ_t^G is a diffeomorphism on $B(G)$, close to the identity, and $\Phi_T^G \Phi_t^G B(G) = \Phi_t^G \Phi_T^G B(G) = \Phi_t^G B(G)$. Therefore $(\Phi_T^G)^n \Phi_t^G B(G) = \Phi_t^G B(G)$ and, then, the uniform asymptotic stability of $B(G)$ under $(\Phi_T^G)^n$ implies $\Phi_t^G B(G) \subset B(G)$. Thus, $B(G)$ is positively invariant under the flow Φ_t^G, $t \geq 0$. To prove that $B(G)$ is invariant for the RFDE(G), we need to extend the flow of G on $B(G)$ to $t < 0$. Let $\varphi \in B(G)$ and consider the curve $s \to (\Phi_s^G)^{-1} \varphi$ defined for the values $s \geq 0$ for which this curve lies in $B(G)$. Fix $s_0 \leq 0$ in the domain of this curve, choose $t_0 > -s_0 + 2r$ and consider the solution curve $t \to \Phi_t^G [(\Phi_{t_0}^G)^{-1} \varphi]$, $t \geq 0$. For $s \in [-2r + s_0, 0]$ and $t = t_0 + s$ we have

$$\Phi_t^G [(\Phi_{t_0}^G)^{-1} \varphi] = (\Phi_{-s}^G)^{-1} \varphi.$$

If

$$y(s + \theta) = \Phi_{t_0 + s}^G (\Phi_{t_0}^G)^{-1} \varphi(\theta), \quad s \in [-r + s_0, 0], \quad \theta \in [-r, 0],$$

then

$$\dot{y}(s) = G(y_s).$$

This shows that $B(G)$ is invariant under the RFDE(G). Thus $B(G) \subset A(G)$.

On the other hand, $(\Phi_T^G)^n B(G) \subset B(G) \subset A(G)$ together with the uniform asymptotic stability of $B(G)$, the upper semicontinuity of $A(G)$ in G (Theorem 3.1.11) and the fact that $B(G) \to A(F)$ as $G \to F$, imply that $A(G) \subset B(G)$. Thus $A(G) = B(G)$ and the rest of the statement follows from Lemma 9.0.6. ∎

As mentioned before, if $F \in \mathcal{X}^1$ is an RFDE defined by an ordinary differential equation on a manifold M, as in Example 3.2.2, the attractor set $A(F)$ is also given by a C^1-retraction. Therefore, the preceding ideas can be applied to establish another class of RFDE whose attractors are C^1-manifolds, namely the RFDE close to ordinary differential equations.

Theorem 9.0.8. *(Kurzweil [116], [117], [115]) Let X be a C^1-vector field defined on a compact manifold M. There is a neighborhood V of $F = X \circ \rho$ in $\mathcal{X}^1$ such that $A(G)$ is a C^1-manifold diffeomorphic to M for $G \in V$, $A(G) \to A(F)$ as $G \to F$ and the restriction of Φ_t^G, $t \geq 0$ to $A(G)$ is a one-parameter family of diffeomorphisms.*

Proof: Let $\Sigma_X : M \to C^0$ be the map such that $\Sigma_X(\varphi)$ is the restriction of the solution of X through φ at $t = 0$, to the interval $I = [-r, 0]$. The map $\gamma = \Sigma_X \rho$ is a C^1-retraction which commutes with the flow of F, and $A(F) = \gamma(C^0)$. Given arbitrary $\bar{R}, \bar{L} > 0$, there is a neighborhood V of F in $\mathcal{X}^1$ such that $\Phi_r^G \in \mathcal{L}_{\Phi_r^F}^1(\bar{R}, \bar{L})$. Lemma 9.0.6 can now be applied with $w_0 = \Phi_r^F$ and $w = \Phi_r^G$. The rest of the proof is identical to the second part of the proof of the preceding theorem.   ∎

Remark 9.0.9. One can obtain higher order of smoothness for the manifolds obtained in the preceding results. In fact, the manifold B_w of Lemma 9.0.6 will be C^k if w^0, w and γ are of class C^k, $k \geq 1$, and the condition (iii) is replaced by

(iii)' $\left\| d\left(w_0 | \gamma^{-1}(p)\right)(p)\right\| \left\| d\left(w_0 | A(F)\right)(p)^{-1}\right\|^k \leq \xi < 1$ for all $p \in A(F)$.

This last condition holds trivially in Theorem 9.0.8, since

$$\left\| d\left(w_0 | \gamma^{-1}(p)\right)(p)\right\| = 0$$

for all $p \in A(F)$.

Since in persistence Kurzweil theorem 9.0.8 $A(F)$ is a compact manifold, the statement follows from general results on (attracting) normally hyperbolic manifolds (see for instance Theorem 14.2 in Ruelle [176] p. 88 to 93). The persistence and smoothness of hyperbolic invariant manifolds for RFDE was also considered in Magalhães [132] with basis on the spectral properties of skew-product semiflows.

For the case when $M = \mathbb{R}^n$ and F is given by an ordinary differential equation, a result somewhat similar to the preceding theorem, was announced by Kurzweil [116]. The proof given here uses considerations different from the above and having some independent interest. The main idea is to look for the manifold of global orbits by finding the ordinary differential equation defining the flow on that manifold. This is accomplished by using a nonlinear variation of constants formula in such a way that one finds the perturbed invariant manifold by finding first the dynamics on it. So, the following theorem is due to Kurzweil [116] but the proof in the text is new (see Magalhães [131] for more details and for an extension of the given result and related facts on smoothness, stability and global solutions). See also [118].

Theorem 9.0.10. *Let $f : \mathbb{R}^n \to \mathbb{R}^n$ be a C^2 function which is bounded and has bounded derivatives, and define $F : C^0 = C^0(I, \mathbb{R}^n) \to \mathbb{R}^n$ by $F(\varphi) = f(\varphi(0))$. For $G \in \mathcal{X}^1(C^0(I, \mathbb{R}^n), \mathbb{R}^n)$, consider the RFDE given by*

$$\dot{x}(t) = G(x_t). \tag{9.2}$$

There exists a neighborhood V of F in $\mathcal{X}^1(C^0(I, \mathbb{R}^n), \mathbb{R}^n)$ such that, for $G \in V$, the set $B(G)$ of all points belonging to orbits of global solutions of (9.1) is diffeomorphic to $\mathbb{R}^n$, depends continuously on G, and the flow of (9.1) in $B(G)$ is given by a one-parameter group of diffeomorphisms, i.e., there exists $g : \mathbb{R}^n \to \mathbb{R}^n$ such that the solutions of $\dot{x}(t) = g(x(t))$ and the global solutions of (9.1) coincide.

Proof: Let us denote by $\xi(t; a, g)$ the value at t of the solution of the ODE $\dot{x}(t) = g\big(x(t)\big)$ which satisfies the initial condition $x(0) = a$, and set $H(t, a) = \frac{\partial \xi}{\partial a}(t; a, f)$. We can write $\xi(\cdot; a, g) : I \to \mathbb{R}^n$ and

$$B(F) = \big\{\xi(\cdot; a, f) : a \in \mathbb{R}^n\big\}.$$

On the other hand, equation (9.1) can be written as

$$\dot{x}(t) = f\big(x(t)\big) + \big[G(x_t) - f\big(x(t)\big)\big].$$

Let $\mathcal{G}^{0,1}(L)$ denote the set

$$\mathcal{G}^{0,1}(L) = \big\{s \in \mathcal{X}^1(\mathbb{R}^n, C^0) : |s(a)| \leq L,\ |s(a) - s(b)| \leq L|a - b|$$
$$\text{for all } a, b \in \mathbb{R}^n\big\}.$$

This set is a complete metric space with distance

$$d(s, s') = \sup\{|s(a) - s'(a)| : a \in \mathbb{R}^n\}.$$

If V is a sufficiently small neighborhood of F we consider the function $\mathcal{H}$ defined in $\mathcal{X}^1(\mathbb{R}^n, C^0) \times \mathcal{X}^1(C^0, \mathbb{R}^n)$, $\mathcal{H}(s, G) \in \mathcal{G}^{0,1}(L)$, by

$$\mathcal{H}(s, G)(a)[\theta] := \mathcal{H}(s, G)(a, \theta) = \int_0^\theta H\big[\theta - \tau, \xi(\tau; a, f) + s(a)(\tau)\big] \cdot$$
$$\cdot \big\{G\big[\xi(\cdot; \xi(\tau; a, f) + s(a)(\tau), f) + s(\xi(\tau; a, f) + s(a)(\tau))\big]$$
$$- f\big[\xi(\tau; a, f) + s(a)(\tau)\big]\big\} d\tau, \theta \in I.$$

Reducing V, if necessary, it can be shown after some computations that, if V is a sufficiently small neighborhood of F in $\mathcal{X}^1\big(C^0(I, \mathbb{R}^n), \mathbb{R}^n\big)$, then $\mathcal{H}(\cdot, G)$ is a uniform contraction from $\mathcal{G}^{0,1}(L)$ into $\mathcal{G}^{0,1}(L)$, for $G \in V$. Applying the uniform contraction principle we obtain, for each $G \in V$, a unique fixed point $\bar{s} = \bar{s}(G) \in \mathcal{G}^{0,1}(L)$ of $\mathcal{H}(\cdot, G)$, which depends continuously on $G \in V$ and satisfies $\bar{s}(F) = 0$. By formally differentiating relative to a the equation $\bar{s}(a)(\theta) = \mathcal{H}(\bar{s}, G)(a, \theta)$, and using the definition of differentiability, we can prove the existence of a function which turns out to be continuous and equal the derivative of $\bar{s}$ relative to a. This establishes that $\bar{s}$ is C^1 in a.

Define $g : \mathbb{R}^n \to \mathbb{R}^n$ by $g(b) = G\big[\xi(\cdot; b, f) + \bar{s}(G)(b)\big]$ and consider the set $S(G) = \{\varphi \in C^0 : \varphi(\theta) = \xi(\theta; a, g),\ a \in \mathbb{R}\}$. Clearly, $S(G)$ is diffeomorphic to $\mathbb{R}^n$, depends continuously on G and the flow of (9.1) on $S(G)$ is given by the ODE $\dot{x}(t) = g\big(x(t)\big)$. Consequently, in order to finish the proof, we need to show $B(G) = S(G)$.

Let

$$y(\theta) = \xi(\theta; a, f) + \bar{s}(G)(a)(\theta), \quad \theta \in [-r, 0].$$

We have

$$y(\theta) = \xi(\theta; a, f) + \int_0^\theta (\theta - \tau, y(\tau)) \left[g\big(y(\tau)\big) - f\big(y(\tau)\big) \right] d\tau$$

which is the nonlinear variation of constants formula for $\dot{x}(t) = f[x(t)] + [g(x(t)) - f(x(t))]$, and, therefore, we have

$$\xi(\theta; a, g) = \xi(\theta; a, f) + \bar{s}(G)(a)(\theta), \quad \theta \in [-r, 0].$$

Also, using the last identity and the definition of g,

$$\begin{aligned}
\xi(t; a, g) = g\big(\xi(t; a, g)\big) &= G\big(\xi(\cdot; \xi(t; a, g), g)\big) \\
&= G\big(\xi(t + \cdot; a, g)\big).
\end{aligned}$$

Therefore, $\xi(t; a, g)$ is a global solution of (9.1) and $S(G)$ is invariant under (9.1), proving that $S(G) \subset B(G)$.

The rest of the proof is similar to the argument used for the analogous situation in Theorem 9.0.7. We begin with the proof that $S(G)$ is uniformly asymptotically stable under (9.1), by showing that there exist $\sigma, \beta > 0$ such that

$$\inf_{\psi \in S(G)} |x_t(\varphi, G) - \psi| \le \sigma e^{-\beta t}, \quad t \ge r, \ \varphi \in C^0.$$

Using this and the fact that $B(G)$ is the set of points in the global orbits of (9.1), we get $B(G) \subset S(G)$, and consequently, $B(G) = S(G)$.   ∎

Remark 9.0.11. Under certain conditions, the preceding proof can be generalized to situations where the unperturbed RFDE is not given by an ordinary differential equation, but there exists a submanifold S of the phase space where the flow is given by a C^1 ODE in $\mathbb{R}^n$, in the sense that there exists a C^1 function $h : \mathbb{R}^n \to \mathbb{R}^n$ such that the ODE $\dot{x}(t) = h\big(x(t)\big)$ has unique solutions for each arbitrary initial condition $x(0) = a \in \mathbb{R}^n$ and that its solutions coincide with the solutions of the unperturbed RFDE $\dot{x}(t) = F(x_t)$ which have initial data on S (such manifolds are necessarily diffeomorphic to $\mathbb{R}^n$).

The method of proof of Theorem 9.0.10, developed in Magalhães [131], was applied to retarded semilinear wave equations by Taboada and You [196], in situations that include the Sine-Gordon equation with a retarded perturbation

$$\frac{\partial^2 u}{\partial t^2} - \Delta u + \beta sinu = M(u_t), \quad in \ \mathbb{R}^+ \times I, \tag{9.3}$$

where $I = (a, b)$ is a bounded interval, $\beta \in \mathbb{R}$, and u satisfies the boundary and initial conditions

$$u = 0 \quad on \quad \mathbb{R}^+ \times (a, b) \tag{9.4}$$

$$(u(\theta, x), \frac{\partial u}{\partial t}(\theta, x)) = w_0(\theta, x), \quad (\theta, x) \in [-h, 0] \times [a, b], \tag{9.5}$$

and also the dissipative Klein-Gordon equation with a retarded perturbation:

$$\frac{\partial^2 u}{\partial t^2} + \beta \frac{\partial u}{\partial t} - \Delta u + mu + |u|^2 u = M(u_t), \quad in \ \mathbb{R}^+ \times \Omega,$$

$$u = 0 \quad on \ \mathbb{R}^+ \times \partial\Omega,$$

$$u(x,0) = u_0(x), \quad \frac{\partial u}{\partial t}(0,x) = u_1(x), \quad in \ \Omega \quad (9.6)$$

where $\beta > 0$ and $m > 0$ are constants and Ω is a bounded region of $\mathbb{R}^3$ with a sufficiently regular boundary.

In fact, Taboada and You follow the constructions of Magalhães [131], applying them to perturbations of hyperbolic equations in a suitable Hilbert space H:

$$\frac{d^2 u}{dt^2} + Au = R(u), \quad u(0) = u_0, \quad \frac{\partial u}{\partial t}(0) = u_1, \quad (9.7)$$

where it is assumed:

- (H1)- the linear operator $A : D(A) \to H$ is a densely defined, positive self-adjoint linear operator on a real Hilbert space H, and has compact resolvent;
- (H2)- the nonlinear operator $R(.)$ is a mapping from $U = D(A^{1/2})$ into H, which is locally bounded and locally Lipschitz continuous. Moreover, $R(.)$ is Fréchet differentiable on U, and its Fréchet derivative, is locally Lipschitz continuous on U;
- (H3)- let $h > 0$ be given. For any b such that $0 < b < \infty$, and any function $u \in C([-h, b); U) \cap C^1([h, b), H)$, the following holds

$$\int_0^t \langle R(u(s)), \frac{\partial u}{\partial t}(s)\rangle_H ds \leq C(u_0, u_1) < \infty,$$

where $\langle ., .\rangle_H$ denotes the inner product in H and $C(u_0, u_1)$ is a constant independent of b, but which may depend on the initial data $u_0 = u(0)$ and $u_1 = (\frac{\partial u}{\partial t})(0)$;
- (H4)- assume the mapping $g : C_X \to X$ is locally bounded and locally Lipschitz continuous, where $X = U \times H$ and $C_X = C([-h, 0]X)$.

Together with the evolution equation (9.7), one considers the perturbed equation

$$\frac{d^2 u}{dt^2} + Au = R(u(t)) + M(u_t), \quad t \geq 0, \quad u_0(\theta) = \phi(\theta), \quad (\frac{\partial u}{\partial t})_0(\theta) = u_1(\theta), \quad (9.8)$$

where $u_t(\theta) = u(t + \theta)$, $\theta \in [-h, 0]$, and $(\phi, u_1) \in C([-h, 0], X) = C_X$. Then, with the methods developed in [131], Taboada and You [196] proved the Theorem 9.0.12 below.

Start by defining $g(\begin{pmatrix} \phi \\ \psi \end{pmatrix}) = \begin{pmatrix} 0 \\ M(\phi) \end{pmatrix}$ with $\begin{pmatrix} \phi \\ \psi \end{pmatrix} \in C([-h, 0], X)$ and

$w_0 = \begin{pmatrix} \phi \\ u_1 \end{pmatrix}$, $w(t) = \begin{pmatrix} u(t) \\ \frac{\partial u}{\partial t}(t) \end{pmatrix}$, then we can rewrite (9.8) as a first order retarded evolution equation

$$\frac{dw}{dt} = \tilde{A}w(t) + f(w(t)) + g(w_t) \quad t \geq 0, \tag{9.9}$$

where

$$\tilde{A} = \begin{pmatrix} 0 & I \\ -A & 0 \end{pmatrix} : D(\tilde{A}) = D(A) \times U \to X,$$

$$f(w) = \begin{pmatrix} 0 \\ R(u) \end{pmatrix}, \quad for \ w = \begin{pmatrix} u \\ v \end{pmatrix} \in X.$$

Theorem 9.0.12. *([196]) Under the hypotheses (H1) - (H4), the retarded evolution equation (9.9) is such that there exists a strongly continuous mapping $K_g : \Omega \subset X \to X$ such that*

$$K_g(\xi) = g[\phi(.,\xi, f + K_g)], \quad for \ \xi \in \Omega, \tag{9.10}$$

where $\phi(.,\xi, f + K_g)$ is the segment of the solution of

$$\frac{dw}{dt} = \tilde{A}w(t) + f(w(t)) + K_g(w(t))$$

on $[-h, 0]$ through the initial data ξ, provided the delay $h > 0$ is sufficiently small. More precisely, for any $d > 0$, there exists an h_0 such that if $0 < h < h_0$, there is a continuous mapping

$$K_g : \Omega_0 := \{\xi \in X :\| \xi \| < d\} \to X$$

that satisfies (9.10). The same result holds if instead of assuming that h is small, one assumes that the magnitude $B(g)$ (see [196] p.347 for the definition) and the Lipschitz constant $L(g)$ of the perturbation g are sufficiently small.

As mentioned before, this result can be applied to the Sine-Gordon equation (9.3) with boundary and initial conditions (9.4) and (9.5). Set $H = L^2(\Omega)$ and $U = H_0^1(\Omega)$, $X = U \times H$, $C_U = C([-h, 0], U)$ and $C_X = C([-h, 0], X)$. Assume that $M : C_U \to H$ is C^1, globally bounded and globally Lipschitz continuous. A typical example of such a mapping M is

$$M(\psi) = \mu_1 \psi(-h_1) + \ldots + \mu_m \psi(-h_m),$$

where $0 \leq h_1 < \ldots < h_m = h$ and the μ_i are the hysteresis functions appearing in the theory of shape memory alloys. Define the operators A and R by

$$A(u) = -\Delta u, \quad D(A) = H^2(\Omega) \cap H_0^1(\Omega) \quad and \quad R(u) = -\beta \ sin u,$$

and notice that $R : U \to H$ is uniformly bounded and uniformly Lipschitz continuous. One can verify that the hypotheses (H1) - (H4) are satisfied. Therefore the conclusions of Theorem 9.0.12 apply and we have the following result:

Theorem 9.0.13. *([196]) For suitable small $h > 0$, the retarded evolution equation* (9.9)

$$\frac{dw}{dt} = \tilde{A}w(t) + f(w(t)) + g(w_t) \quad t \geq 0$$

associated to the retarded Sine-Gordon equation (9.3)-(9.4) has a globally invariant manifold

$$\mathcal{M}_g = \{\phi(.,\xi, f + K_g) : \xi \in U \times H\},$$

where K_g is a mapping from $U \times H$ into itself, determined by g as in Theorem 9.0.12.

We also mentioned above that Theorem 9.0.12 can be applied to the dissipative Klein-Gordon equation with a retarded perturbation and boundary conditions (9.6). Let

$$A = -\Delta : H^2(\Omega) \cap H_0^1(\Omega) \to H = L^2(\Omega),$$

$$R(u) = -(m + |u|^2)u : U = H_0^1(\Omega) \to H.$$

Note that, since $\Omega \subset \mathbb{R}^3$, $H_0^1(\Omega)$ is continuously embedded in $L^6(\Omega)$. Assume that $M : C_U = C([-h, 0], U) \to H$ is a C^1 mapping and is globally bounded and locally Lipschitz continuous. Define

$$\tilde{A} = \begin{pmatrix} 0 & I \\ -A & \beta I \end{pmatrix} : (H^2(\Omega) \cap H_0^1(\Omega)) \times U \to X = U \times H,$$

and

$$f(w) = \begin{pmatrix} 0 \\ R(u) \end{pmatrix}, \quad for \quad \begin{pmatrix} u \\ v \end{pmatrix} \in X,$$

$$g\begin{pmatrix} \psi \\ \rho \end{pmatrix} = \begin{pmatrix} 0 \\ M(\psi) \end{pmatrix}, \quad for \quad \begin{pmatrix} \psi \\ \rho \end{pmatrix} \in C_X = C([-h, 0], X).$$

It can be verified that hypotheses $(H1)$ - $(H4)$ hold, so that Theorem 9.0.12 yields the following result:

Theorem 9.0.14. *([196]) For suitable small $h > 0$, the retarded evolution equation* (9.9)

$$\frac{dw}{dt} = \tilde{A}w(t) + f(w(t)) + g(w_t) \quad t \geq 0$$

associated with the dissipative Klein-Gordon equation with a retarded perturbation and boundary conditions (9.6) has a globally invariant manifold in C_X

$$\mathcal{M}_g = \{\phi(.,\xi, f + K_g) : t \geq 0, \| \xi \| \leq r\}$$

where $r > 0$ is suitably chosen and K_g is a mapping determined by g as in Theorem 9.0.12 (see [196] p.368).

The above results, Theorems 9.0.13 and 9.0.14, establish that for the retarded Sine-Gordon equation and the dissipative Klein-Gordon equation with a retarded perturbation, "small delays do not matter", as in the expression of Kurzweil in [118]. These results of Taboada and You [196] show the power of the methods developed in [131] as they can be applied to retarded semilinear wave equations.

10 Monotonicity

In the last years several papers have appeared which deal with global properties of monotone dynamical systems, that is, maps or semiflows admitting a discrete Lyapunov functional. We refer also to the book [192] for some aspects of the theory as well as extensive references. Our objective in this chapter is to concentrate on certain properties of monotonicity. Between these properties we mention: transversality, Morse–Smale structure, connections between critical elements, Morse decomposition, etc. All of them are used to clarify the structure of the flow including the determination of stability under perturbation of parameters.

Let E be a real normed vector space. A subset $M \subset E$ *generates* E if its linear hull is dense in E. The smallest cardinality of all sets generating E is called the *dimension* of E ($\dim E$). If E is finite dimensional, $\dim E$ is the number of elements of a basis; if, otherwise, E is infinite dimensional, $\dim E$ is the minimal cardinality of dense subsets of E. The notion of cone plays a special role in the concept of monotonicity.

10.1 Usual cones

Definition 10.1.1. *A usual cone in a (finite or infinite dimensional) real normed vector space E is a non empty closed subset C of E such that:*

(1) $x \in C$ and $\alpha \in [0, \infty)$ imply $\alpha x \in C$;
(2) $x, y \in C$ imply $(x + y) \in C$;
(3) $C \cap (-C) = \{0\}$.

As we see, usual cones are convex sets and they do not contain any linear subspace, except $\{0\}$.

Definition 10.1.2. *. Any usual cone C defines on E a partial order by setting $x \leq y \iff (y - x) \in C$. It is clear that $x \leq x$; also, $x \leq y$ and $y \leq x$ imply $x = y$; and, finally, $x \leq y$ and $y \leq z$ imply $x \leq z$.*

Example 10.1.3. The set $P = \{(x_i) \in \mathbb{R}^N \mid x_i \geq 0, i = 1, \ldots, N\}$ is a usual cone and it defines a natural partial order in $\mathbb{R}^N$. In the Banach spaces C, C^1 and L_p, the set K_+ of all non negative functions is a usual cone and it generates the natural partial order in each of these spaces. The set of all non-negative definite self-adjoint operators in a Hilbert space is a usual cone in the space of all bounded self-adjoint operators. The usual cone P and also K_+ in C, C^1 have non empty interior but K_+ in L_p has empty interior.

In 1907 Perron published his famous theorem on positive matrices (matrices with positive entries) that can be stated as follows:

Theorem 10.1.4. *(Perron) Let A be a positive $N \times N$ matrix (so we necessarily have $A(P - \{0\}) \subset intF$). Then*

(i) *the spectral radius $\rho(A)$ of A is positive.*
(ii) *$\rho(A)$ is a simple eigenvalue of A.*
(iii) *if $\mu \neq \rho(A)$ is an eigenvalue of A then $|\mu| < \rho(A)$.*
(iv) *the eigenvector v associated to $\rho(A)$ can be taken in the interior of the cone P.*
(v) *the eigenvector v is unique up to a multiplicative constant.*

Definition 10.1.5. *Let $C \subset E$ be a usual cone with non empty interior and $A : E \to E$ a linear operator. A is said to be positive with respect to C if $A(C \setminus \{0\}) \subset intC$.*

In 1948 Krein and Rutman published a remarkable extension of Perron's theorem that covers the case where A is compact and $\dim E = \infty$.

Theorem 10.1.6. *(Krein-Rutman) Let $C \subset E$ be a usual cone with non empty interior and $A : E \to E$ be a compact linear operator such that A is positive with respect to C. Then the same conclusions of Perron's theorem are true.*

An elementary proof for the last theorem that uses dynamical systems tools only can be seen in [3] where it is also proved the following:

Theorem 10.1.7. *Let E be a finite dimensional normed space and $C \subset E$ a closed set with non empty interior such that*

(1) $x \in C$, $-x \in C \implies x = 0$
(2) $\alpha \in [0, \infty)$, $x \in C \implies \alpha x \in C$.

Assume also that A is a linear operator such that $A(C \setminus \{0\}) \subset intC$. Then the same conclusions of Perron's theorem hold. Moreover, the convex hull $\hat{C}$ of C is a usual cone with non empty interior.

10.2 Cones of rank k

Definition 10.2.1. *A cone of rank k (finite or infinite) is a non empty and closed set $K \subset E$ such that:*

K_1) $x \in K, t \in \mathbb{R} \Longrightarrow tx \in K$,

K_2) K contains at least one (linear) subspace of dimension k but no subspaces of higher dimension.

A trivial example of a cone of rank k is a k-dimensional linear subspace of E.

Remark 10.2.2. If C is a usual cone then $K = C \cup -C$ is a cone of rank 1.

From the proof of Theorem 10.1.7 it follows the following corollary which also implies Perron's result:

Corollary 10.2.3. *Let K be a cone of rank 1 with non empty interior, contained in a finite dimensional normed space E, $dimE = n$. Assume that A is a linear operator such that $A(K \setminus \{0\}) \subset intK$. Then there exist linear subspaces W_1, W_2, $W_1 \cap W_2 = \{0\}$, $\dim W_1 = 1$, $\dim W_2 = n-1$, which are invariant under A, and such that $W_1 \subset \{0\} \cup intK$, $W_2 \cap K = \{0\}$. Moreover if λ is an eigenvalue of A corresponding to W_1 and $\mu \in \sigma(A)$, $\mu \neq \lambda$, then $\rho(A) = |\lambda| > |\mu|$.*

The last corollary was extended to cones of rank k (see [65]) and is very important in the theory of finite dimensional monotone systems, as we will see below.

Theorem 10.2.4. *Let E be a finite n-dimensional normed space, K a cone of rank $k < n$ with non empty interior and A a linear operator of E such that $A(K \setminus \{0\}) \subset intK$. Then there exist (unique) subspaces W_1, W_2 such that:*

(1) $W_1 \cap W_2 = \{0\}$, $dimW_1 = k$, $dimW_2 = n-k$;
(2) $AW_j \subset W_j$, $j = 1, 2$;
(3) $W_1 \subset (\{0\} \cup intK)$, $W_2 \cap K = \{0\}$.

Moreover, if $\sigma_1(A)$, $\sigma_2(A)$ are the spectra of A restricted to W_1 and W_2, then between $\sigma_1(A)$ and $\sigma_2(A)$ there is a gap:

$$\lambda \in \sigma_1(A), \mu \in \sigma_2(A) \Longrightarrow |\lambda| > |\mu|.$$

Theorem 10.2.4 can be used in many applications of the finite dimensional theory of monotone systems, as we will see in the sequel. It can also be inverted and we obtain (see also [65]):

Proposition 10.2.5. *Let A be a linear operator of a finite n-dimensional normed space E, $W_1, W_2 \subset E$ subspaces of dimensions k and $n-k$, such that $W_1 \cap W_2 = \{0\}$. Assume that:*

a) $AW_j \subset W_j, j = 1, 2$;
b) $\lambda \in \sigma(A_{|W_1}), \mu \in (A_{|W_2}) \Longrightarrow |\lambda| > |\mu|$.

Then, there is a subset $K \subset E$ with non empty interior which is a cone of rank k and $A(K \setminus \{0\}) \subset intK$.

10.3 Monotonicity in finite dimensions

Let be given a cone K of rank k with non empty interior, contained in $\mathbb{R}^n$. Denote by d_k the rank k of K and set $\overset{\circ}{K} = \{0\} \cup intK$.

Definition 10.3.1. *Let $\Omega \subset \mathbb{R}^n$ be a connected open set and $\varphi : \Omega \to \mathbb{R}^n$ an injective and differentiable map with derivative $\varphi'(x) : \mathbb{R}^n \to \mathbb{R}^n$ injective for all $x \in \mathbb{R}^n$. Then φ is said to be monotone if there exists a family of nested cones of finite rank: $K_1 \subset K_2 \subset \ldots K_{r+1} = \mathbb{R}^n$, such that*

$$d_{K_i} < d_{K_{i+1}}$$

and

$$x \in \Omega, \quad y \in K_i \Longrightarrow \varphi'(x)y \in \overset{\circ}{K_i}, \quad i = 1, \ldots, r.$$

A vector field $(\dot{x} = X(x))$ is monotone if, for each fixed $t \in \mathbb{R}$, its flow map φ_t is a monotone map. A discrete Lyapunov functional associated to this family of cones is the map $N : \mathcal{N} \to \{d_{K_1}, \ldots, d_{K_r}, n\}$, where $\mathcal{N} = \mathbb{R}^n \setminus \bigcup_{i=1}^{r} \partial K_i$, and one defines $N(y) = d_{K_i}$ for $y \in (\overset{\circ}{K_i} \setminus K_{i-1})$, $i = 1, \ldots, r + 1$ (here $K_0 = \{0\}$).

Remark 10.3.2. The discrete Lyapunov functional N is non-increasing along the trajectories of each tangent map.

Example 10.3.3. (Jacobi systems, see [63]) Let us consider the system of ordinary differential equations:

$$\dot{x}_1(t) = f_1(x_1, x_2)$$
$$\dot{x}_i(t) = f_i(x_{i-1}, x_i, x_{i+1}), \quad i = 2, \ldots, n-1,$$
$$\dot{x}_n(t) = f_n(x_{n-1}, x_n),$$

where, for each $x \in \mathbb{R}^n$, the Jacobian matrix of the system at x is a positive Jacobi matrix, that is, its off-diagonal terms are positive numbers. Consider now the following family of nested cones: take $y = (y_1, \ldots, y_n) \in \mathbb{R}^n$, all $y_i \neq 0$, and call $N(y)$ the integer 1 plus the number of sign changes in y. Define, also, for an arbitrary $y \in \mathbb{R}^n$, $N_M(y)$ as the maximum of $N(z)$ for z with non zero components in a small neighborhood of y. Let us set K_i equal to the closure of the set $\{y \in \mathbb{R}^n | N_M(y) \leq i\}$. It can be proved that the K_i have non empty interior and are cones of rank $d_{K_i} = i$ and

$$K_1 \subset K_2 \subset \ldots \subset K_n = \mathbb{R}^n.$$

In [63] it is proved that any Jacobi system is monotone with respect to this family of cones.

Example 10.3.4. (Cyclic systems, see [64]) Consider the system of ordinary differential equations:

$$\dot{x}_1(t) = f_1(x_1, x_2, x_n)$$
$$\dot{x}_i(t) = f_i(x_{i-1}, x_i, x_{i+1}), \quad i = 2, \ldots, n-1,$$
$$\dot{x}_n(t) = f_n(x_1, x_{n-1}, x_n),$$

where, for each $x \in \mathbb{R}^n$, the Jacobian matrix of the system at x has all its off-diagonal elements, $a_{j,k}(x)$, greater or equal to zero and

$$\prod_{j=1}^{n} a_{j,j-1}(x) + \prod_{j=1}^{n} a_{j,j+1}(x) > 0, j = 1, \ldots, n \quad (x_0 = x_n; x_{n+1} = x_1).$$

A family of nested cones for cyclic systems is defined by introducing N, N_M, as in the previous example, but using the cyclic sequence $(x_n, x_1, \ldots, x_n)$ instead of $(x_1, \ldots, x_n)$, and get $N : \mathcal{N} \to \{1, 3, 5, \ldots\}$. Define now $K_i =$ closure of $\{y \in \mathbb{R}^n | N_M(y) \leq (2i-1)\}$. The set K_i is a cone of rank $d_{K_i} = (2i-1)$, $i = 1, 2, \ldots$, with non empty interior. In [64] it is proved that cyclic systems are monotone with respect to the family of cones

$$K_1 \subset K_2 \subset \ldots \subset K_{(\bar{n}+3)/2} = \mathbb{R}^n,$$

where $\bar{n} = n$ if n is odd and $\bar{n} = n + 1$ if n is even.

Example 10.3.5. (Quadratic cones) Let $P_i = diag(\delta_1, \ldots, \delta_n)$ be a $n \times n$ diagonal matrix with $\delta_k = 1$ (if $k = 1, \ldots, i$) and $\delta_k = -1$ (if $k = i+1, \ldots, n$). Define the quadratic cones

$$K_i = \{y \in \mathbb{R}^n | (y, P_i y) \geq 0\}, i = 1, \ldots, (n-1),$$

where $(\, , \,)$ means the usual inner product of $\mathbb{R}^n$. We can easily see that $K_1 \subset K_2 \subset \ldots \subset K_n = \mathbb{R}^n$ and K_i is a cone of rank $d_{K_i} = i$, with non empty interior (see [203] for an example where, for other reasons, it is also used the notion of quadratic cone). Let us consider the following class $\mathcal{L}$ of $n \times n$ matrices A such that :

$$\forall y \neq 0, (y, P_i y) = 0 \Rightarrow (Ay, P_i y) > 0, \quad i = 1, \ldots, n-1. \tag{10.1}$$

Let

$$\dot{x} = f(x) \tag{10.2}$$

be such that $f'(x) \in \mathcal{L}$, for all $x \in \mathbb{R}^n$. System (10.2) is monotone with respect to the family of nested quadratic cones considered above in this example.

Monotone systems have special spectral properties. In fact, using Theorem 10.2.4 we obtain the following result:

Theorem 10.3.6. *Let $x_0 \in \Omega$ be a critical point of a monotone system $\dot{x} = f(x)$ and A be the Jacobian matrix of f at x_0 : $A = f'(x_0)$. Then $\mathbb{R}^n = W_1 \oplus \ldots \oplus W_{r+1}$ is the direct sum of $(r+1)$ subspaces, such that:*

(i) $W_i \backslash \{0\} \subset \overset{\circ}{K}_i \backslash K_{i-1}$, *dim* $W_i = d_{K_i} - d_{K_{i-1}}$, $d_{K_0} = 0$ *and* $y \in W_i \backslash \{0\} \Rightarrow$ $N(y) = d_{K_i}$, $i = 1, \ldots, (r+1)$.
(ii) $AW_i \subset W_i$.
(iii) *(Gap condition) If* μ_i, λ_i *are the maximum and minimum of the real parts of the eigenvalues of* $A|_{W_i}$*, then* $\mu_1 \geq \lambda_1 > \mu_2 \geq \lambda_2 > \ldots > \mu_{r+1} \geq \lambda_{r+1}$.

We will see now that monotone systems with an additional condition have very special transversality properties. In fact, given a map $\varphi : \Omega \to \mathbb{R}^n$ which is monotone with respect to a family $(K_i)_i$ of nested cones, assume that, in addition, φ satisfies:

$$x_1, x_2 \in \Omega, x_1 \neq x_2 \quad \text{and} \quad x_1 - x_2 \in K_i \quad \text{imply} \quad \varphi(x_1) - \varphi(x_2) \in \overset{\circ}{K}_i. \quad (10.3)$$

We remark that the flow maps of Jacobi and cyclic systems satisfy condition (10.3). Moreover, we also mention that *for Jacobi systems, given two hyperbolic critical points P_1 and P_2 with a heteroclinic connection then the unstable manifold $W^u(P_1)$ intersects transversaly the stable manifold $W^s(P_2)$. If Γ_1 is a hyperbolic periodic orbit or a hyperbolic critical point and Γ_2 is a hyperbolic periodic orbit of a cyclic system, with a heteroclinic connection, then the invariant manifolds $W^u(\Gamma_1)$ and $W^s(\Gamma_2)$ intersect transversaly (see [63] and [64]).*

The automatic transversality along a heteroclinic connection between two hyperbolic critical points is not the only fundamental property of a Jacobi system; its flow also puts restrictions on the structure of the nonwandering set (see [160] for the definition of nonwandering set of a flow). We have, in fact, the following theorem.

Theorem 10.3.7. *(see [63]) The nonwandering set of a Jacobi system (see Example10.3.3) coincides with the set of its critical points. In particular if the critical points are hyperbolic and finite in number, the system is Morse-Smale.*

Oscillatory matrices were studied extensively by Gantmacher and Krein (see [66]) and have very interesting spectral properties. For historical notes and a long list of references see the books [109] and [113]. The main properties of oscillatory matrices were revisited recently(see [156]). Just to remember, a $n \times n$ matrix is said to be oscillatory if all minors of all orders are non negative and there exists a power of it such that all minors of all orders are (strictly) positive numbers.

Theorem 10.3.8. *(see [156]) Let φ be a smooth diffeomorphism $\varphi : \Omega \to \mathbb{R}^n$ such that φ is monotone with respect to the family of nested cones of Example*

10.3.3.Assume also that $\varphi'(x)$ is an oscillatory matrix for all $x \in \Omega$ and that φ satisfies (10.3). Then, given two hyperbolic fixed points P_1 and P_2 of φ with a heteroclinic connection, one has that $W^u(P_1)$ and $W^s(P_2)$ intersect transversaly.

Example 10.3.9. The present example fits in the hypothesis of Theorem 10.3.8 and is completely developed in [156]. It deals with a double discretization of the problem

$$u_t = u_{xx} + f(u)$$
$$u(0,t) = u(L,t) = 0, \quad t > 0$$
$$u(x,0) = u_0(x),$$

where $f : \mathbb{R} \to \mathbb{R}$ is smooth and $|f'(x)| \leq k$. The interval $[0,L] = [x_0, x_{N+1}]$ is divided in $(N+1)$ pieces of length $D = \frac{L}{(N+1)}$ and let $h > 0$ be the time step. One can approximate u_t and $u_{xx} + f(u)$ by

$$\frac{u(x_i, t+h) - u(x_i, t)}{h}$$

and

$$\frac{u(x_{i-1}, t+h) - 2u(x_i, t+h) + u(x_{i+1}, t+h)}{D^2} + f(u(x_i, t)),$$

respectively, with $u(x_0, t) = u(x_{N+1}, t) = 0$, or, equivalently,

$$J_{h,D}u(x_i, t+h) = \frac{u(x_i,t)}{h} + f(u(x_i,t)), \quad i = 1,\ldots,N, \tag{10.4}$$

where $J_{h,D}$ is the matrix

$$\begin{bmatrix} \frac{1}{h}+\frac{2}{D^2} & -\frac{1}{D^2} & \cdots & 0 & 0 \\ -\frac{1}{D^2} & \frac{1}{h}+\frac{2}{D^2} & \cdots & 0 & 0 \\ \cdots & \cdots & \cdots & \cdots & \cdots \\ 0 & 0 & \cdots & \frac{1}{h}+\frac{2}{D^2} & -\frac{1}{D^2} \\ 0 & 0 & \cdots & -\frac{1}{D^2} & \frac{1}{h}+\frac{2}{D^2} \end{bmatrix}.$$

It can be proved that for any h, D, the matrix $J_{h,D}$ is nonsingular and its inverse is oscillatory. One defines the map $\phi_{h,D} : \mathbb{R}^N \to \mathbb{R}^N$ by

$$\phi_{h,D}(x) = J_{h,D}^{-1}[\frac{1}{h} + \tilde{f}(x)]$$

where $x = (x_i)_i$ and $\tilde{f}(x) = (f(x_i))_i$, $i = 1,\ldots,N$.

Theorem 10.3.10. *(see[156])* If $hk < 1$, then $\phi_{h,D}(x)$ is a diffeomorphism such that:

(i) $x_1 \neq x_2 \in \mathbb{R}^N \implies N_M(\phi_{h,D}(x_1) - \phi_{h,D}(x_2)) \leq N_M(x_1 - x_2)$, that is, $\phi_{h,D}$ satisfies 10.3.

(ii) $\forall x \in \mathbb{R}^N$, *the Jacobian matrix of* $\phi_{h,D}$ *is oscillatory.*
(iii) The nonwandering set of $\phi_{h,D}$ *is the set of its fixed points.*
(iv) $\phi_{h,D}$ *has a Lyapunov function:*

$$V(x) = \sum_{i=0}^{N} [\frac{1}{D^2}(x_{i+1} - x_i)^2 - \int_0^{x_i} f(s)ds], \quad x_{N+1} = x_0 = 0,$$

that is, $V(\phi_{h,D}(x)) \leq V(x)$ *(equality means that* x *is a fixed point of* $\phi_{h,D}$*).*

Corollary 10.3.11. *If all the fixed points of* $\phi_{h,D}$ *are hyperbolic, then* $\phi_{h,D}$
is a Morse–Smale diffeomorphism.

10.4 Monotonicity in infinite dimensions

In order to deal with monotone systems in the infinite dimensional setting, we have to look for results under the hypothesis that the normed vector space E is such that dim$E$ is not necessarily finite.

Just to motivate the reader, we start with an example. For this, let us consider a bounded projection $P : E \to E$ on a subspace $E_0 \subset E$, that is $P^2 = P$ and $P(E) = E_0$. Define the function

$$\mu(x, \alpha) := \parallel Px \parallel^2 - \alpha \parallel x - Px \parallel^2, \alpha > 0.$$

The set

$$T(\alpha) := \{x | \mu(x, \alpha) \geq 0\}$$

is called the *nonnegative part of the space* E (see [113]). Observe that $T(\beta)$ is a proper subset of $T(\alpha)$ if $\beta > \alpha$.

Proposition 10.4.1. *Every set* $T(\alpha)$ *is a cone of rank* $k = dimE_0$ *and all non zero elements of* E_0 *are interior points of the cone.*

The sets $T(\alpha)$ belong to a wider class the so called class of *normal* cones of rank k with non empty interior (see [113]). An extension of Theorem 10.2.4 can be obtained, for this kind of infinite dimensional cones, when we assume, for the cones $T(\alpha)$, that E is a Hilbert space, $k = dimE_0 < \infty$ and that P is an orthogonal projection. For each $\alpha \neq 0$ the quadratic form $\mu(x, \alpha)$ is an indefinite metric in E and gives an extension for the quadratic cones considered in Example 10.3.5. If $AT(\alpha) \subset T(\alpha)$, $(\alpha > 0)$, holds for some continuous linear operator A, then A is called an *indefinitely positive operator.*

Theorem 10.4.2. *Suppose that a continuous linear operator* A *satisfies the condition* $AT(\alpha) \subset T(\beta)$ *for some* $\beta > \alpha$, $\alpha > 0$ *(so* $AT(\alpha) \subset T(\alpha)$*) and* $A(x) \neq 0$ *for* $x \in T(\beta)$, $x \neq 0$. *Then the operator* A *has a* k-*dimensional invariant linear subspace* $M \subset T(\beta)$ *and is invertible on this subspace. The* *points in the spectrum $\sigma(A)$ of A which are different from the eigenvalues $\lambda_1, \ldots, \lambda_r$ of its restriction to M lie in the disc*

$$|\lambda| \leq \rho.min\{|\lambda_1|, \ldots, |\lambda_r|\}, \quad \rho = \rho(A, T(\alpha)), \quad 0 < \rho < 1.$$

Remark that, in fact, in the last theorem, $\rho = \frac{\chi(A,T(\alpha))-1}{\chi(A,T(\alpha))+1}$, where $\chi(A, T(\alpha)) > 1$ is called the *focusing constant* of A with respect to $T(\alpha)$. The operator A fits in the class of the so called *focusing operators* with respect to $T(\alpha)$. For a quite general statement, dealing with normal cones and focusing operators, see also Theorem 14.4 of [113].

In what follows we will see two more examples in infinite dimensions that present discrete Lyapunov functionals, a fundamental tool for transversality as we already saw in finite dimensions.

10.4.1 The Chafee–Infante problem

As mentioned by Henry in 1985 (see [100]), this problem is certainly the best understood example for the global and geometric theory of semi-linear parabolic equations (see example 2.0.9). The main result of the Chafee–Infante problem, a very surprising success story, essentially states that the corresponding flow has strong properties such as

(a) gradient structure;
(b) compact attractor;
(c) existence of a discrete Lyapunov functional, the so called *zero number* or *lap number* (see [139] and [140]) implying transversality between invariant manifolds of critical points with a heteroclinic connection.

Theorem 10.4.3. *(Henry [100]) Let $f : \mathbb{R} \to \mathbb{R}$ be C^2, $f(0) = 0$, $f'(0) = 1$, $uf''(u) < 0$ for $u \neq 0$ and*

$$limsup_{|u| \to \infty} \frac{f(u)}{u} \leq 0,$$

and consider the Chafee–Infante problem

$$u_t = u_{xx} + \lambda f(u), \quad on \quad 0 < x < \pi$$

$$u = 0 \quad at \quad x = 0 \quad and \quad x = \pi$$

with constant $\lambda \geq 0$. Let

$$F_\lambda : H_0^1(0, \pi) \to H_0^1(0, \pi)$$

be the time one map $u_{|t=0} \mapsto u_{|t=1}$. Then for every $\lambda \notin \{1^2, 2^2, 3^2, \ldots\}$, F_λ is a C^2 Morse-Smale map in the sense of Chapter 6.

Except for transversality that was, in fact, conjectured by Hale in 1981 (see [74]), the other conditions for the map F_λ be Morse–Smale were already proved in [99]. The key argument is an appropriate analysis for the asymptotic behavior of solutions of the linear variational equation along a heteroclinic connection between two equilibria.

As a matter of fact, in [100] it is constructed a series of Morse–Smale systems generalizing the Chafee–Infante problem, allowing, also, non linear boundary conditions; the transversality condition hold for a large class of one-space dimensional parabolic equations, even when the equilibria are not necessarily hyperbolic; as commented by Henry, transversality is somehow automatic and unavoidable. The key argument for transversality appears in the next theorem, expressing a general monotonic property of linear second order parabolic equations in one space variable: roughly speaking, the number of zeros of the solution decreases with time, defining the discrete Lyapunov functional mentioned in item (c) above.

Theorem 10.4.4. *(Henry [100]) Let p, q, r, W be continuous with $p > 0$ on $0 \leq x \leq 1$, $t_0 \leq t \leq t_1$, let the derivative W_x be continuous on $[0,1] \times (t_0, t_1]$ and let W_t, W_{xx} be continuous on $(0,1) \times (t_0, t_1]$, and*

$$W_t = p(x,t)W_{xx} + q(x,t)W_x + r(s,t)W$$

on $[0,1] \times (t_0, t_1]$. Also suppose $\beta_0(t)$, $\beta_1(t)$ are continuous on $[t_0, t_1]$ and the products

$$W.(W_x + \beta_0(t)W) \geq 0 \quad at \quad x = 0$$
$$W.(W_x + \beta_0(t)W) \leq 0 \quad at \quad x = 1.$$

The number of components of

$$\{x \mid \ 0 < x < 1, W(x,t) \neq 0\}$$

decreases with time (is a nonincreasing function of t) on $t_0 \leq t \leq t_1$.

This last result is then applied to the linear variational equation along heteroclinic connections of the given non linear problem to obtain the conclusions on transversality between stable and unstable manifolds of equilibria leading to various examples of Morse–Smale semiflows.

Using refined arguments which include center manifolds, Henry was also able to track the changes in connecting orbits which occur at the pitchfork bifurcations when $\lambda = n\pi$ in the Chafee–Infante problem, founding all the referred connecting orbits.

We have to mention that, in 1986, Theorem 10.4.3 was again proved, now by Angenent, who also worked on the zero set of a solution of a parabolic equation (see [4], [5]).

The determination of all heteroclinic connections that appear in certain semi-linear parabolic equations, as well as the theory of orbit equivalence of their attractors were published more recently. We mention the papers [57] and [58] where the reader will find the basic concepts, results and references.

10.4.2 An infinite dimensional Morse–Smale map

The time-one maps $T(1) : H_0^1(\Omega) \to H_0^1(\Omega)$ corresponding to the flow of some semilinear parabolic equations $u_t = \Delta u + f(u, x)$, $x \in \Omega \subset \mathbb{R}^n$ with $n = 1$ and Dirichlet boundary conditions, have the same strong properties such as the ones mentioned for the Chafee–Infante problem. These maps as well as the time-one maps of other more complicated but similar problems, allowing the dependence of f on u_x and more general boundary conditions, have been also mentioned, under this point of view, in the last section. A nonautonomous problem was considered in [27, 28] where $f = f(t, x, u)$ is one-periodic in time and $\dim \Omega = 1$; the strong properties already mentioned also hold for the time-one map (Poincaré map) associated to this periodic system.

In the present subsection we will talk about the dynamics of a map $\Phi_h :$ $X \to X$, parametrized by $h > 0$, where X is the Hilbert space $H_0^1 = \{u :$ $[0, 1] \to \mathbb{R} \mid u$ *is absolutely continuous*, $u(0) = u(1) = 0$ *and* $u' \in L_2[0, 1]\}$, with the inner product given by

$$(u, v) = \int_0^1 [u'(x)v'(x) + h^{-1}u(x)v(x)]dx.$$

The map Φ_h is defined by

$$\Phi_h(u) = (I - h\Delta)^{-1}(u + hf \circ u), u \in X \qquad (10.5)$$

where $\Delta u = u_{xx}$ and $f : \mathbb{R} \to \mathbb{R}$ is a given C^1 function satisfying

$$K = sup_{y \in \mathbb{R}} |f'(y)| < \pi \quad and \quad hK < 1.$$

Therefore, $\Phi_h(u)$ is the solution $v = v(x)$ of the Poisson equation

$$v(x) - hv''(x) = u(x) + hf(u(x)), x \in (0, 1) \qquad (10.6)$$

with the boundary conditions $v(0) = v(1) = 0$.

Equation (10.5) is obtained as a semi-implicit discretization on time of the one-dimension semi-linear scalar parabolic equation

$$u_t = u_{xx} + f(u), (x, t) \in (0, 1) \times (0, \infty) \qquad (10.7)$$

with Dirichlet boundary conditions $u(0, t) = u(1, t) = 0$, $t > 0$. More precisely, we start with

$$(u_{n+1} - u_n)h^{-1} = \Delta u_{n+1} + f \circ u_n$$

and obtain $u_{n+1} = \Phi_h(u_n)$. We will show that the same strong properties (a), (b) and (c), mentioned above in Example 10.4.1, hold for the discrete dynamics of Φ_h, that is, Φ_h is a Morse–Smale map in the sense of Section 6, provided that it has a finite nonwandering set equal to the totality of its hyperbolic fixed points; moreover, Φ_h is stable relatively to the attractor $A(f)$, for f in a certain class (see[155] for a complete presentation). The map Φ_h is not the time-one map of equation (10.7), although both have similar properties and the fixed points of Φ_h (that do not depend on h) are precisely the equilibrium points of that parabolic equation. It is interesting to remark that if we discretize equation (10.7) not only in time but also in space by dividing $[0, 1]$ into $N + 1$ equal pieces, we obtain the Morse–Smale map that is a diffeomorphism of $\mathbb{R}^n$ whose Jacobian at each point is an oscillatory matrix, already mentioned in Example 10.3.9.

We recall now some of the results proved in [155]:

1) The map $\Phi_h : H_0^1 \to H_0^1$ is compact and its range is contained in $H^3(0, 1)$.
2) For any $\varphi \in H_0^1$, the maps Φ_h and $D\Phi_h(\varphi)$ are injective in H_0^1.
3) Let φ be a fixed point of Φ_h. Then, $D\Phi_h(\varphi)$ is self-adjoint.

For a given $\varphi \in C[0, 1]$ one defines the number of sign changes of φ (lap number) by $S(\varphi) = max\{0, sup\{k \in N \,|\, there\, exists\, 0 < x_0 < \ldots < x_k < 1\, such\, that\, \varphi(x_j)\varphi(x_{j-1}) < 0\, for\, j = 1, \ldots, k\}\}$; we allow $S(\varphi) = \infty$.

4) For each given $\varphi \in C[0, 1]$, let $v \in X$ be such that $v = L\varphi$ where L is the linear operator $(I - h\Delta)^{-1}$. Then $S(v) \leq S(\varphi)$; in other words, the operator L with Dirichlet boundary conditions does not increase the number of sign changes.
5) If $v^1, v^0, u \in H_0^1$ satisfy $v^1 = D\Phi_h(u)v^0$, then $S(v^1) \leq S(v^0)$.
6) If u, v are in H_0^1 then $S(\Phi_h(u) - \Phi_h(v)) \leq S(u - v)$, that is, Φ_h satisfies condition 10.3.
7) Let p be a hyperbolic fixed point of Φ_h and $u_0 \in H_0^1$, $u_0 \neq p$. Assume that $u_{n+1} = \Phi_h^n(u_0)$, $n \geq 1$, satisfies $p = lim_{n\to\infty} u_n$. Then we have: (i)- the sequence $x_n = u_{n+1} - u_n$ is such that $x_n/\|x_n\| \to v$ as $n \to \infty$ where v is an eigenfunction of $A = D\Phi_h(p)$. (ii)- if $w_0 \neq 0$ is given in H_0^1 and $w_{n+1} = D\Phi_h(u_n)w_n$, $n \geq 0$, then $w_n/\|w_n\| \to w$ as $n \to \infty$, where w is an eigenfunction of A.

Using these results one can prove the following two basic statements for transversality:

Proposition 10.4.5. *Let $(u_n)_{n\in Z}$ be a heteroclinic connection between two hyperbolic fixed points ψ and φ of the map Φ_h ($u_n \to \psi$ as $n \to -\infty$ and $u_n \to \varphi$ as $n \to \infty$). Then the dimension of the unstable manifold of ψ is greater than the dimension of the unstable manifold of φ.*

Theorem 10.4.6. *Under the hypotheses of Proposition 10.4.5 the unstable manifold of ψ meets transversaly the local stable manifold of φ.*

These two results imply that Φ_h is a Morse–Smale map. For the proofs and a conclusion about the A-stability of that infinite dimensional map, we refer the reader to [155].

10.5 Negative feedback: Morse decomposition

This example illustrates another important reason for restricting the flow of an infinite dimensional system to the compact global attractor.

Suppose that $T(t)$, $t \geq 0$, is a C^0-semigroup on a Banach manifold X for which there is a compact global attractor $\mathcal{A}$. A *Morse decomposition* of the attractor $\mathcal{A}$ is a finite ordered collection $\mathcal{A}_1 < \mathcal{A}_2 < \cdots < \mathcal{A}_M$ of disjoint compact invariant subsets of $\mathcal{A}$ (called *Morse sets*) such that, for any $\varphi \in \mathcal{A}$, there are positive integers N and K, $N \geq K$, such that $\alpha(\varphi) \subset \mathcal{A}_N$ and $\omega(\varphi) \subset \mathcal{A}_K$ and $N = K$ implies that $\varphi \in \mathcal{A}_N$. In the case $N = K$, we have $T(t)\varphi \in \mathcal{A}_N$ for $t \in \mathbb{R}$.

The union of the Morse sets and the *connecting orbits*

$$C_K^N \equiv \{ \varphi \in \mathcal{A} : \alpha(\varphi) \subset \mathcal{A}_N, \omega(\varphi) \subset \mathcal{A}_K \}$$

for $N > K$ is the compact global attractor $\mathcal{A}$.

There are two obvious Morse decompositions; namely, the set $\mathcal{A}$ itself or the empty set. Neither of these decompositions are interesting. A Morse decomposition becomes important when it gives some additional information about the flow defined by the semigroup restricted to the attractor. In this section, we consider in some detail a Morse decomposition for a special class of differential difference equations. We only give some ideas of the proofs and the reader may consult the references mentioned in the presentation. Here we recall that Conley [34] defined the notion of Morse decomposition for a compact metric space with a flow (see [137]).

Consider the equation

$$\dot{x}(t) = -\beta x(t) - g(x(t-1)) \tag{10.8}$$

where $\beta \geq 0$ and the following hypotheses are satisfied:

1. (H_1) $-g \in C^\infty(\mathbb{R}, \mathbb{R})$ and has *negative feedback*; that is, $xg(x) > 0$ for $x \neq 0$ and $g'(0) > 0$,
2. (H_2) there is a constant K such that $g(x) \geq -K$ for all x,
3. (H_3) the zero solution of (10.8) is hyperbolic.

There are many applications of equation (10.8) to such diverse fields as laser optics, mathematical biology and ecology. It also includes the famous equation of Wright (see[204]),

$$\dot{x}(t) = -\alpha x(t)(1 + x(t)),$$

encountered in the theory of prime numbers. The interesting situation is $\alpha > 0$ and initial values φ with $\varphi(0) > -1$. Under the transformation $1 + x(t) = e^{y(t)}$, we obtain the equation

$$\dot{y}(t) = -\alpha(e^{y(t-1)} - 1),$$

which is a special case of (10.8) with negative feedback.

Wright's equation played an important role in the early development of the theory of functional differential equations. In fact, for $\alpha > \pi/2$, there is a periodic solution of (10.8) with the distance between zeros strictly greater than 1, a *slowly oscillating solution*. The method developed for the proof of this result relied heavily upon the consideration of delay equations as defining a flow in the function space $C([-1,0],\mathbb{R})$. The analysis also was the motivation for new fixed point theorems of mappings which are now referred to as *ejective fixed point theorems*. We refer the reader to Hale and Verduyn-Lunel [86] for a discussion and detailed references to this subject.

The results to be stated below actually hold in situations more general that (10.8). In fact, the right hand side of (10.8) can be replaced by $f(x(t), x(t-1))$ with a modified definition of negative feedback. Hypothesis (H_2) can be replaced by the assumption that the semigroup associated to (10.8) is point dissipative. We need only the existence of a compact global attractor and the negative feedback restriction. We prove below that (H_1) and (H_2) imply the existence of the compact global attractor. Hypothesis (H_3) is not necessary, but to eliminate it will introduce additional complications and modified definitions.

Lemma 10.5.1. . *Under hypotheses (H_1) and (H_2), the semigroup $T(t)$ generated by (10.8) is a bounded map and is point dissipative. Thus, there is a compact global attractor $\mathcal{A}_\beta$. Furthermore, there is a bounded set $B \subset C$ such that $\mathcal{A}_\beta \subset B$ for all $\beta \geq 0$.*

Proof: By integrating (10.8) over $[0, 1]$, one deduces that $T(1)$ is a bounded map. Thus, $T(t)$ is a bounded map for any $t > 0$.

Let us first show that $T(t)$ is point dissipative if $\beta > 0$. If we observe that

$$\frac{d}{dt}(e^{\beta t} x(t)) \leq e^{\beta t} K$$

for all $t \geq 0$, then

$$x(t) \leq x(0)e^{-\beta t} + \frac{K}{\beta}(1 - e^{-\beta t})$$

and $\limsup_{t \to \infty} x(t) \leq 2K/\beta$. Since $x(t)$ is bounded above, it follows that $-g(x(t-1))$ is bounded below by a constant K_1. Therefore, arguing as above, one obtains that $\liminf_{t \to \infty} x(t) \geq -2K_1/\beta$. This shows that $T(t)$ is point dissipative and, from Theorem 2.0.14, we deduce that the compact global attractor exists.

To obtain the uniform bound on $\mathcal{A}_\beta$ for all $\beta > 0$, we first note that

$$x(t) \leq x(t-1)e^{-\beta} + \frac{k}{\beta}(1 - e^{-\beta}).$$

If $x(t)$ is a bounded solution on $(-\infty, \infty)$ and there is a τ such that $x(\tau) > (k/\beta)(1 - e^{-\beta})$, then $\dot{x}(\tau) < 0$ and $x(t)$ is decreasing on $(-\infty, \tau]$. This implies that $x(t)$ approaches a positive equilibrium point as $t \to -\infty$, which is a contradiction. This completes the proof of the lemma for the case $\beta > 0$.

When $\beta = 0$, we obtain the estimate $x(t) \leq x(t-1) + K$, for $t \geq 0$ and the proof proceeds as before. This completes the proof of the lemma. ∎

To simplify notation, we write $\mathcal{A}$ for $\mathcal{A}_\beta$. We now describe an interesting Morse decomposition, due to Mallet-Paret, of the flow on the attractor $\mathcal{A}$ defined by the semigroup $T(t)$ corresponding to (10.8). To do this, it is convenient to think of the flow on $\mathcal{A}$ in the following way. For any $\varphi \in \mathcal{A}$, we know that $T(t)\varphi \in \mathcal{A}$ for all $t \in \mathbb{R}$. Since $(T(t)\varphi)(\theta) = (T(t+\theta)\varphi)(0)$ for $\theta \in [-1, 0]$, the orbit $T(t)\varphi$, $t \in \mathbb{R}$, can be identified with the function $x(t, \varphi) = (T(t)\varphi)(0)$, $t \in \mathbb{R}$. With this observation, we define

$$\tilde{A} = \{x \in C(\mathbb{R}, \mathbb{R}) : x_t \in \mathcal{A}, \; t \in \mathbb{R}\} \tag{10.9}$$

and endow $\tilde{A}$ with the compact open topology and it has the structure of a compact metric space (see [137]).

For any $x \in \tilde{A}$ and any $t \in \mathbb{R}$ the function $(x \cdot t)(\theta) \equiv x(t + \theta), \theta \in \mathbb{R}$ belongs to $\tilde{A}$. We refer to $x \cdot t$ as the *translational flow* on $\tilde{A}$ and use this particular notation in order not to confuse it with the notation x_t for elements of C.

Definition 10.5.2. *If $x \in \tilde{A} \setminus \{0\}$, define*

$$V(x) = \begin{cases} \text{the number of zeros, counting multiplicity, of } x(t) \text{ in } (\sigma - 1, \sigma] \\ 1 \text{ if no } \sigma \text{ exists} \end{cases}$$

where

$$\sigma = \inf\{t \geq 0 : x(t) = 0\}.$$

We will refer to V either as the *Lyapunov function* for (10.8) or as the *zero number of* $x \in \tilde{A} \setminus \{0\}$. The following important result is due to Mallet-Paret [137].

Theorem 10.5.3.

(i) $V(x \cdot t)$ is nonincreasing in t for each $x \in \tilde{A} \setminus \{0\}$.

(ii) $V(x)$ is an odd integer for each $x \in \tilde{A} \setminus \{0\}$.

(iii) There is a constant K such that $V(x) \leq K$ for all $x \in \tilde{A} \setminus \{0\}$.

We give only some of the intuitive ideas of why Theorem 10.5.3 is true. The complete proof requires several nontrivial observations and the interested reader should consult [137]. Let us first suppose that $x \in \tilde{A}$ has only simple zeros. Let $\sigma_0 < \sigma_1$ be consecutive zeros of $x(t)$ with $x(t) > 0$ for $\sigma_0 < t < \sigma_1$. Then $\dot{x}(\sigma_0) > 0$ and $\dot{x}(\sigma_1) < 0$. From the negative feedback condition (H₁), it follows that $x(\sigma_0 - 1) < 0$ and $x(\sigma_1 - 1) > 0$. Thus, $x(t) = 0$ at some point in $(\sigma_0 - 1, \sigma_1 - 1)$; that is, $x(t)$ can have no more zeros in $(\sigma_1 - 1, \sigma_1]$ than it does in $(\sigma_0 - 1, \sigma_0]$. This shows that $V(x)$ is nonincreasing in t. Again, if we assume that the zeros of $x(t)$ are simple, then $x(\sigma) = 0$, $\dot{x}(\sigma) > 0$ (resp. $\dot{x}(\sigma) < 0$) imply that $x(\sigma - 1) < 0$ (resp. $x(\sigma - 1) > 0$), which in turn implies that the number of zeros of $x(t)$ in $(\sigma - 1, \sigma]$ is odd.

If the zeros of $x(t)$ are not simple, one first proves that $V(x) < \infty$ for $x \in \tilde{A} \setminus \{0\}$. This requires several technical estimates. Also, if $x \in \tilde{A} \setminus \{0\}$ has a zero of order exactly k at $t = \sigma$, then it is easy to see that $t = \sigma - 1$ is a zero of exactly order $k - 1$ and $D^{k-1}x(\sigma - 1)D^k x(\sigma) < 0$. The proofs of (i) and (ii) are completed by noting sign changes near the zeros.

The proof of property (iii) in the theorem is more difficult. We have already remarked that $V(x) < \infty$ for any $x \in \tilde{A} \setminus \{0\}$. By a careful analysis of sign changes, one can show that, if $x^n \in \tilde{A} \setminus \{0\}$ and $V(x^n) \geq N$ for all n, then $x^n \to x \neq 0$ in $\tilde{A}$ implies that $V(x) \geq N$. From this fact, we see that V is bounded in the exterior of any neighborhood of 0.

The most difficult part of the proof of the theorem is to analyze the behavior of the solutions in $\tilde{A}$ near the origin. Since we are assuming that the origin is hyperbolic (H_3), each solution of the characteristic equation

$$\lambda + \beta + g'(0)e^{-\lambda} = 0, \tag{10.10}$$

has nonzero real parts.

A detailed analysis of the solutions of (10.10) is necessary to complete the proof. It can be shown that the eigenvalues of (10.10) occur in complex conjugate pairs $\mu_j \pm i\nu_j$, $j \geq 1$, with $\mu_{j+1} < \mu_j$ and $2j\pi < \nu_k < (2j + 1)\pi$ for all j. This implies that the number of zeros of the eigenfunctions in a unit interval increases as the real part of the eigenvalue decreases. It also implies that the number of eigenvalues with positive real parts is even, say N^*. The dimension of the unstable manifold $W^u(0) \subset C$ of the origin is N^*. The stable manifold $W^s(0) \subset C$ of 0 has codimension N^*.

From the above mentioned properties of the eigenfunctions of (10.10), the finite dimensionality of $W^u(0)$ implies that, for any $x \in \tilde{A}$ for which $x_t \in C$ remains in a small neighborhood of the origin for $t \leq 0$, we must have $x_t \in W^u(0)$ for $t \leq 0$. Furthermore, $\dot{x}_t$ should approach the origin as $t \to -\infty$ along an eigenspace of the linear equation. Therefore, the number of sign changes in any unit interval should be $\leq N^*$; that is, $V(x) \leq N^*$. Since $V(x)$ is odd for $x \in \tilde{A}$, we must have $V(x) > N^*$.

Analogously, if $x \in \tilde{A}$ and $x_t \in C$, $t \geq 0$, remains in a small neighborhood of the origin, then we must have $x_t \in W^s(0)$ for $t \geq 0$. Again, one would expect that the tangent to the orbit of this solution would approach the origin along one of the eigenspaces of the linear equation. If this were the case, then we would have $V(x) \geq N^*$. Since $V(x)$ is odd, we would have $V(x) > N^*$. However, since $W^s(0)$ is infinite dimensional, this is not obvious and it is conceivable that there is a solution which approaches zero faster than any exponential (the so called small solutions). It is true that no such solutions exists, but the proof is far from trivial.

We are now in a position to define a Morse decomposition of the space $\tilde{A}$. It is tempting to consider, for each odd integer N, the following sets as part of a Morse decomposition: $\{x \in \tilde{A} : V(x \cdot t) = N$ for all $t \in \mathbb{R}\}$. However, this will not work because the function V is not defined at the origin and these sets in general are not closed. In fact, several of them may contain the point 0 in their closure. The definition must be refined to keep the orbits away from the origin.

The fact that we are assuming that the origin is hyperbolic allows us to define the Morse decomposition in the following way. Let N^* be the dimension of the unstable manifold of the origin and define $\mathcal{A}_{N^*} = \{0\}$. For any odd integer N, define

$$\mathcal{A}_N = \{x \in \tilde{A} : V(x \cdot t) = N \text{ for } t \in \mathbb{R} \text{ and } 0 \notin \alpha(x) \cup \omega(x)\}.$$

With this definition, the sets $\mathcal{A}_N$ for N odd are compact and do not contain the origin. We remark that $\mathcal{A}_N = \emptyset$ for large N by (iii) of 10.5.3.

In [137] it is proved the following result.

Theorem 10.5.4. . *Assume the origin is hyperbolic. Then the sets $\mathcal{A}_N$, $N \in \{N^*, 1, 3, 5, \ldots\}$, form a Morse decomposition of $\tilde{A}$ with the ordering $\mathcal{A}_N < \mathcal{A}_K$ if and only if $K < N$.*

Further properties also are known about the Morse sets $\mathcal{A}_N$. In particular, for N an odd integer, if $x \in \tilde{A}$, then the zeros of x are simple. This allows one to prove that each $\mathcal{A}_N$ for $N < N^*$ is not empty and contains a periodic orbit $x_N(t)$ with least period τ satisfying $2/N < \tau < 2/(N-1)$ and $x_N(t)$ has exactly two zeros in $[0, \tau)$. In particular, if $N^* \geq 2$, then $\mathcal{A}_1$ contains a periodic orbit with exactly two zeros in $[0, \tau)$. Compare this remark with the remarks above about Wright's equation.

The proof of this last fact uses a special type of Poincaré map. Consider the map $\Theta : \mathcal{A}_N \to S^1$ from the Morse set $\mathcal{A}_N$ to the unit circle S^1 in the plane with center 0, induced by the map

$$x \in \mathcal{A}_N \mapsto (x(0), \dot{x}(0)) \in \mathbb{R}^2 \setminus \{0\}.$$

From the properties mentioned above, the image of the orbit winds around the circle infinitely often as $t \to \pm\infty$. In particular, it has a transversal cross section; namely, the half line $x = 0$, $\dot{x} > 0$ in $\mathbb{R}^2 \setminus \{0\}$ and has a corresponding Poincaré map. It is this map that is used to prove the existence of the periodic solutions mentioned before.

The existence of the Morse decomposition in Theorem 10.5.4 does not imply that the flow on the attractor $\mathcal{A}$ is simple. In fact, numerical evidence and some theoretical results indicate that there can be chaotic dynamics in some of the Morse sets.

In spite of the complexity that may exist in the dynamics of equations with negative feedback, using the Conley index and concepts of semiconjugacy of flows, McCord and Mischaikow [141] proved the following result about orbits connecting the Morse sets.

Theorem 10.5.5. $C_K^N \neq \emptyset$ *for any* $N > K$.

The main idea in [141] is to extend the results above to flows for which the dynamics on the global attractor is adequately described, via semiconjugacy, onto the dynamics of a Morse–Smale system of ordinary differential equations (see [160]) which is common to some monotone cyclic feedback systems. The exact flow considered is a model flow on the disk D^{2P}, the closed unit ball in $\mathbf{R}^{2P}$. Let $z = (z_0, \ldots, z_{2P-1}) \in \mathbf{R}^{2P}$ be written in polar coordinates $z = r\zeta$, where $r \geq 0$ and $\zeta \in S^{2P-1}$ the unit sphere in $\mathbf{R}^{2P}$. Then, the model flow $\psi : \mathbf{R} \times D^{2P} \rightarrow D^{2P}$ is generated by the equations $\dot{r} = r(r - 1)$, and $\dot{\zeta} = A\zeta - \langle A\zeta, \zeta \rangle \zeta$ (the projection of the linear flow $\dot{z} = Az$ onto the unit sphere), where $A : \mathbf{R}^{2P} \rightarrow \mathbf{R}^{2P}$ is given by a matrix of the form $A = \mathrm{diag}(A_1, \ldots, A_P)$ with $A_p = \begin{bmatrix} p^{-1} & 2\pi \\ -2\pi & p^{-1} \end{bmatrix}$, $p = 1, 2, \ldots, P$. This flow is Morse–Smale with the following properties:

(i) the origin $0 = \Pi(P)$ is a fixed point with a $2P$ dimensional unstable manifold $W^u(0)$ and $\mathrm{cl}(W^u(0)) = D^{2P}$;

(ii) for each $p = 0, \ldots, P - 1$, the set $\Pi(p) := \{z \in D^{2P} | z_{2p}^2 + z_{2p+1}^2 = 1\}$ is a periodic orbit with period 1 and $\mathrm{cl}(W^u(\Pi(p)))$ is the $(2p + 1)$-sphere $\{z | \sum_{i=0}^{2p+1} z_i^2 = 1\}$; and,

(iii) $\mathcal{M}(D^{2P}) := \{\Pi(p) | p = 0, \ldots, P\}$ is a Morse decomposition of D^{2P} with ordering $0 < 1 < \cdots < P$.

It is proved that any flow satisfying a set of five assumptions given below (shown to be naturally satisfied by the scalar delay differential equation with negative feedback, considered above) must posses as minimal dynamics on the attractor the same structure of the flow ψ.

Let $\mathcal{A}$ denote the global attractor of a semi-flow Φ on a Banach space. Then, the five assumptions are the following:

(A1) the restriction φ of the semi-flow Φ to $\mathcal{A}$ is a flow on $\mathcal{A}$;

(A2) The collection $\mathcal{M}(\mathcal{A}) = \{M_p | p = 0, \ldots, P\}$ with ordering $0 < 1 < \cdots < P$ is a Morse decomposition of $\mathcal{A}$, under the flow $\varphi : \mathbf{R} \times \mathcal{A} \rightarrow \mathcal{A}$;

(A3) for each $p = 0, \ldots, P - 1$, M_p has a Poincaré section $\Pi(p)$ defined on a neighborhood of M_p;

(A4) the cohomology Conley indices of the Morse sets are

$$C\check{H}^k(M_P, \mathbf{Z}) \approx \begin{cases} \mathbf{Z} \text{ if } k = 2P \\ 0 \text{ otherwise} \end{cases}$$

and for $p = 0, \ldots, P - 1$

$$C\check{H}^k(M_p, \mathbf{Z}) \approx \begin{cases} \mathbf{Z} \text{ if } k = 2p, 2p + 1 \\ 0 \text{ otherwise} \end{cases} \quad ;$$

(A5) for each M_p, $p < P$, there is a continuation of the flow in a neighborhood of M_p (preserving the Poincaré section) to an isolated invariant set which consists of the disjoint union of a hyperbolic periodic orbit and a set with trivial Conley index.

Then, under the above assumptions for the flow φ on $\mathcal{A}$, there exist a continuous surjective function $f : \mathcal{A} \to D^{2P}$ for which $M_p = f^{-1}(\Pi(p))$, $p = 0, \ldots, P$, and a continuous flow $\tilde{\varphi} : \mathbf{R} \times \mathcal{A} \to \mathcal{A}$ obtained by an order preserving time reparametrization of φ such that we have the following semiconjugacy:

$$\psi \circ (id \times f) = f \circ \tilde{\psi}.$$

As pointed out by McCord and Mischaikow in [141], this simple picture corresponds to the collapsing of any non-trivial recurrent dynamics in each Morse set onto a simple periodic orbit, which, in this way, is the minimal cohomological complexity of the invariant set.

The proof uses the Conley index theory and proceeds along the natural steps: (a) define a coordinate system on $\mathcal{A}$ reparametrizing the flow φ with the help of a Lyapunov function and construct a map $\tilde{f}$ (not necessarily continuous) between $\mathcal{A}$ and a certain set X; (b) define an equivalence relation $\sim$ on X in such way that the induced quotient space $X/\sim$ is homeomorphic to D^{2P}, the induced map $f : \mathcal{A} \to D^{2P}$ is continuous and the induced flow on D^{2P} is conjugate to the flow ψ; and (c) show that f is surjective. In this last step the information on the connecting homomorphisms of the attractor repeller pairs is used to establish the desired lower bounds on the cohomological complexity of the global attractor.

11 The Kupka–Smale Theorem

The aim of the generic theory of differential equations is to study qualitative properties which are typical of the class of equations considered, in the sense that they hold for all equations defined by functions of a residual set of the function space being considered. More precisely, if X is a complete metric space, then a property $\mathcal{P}$ on the elements $x \in X$ is said to be *generic* if there is a residual set $Y \subset X$ such that each element of Y has property $\mathcal{P}$. Recall that a residual set is a countable intersection of open dense sets. As for ordinary differential equations, the constant and the periodic solutions, and their stable and unstable manifolds, play an important role in the generic theory of RFDE.

Given an RFDE(F) on a manifold M, we say that a constant function $\varphi \in C^0(I, M)$ is a *critical point* or an *equilibrium point* of F, if the solution of F with initial data φ is constant, i.e., $F(\varphi) = 0$. A critical point φ of F is said to be *nondegenerate* if zero is not a characteristic value of the linear variational equation of F at φ; φ is said to be hyperbolic if there is no characteristic value of the linear variational equation having real part equal to zero. Locally, the RFDE(F) on M can be identified with an equation on an Euclidean space, and there exist manifolds $W^s_{\text{loc}}(\varphi)$ and $W^u_{\text{loc}}(\varphi)$—the *local stable manifold* and *local unstable manifold* of F at φ—which have the property that, for some $\varepsilon_0 > 0$ and all balls $B_\varepsilon(\varphi) = \{\psi \in C^0 : \text{dist}(\varphi, \phi) \leq \varepsilon\}$, $0 < \varepsilon < \varepsilon_0$ they consist of the points in orbits of F which stay in $B_\varepsilon(\varphi)$ for all $t \geq 0$ or $t \leq 0$, respectively. The manifolds $W^s_{\text{loc}}(\varphi)$ and $W^u_{\text{loc}}(\varphi)$ are "tangent" at φ to linear manifolds S and U which decompose, as a direct sum, the phase space of the linear variational equation of F. The dimension of $W^u_{\text{loc}}(\varphi)$ and U is finite. The solutions $x(t)$ of F with initial data in W^u_{loc} are defined for all $-\infty < t < 0$, and the union of the complete orbits having initial data in $W^u_{\text{loc}}(\varphi)$ defines in some cases a manifold $W^u(\varphi)$ called the *global unstable manifold of F at φ*. The flow of F on the finite dimensional manifold $W^u_{\text{loc}}(\varphi)$ can be associated with an ordinary differential equation.

The concepts of nondegeneracy and hyperbolicity can also be defined for periodic solutions of an RFDE(F). If $p(t)$ is a nonconstant ω-periodic solution of F, by compactness of the intervals $[t_0, t_0 + T]$, the RFDE(F) on M can be identified, locally around $p(t)$, with an equation on Euclidean space $\mathbb{R}^n$. One can then consider the linear variational equation relative to $p(t)$. This equation is a linear periodic system of period ω, having $p(t)$ as one of its solutions. It follows that $\mu = 1$ is a characteristic multiplier of the linear variational equation relative to $p(t)$. We say the *periodic orbit* $\Gamma = \{p(t), t \in \mathbb{R}\}$ is *nondegenerate* if the characteristic multiplier $\mu = 1$ is simple and we say the *periodic orbit* Γ is *hyperbolic* if it is nondegenerate and $\mu = 1$ is the only characteristic multiplier with $|\mu| = 1$.

The Kupka–Smale theorem for ordinary differential equations, perhaps the most basic result of generic theory, asserts that the property that all critical points and periodic orbits are hyperbolic and the stable and unstable manifolds intersect transversaly is generic in the class of all ordinary differential equations $\dot{x} = f(x)$, $x \in \mathbb{R}^n$ or $x \in M$ (M a compact submanifold of $\mathbb{R}^n$) for which f is smooth in an adequate topology. The complete proof of the Kupka–Smale theorem for RFDE is not presently available, but some results in this direction are known.

The first generic results for RFDE were established for equations defined on a compact manifold M, proving that the sets G_0^k and G_1^k of all RFDE in $\mathcal{X}^k(I, M)$ which have all critical points nondegenerate and hyperbolic, respectively, are open and dense in $\mathcal{X}^k(I, M)$, $k \geq 1$, and the sets $G_{3/2}^k(T)$ and $G_2^k(T)$ of all RFDE in $\mathcal{X}^k(I, M)$ for which all nonconstant periodic solutions with period in $(0, T]$ are nondegenerate and hyperbolic, respectively, are open in $\mathcal{X}^k(I, M)$, $k \geq 1$. For the case of RFDE on $\mathbb{R}^n$ it is known that the set of all RFDE in a convenient class of functions $\mathcal{X}$ which have all critical points and all periodic orbits hyperbolic is a residual set in $\mathcal{X}$. These results are described below in detail, since they illustrate the techniques used in the generic theory of RFDE.

The proof follows the general pattern that was developed for ordinary differential equations. We consider RFDE on $\mathbb{R}$, defined by

$$\dot{x}(t) = f(x_t) \tag{11.1}$$

with $f \in \mathcal{X} = \mathcal{X}^k(I, \mathbb{R}^n)$, $k \geq 2$, and taking $\mathcal{X}^k$ with the C^k-uniform topology. For each compact set $K \subset \mathbb{R}^n$ and each $A > 0$ define the subsets of

$\mathcal{G}_0(K) = \{f : \text{all critical points in } K \text{ are nondegenerate}\}$

$\mathcal{G}_1(K) = \{f : \text{all critical points in } K \text{ are hyperbolic}\}$

$\mathcal{G}_{3/2}(K, A) = \{f : \text{all periodic orbits lying in } K \text{ and with period in } (0, A] \text{ are nondegenerate}\}$

$\mathcal{G}_2(K, A) = \{f \in \mathcal{G}_1 : \text{all periodic orbits lying in } K \text{ and with period in } (0, A] \text{ are hyperbolic}\}$.

Theorem 11.0.1. *The set of all $f \in \mathcal{X}$ such that all critical points and all periodic orbits of (11.1) are hyperbolic is residual in $\mathcal{X}$.*

Proof: We break the proof in several steps:

1) the sets $\mathcal{G}_0(K)$, $\mathcal{G}_1(K)$, $\mathcal{G}_{3/2}(K,A)$, $\mathcal{G}_2(K,A)$ are open.

This is a consequence of general perturbation results associated with the saddle-point property.

2) $\mathcal{G}_0(K)$ is dense in $\mathcal{X}$.

Any $f \in \mathcal{X}$, by restriction to the constant functions in C^0, gives a C^k function $\bar{f} : \mathbb{R}^n \to \mathbb{R}^n$. It is easily seen that $\varphi \in K$ is a critical point of (11.1) if and only if the origin of $\mathbb{R}^n$ is a regular value of the restriction of $\bar{f}$ to a compact set $\bar{K} \subset \mathbb{R}^n$. By Sard's theorem, the set of singular values of $\bar{f}$ has measure zero, so there are regular values $\varepsilon \in \mathbb{R}^n$ of $\bar{f}$ arbitrarily close to zero. Letting $G : \mathbb{R}^n \to \mathbb{R}^n$ be a C^∞-function with compact support and equal to 1 on $\bar{K}$, and $g_\varepsilon : C^0 \to \mathbb{R}^n$ be defined by $g_\varepsilon(\varphi) = f(\varphi) - \varepsilon G(\varphi(0))$, we get $g_\varepsilon \in \mathcal{X}$ and $\bar{g}_\varepsilon = \bar{f} - \varepsilon G$. Consequently, $g_\varepsilon \in \mathcal{G}_0(K)$ and, therefore, $\mathcal{G}_0(K)$ is dense in $\mathcal{X}$.

3) $\mathcal{G}_1(K)$ is dense in $\mathcal{G}_0(K)$.

Take $f \in \mathcal{G}_0(K)$. Each zero of $\bar{f}$ in $\bar{K}$ is isolated and, since $\bar{K}$ is compact, the zeros of $\bar{f}$ are finitely many. By the Implicit Function Theorem, these zeros persist under small perturbations of f and no other new zeros of $\bar{f}$ appear in some neighborhood of $\bar{K}$. If we can perturb f locally around each critical point, by adding to f a function having support in a small neighborhood of the point, in such a way that the associated critical point of the perturbed equation is hyperbolic, then we can construct perturbations of f which have the same number of critical points on K as f does, but with all of them hyperbolic. This would imply that $\mathcal{G}_1(K)$ is dense in $\mathcal{G}_0(K)$.

To show that such local perturbations exist, let $a \in \bar{K}$ be a zero of $\bar{f}$ and change coordinates so that $a = 0$. Let $H : \mathbb{R}^n \times \mathbb{R}^n \to \mathbb{R}^n$ be C^∞ with arbitrarily small compact support and $H(0,0) = 0$, $DH(0,0) = (0,I)$. For $\varepsilon \in \mathbb{R}$, let $L_\varepsilon : C^0 \to \mathbb{R}^n$ be $L_\varepsilon(\varphi) = -\varphi(0) + [(e^\varepsilon - 1)/\varepsilon]f'(0)\varphi$ and define $g(\varphi) = f(\varphi) - \varepsilon H(\varphi(0), L_\varepsilon(\varphi))$. Then, as $\varepsilon \to 0$, $g_\varepsilon \to f$ in $\mathcal{X}$, and the characteristic function $\Delta_\varepsilon(\lambda)$ of the linearized equation at zero satisfy $\Delta_\varepsilon(\lambda) = \Delta_0(\lambda + \varepsilon)$. For all $\varepsilon \neq 0$ small, Δ_0 has no zeros on $\mathrm{Re}\,\Delta = \varepsilon$. Thus 0 is an hyperbolic critical point of $\dot{x}(t) = g_\varepsilon(x_t)$.

4) $\mathcal{G}_{3/2}(K, 3A/2)$ is dense in $\mathcal{G}_2(K,A)$.

The main idea for proving this statement is to consider, for each $f \in \mathcal{G}_2(K,A)$, perturbations on a conveniently chosen finite-dimensional subspace of $\mathcal{X}$. The elements of this subspace are taken from the set

$$\mathcal{Y} = \Big\{ f \in \mathcal{X} : f(\varphi) = F\big(\varphi(0), \varphi(-r/N), \varphi(-2r/N), \ldots, \varphi(-r)\big)$$

$$\text{for some } N \text{ and some } F \in C^k(\mathbb{R}^{n(N+1)}, \mathbb{R}^n) \Big\}.$$

Fix $f \in \mathcal{G}_2(K,A)$ and $g_1, \ldots, g_J \in \mathcal{Y}$. For $\eta = (\eta_1, \ldots, \eta_J) \in \mathbb{R}^J$, let $f_\eta = f + \sum_{j=1}^J \eta_j g_j$, and denote by $x(t; \varphi, \eta)$ the solution of the initial value problem

$$\dot{x}(t) = f_\eta(t), \quad x_0 = \varphi. \tag{11.2}$$

Consider the map $\Psi : (0,\infty) \times C^0 \times \mathbb{R}^J \to C^0$ given by $\Psi(t,\varphi,\eta) = x_t(\varphi,\eta) - \varphi$. Clearly, the zeros of Ψ correspond to initial data of periodic solutions of (11.2). Let $x^*(t)$ be a nonconstant periodic solution of (11.1) lying in K, having period $t^* \in (0, 3A/2]$ and nonconstant initial conditions $\varphi^* = x_0^*$. Then $\Psi(t^*,\varphi^*,0) = 0$. The Implicit Function Theorem cannot be applied to Ψ, since this map may fail to be differentiable at (t,φ,η) if $\varphi \notin C^1$. For this reason, we introduce, for each integer N, the map $\Psi_N(t,\varphi,\eta) = \Psi(Nt,\varphi,\eta)$, which is differentiable at (t,φ,η) provided N is sufficiently large (see Th. 3.1.2). By application of the Implicit Function Theorem to functions Ψ_N, for conveniently chosen N, we get the following lemma.

Lemma 11.0.2. *If $D\Psi(t^*,\varphi^*,0)$ is surjective, then there is a neighborhood U of $(t^*,\varphi^*,0)$ in $(0,\infty) \times C^0 \times \mathbb{R}^J$ such that $M = \Psi^{-1}(0) \cap U$ is a C^2-manifold. At each point $(t,\varphi,\eta) \in M$, $D\Psi(t,\varphi,\eta)$ is surjective, and the tangent space of M is the null space of $D\Psi(t,\varphi,\eta)$.*

Proof: (of Lemma 11.0.2) Let $\Lambda = D_\varphi x_{t^*}^*(\varphi^*,0)$, $\Gamma = D_\eta x_{t^*}^*(\varphi^*,0)$ and notice that Λ can be defined by the solution map of the linear variational equation of (11.1) at the periodic solution $x^*(t)$

$$\dot{y}(t) = f'(x_t^*)y_t, \quad y_0 = \xi \tag{11.3}$$

as

$$\Lambda\xi = y_{t^*}. \tag{11.4}$$

A straightforward computation gives

$$D\Psi_N(t^*,\varphi^*,0)(s,\psi,\sigma) = \left(\sum_{i=1}^{N-1} \Lambda^i\right) (\dot{\varphi}^*s + (\Lambda - I)\psi + \Gamma\sigma). \tag{11.5}$$

Since Λ is defined by (11.3)–(11.4), some power of Λ is a compact operator. There are finitely many points of norm one in the spectrum $\sigma(\Lambda)$ of Λ. Therefore, there exist relatively prime positive integers N_1, N_2 such that $\exp(2\pi ik/N_3) \notin \sigma(\Lambda)$ for all $0 < k < N_3 = N_1 N_2$. Then

$$0 \notin \sigma\left(\sum_{i=0}^{N_j-1} \Lambda^i\right) \quad \text{for } j = 1,2,3$$

and therefore, these operators are isomorphisms. Thus $D\Psi_{N_j}(t^*,\varphi^*,0)$, $j = 1,2,3$ are surjective and the null spaces of $D\Psi_{N_j}(t^*,\varphi^*,0)$, $j = 1,2,3$ and of $D\Psi(t^*,\varphi^*,0)$ are equal. Let P and Q, $P + Q = I$, be the usual spectral projections onto Λ-invariant subspaces, where $\Lambda - I$ is nilpotent on the finite-dimensional P space, and has an inverse L on the Q space. Noting that $\Lambda\dot{\varphi}^* = \dot{\varphi}^*$ and $P\dot{\varphi}^* = \dot{\varphi}^*$, we see that $(s,\psi,\sigma) \in \text{null}(D\Psi)$ is equivalent to the following system

$$\dot\varphi^* s + (\Lambda - I)P\psi + P\Gamma\sigma = 0 \tag{11.6}$$

$$Q\Psi = -LQI\,\sigma. \tag{11.7}$$

The only independent parameters in this system are s, $P\psi$ and σ, which are all finite-dimensional. Thus $\text{null}(D\Psi) = \text{null}(D_{N_j}\Psi)$, $j = 1, 2, 3$ is finite dimensional, and, consequently, has a closed complement. Since $N_j > 2/t^*$, $j = 1, 2, 3$, it follows that Ψ_{N_j} is C^2 at $(t^*, \varphi^*, 0)$. By the Implicit Function Theorem, there is a neighborhood U of $(t^*, w^*, 0)$ such that $M_j = \Psi_{N_j}^{-1}(0) \cap U$ is a C^2-manifold, and at each $(t, \varphi, \eta) \in M_j$ the map $D\Psi_{N_j}$ is surjective, and the tangent space of M_j is the null space of $D\Psi_{N_j}(t, \varphi, \eta)$. Any solution of period $N_j t$ ($j = 1, 2$) is also of period $N_3 t$, so $M_j \subset M_3$, $j = 1, 2$. Furthermore, the tangent space of M_j at $(t^*, \varphi^*, 0)$ is independent of $j = 1, 2, 3$. So, by restriction to a smaller neighborhood U, if necessary, we get $M_1 = M_2 = M_3 \overset{\text{def}}{=} M_0$. Clearly, $\Psi^{-1}(0) \cap U \subset M_0$. On the other hand, $(t, \varphi, \eta) \in M_0$ is associated with a solution of (11.2) with periods $N_1 t$ and $N_2 t$ and, since N_1, N_2 are relatively prime, this solution has period t and $(t, \varphi, \eta) \in \Psi^{-1}(0) \cap U$. This proves $\Psi^{-1}(0) \cap U = M_0$ is a C^2-manifold. The assertions in Lemma 11.0.2 about surjectivity and tangent space follow from (11.5), evaluated at (t, φ, η), by noting that $\sum_{i=0}^{N_j-1}(D_\varphi x_y(\varphi, \eta))^i$, $j = 1, 2$ are isomorphisms for $(t, \varphi, \eta) \in M_0$, provided U is taken sufficiently small. This finishes the proof of Lemma 11.0.2. ∎

On the basis of Lemma 11.0.2, we now need to prove that $D\Psi(t^*, \varphi^*, 0)$ is surjective. If $x^*(t)$ is a nondegenerate periodic solution of (11.1) with period t^*, then, the map $(s, \psi) \to \varphi^* s + (\Lambda - I)\psi$ from $(0, \infty) \times C^0$ into C^0 is surjective. From (11.5) with $N = 1$, it follows that $D\Psi(t^*, \varphi^*, 0)$ is surjective. On the other hand, if $x^*(t)$, with period t^*, is degenerate, then $f \in \mathcal{G}_2(K, A)$ implies $t^* \in [A, 3A/2]$ and, consequently t^* is the least period of $x^*(t)$. The following lemma, whose proof is postponed, guarantees that $D\Psi(t^*, \varphi^*, 0)$ is surjective for some choice of $g_1, \dots, g_J \in \mathcal{Y}$.

Lemma 11.0.3. *If $\bar{t}$ is the least period of a periodic solution of (11.1) through $\bar\varphi$, then there exist $g_1, \dots, g_J \in \mathcal{Y}$ such that $D\Psi(\bar{t}, \bar\varphi, 0)$ is surjective.*

Lemma 11.0.2 can then be applied at each point (t, φ) in the set $F(0)$, where

$$F(\eta) = \{(t, \varphi); 0 < t \le 3A/2, \varphi \text{ is not constant,}$$

$$x(s; \varphi, \eta) \in K \text{ for all } s \in [0, 3A/2] \text{ and } \Psi(t, \varphi, \eta) = 0\}.$$

J and $g_1, \dots, g_J \in \mathcal{Y}$ are possibly different for different points (t, φ). To remove this dependence on (t, φ), notice that, given $f \in \mathcal{G}_0(K)$, there exists a neighborhood $\mathbb{N}$ of f such that the periods of nonconstant periodic solutions of $\dot{x}(t) = g(x_t)$, $g \in \mathbb{N}$ lying in K are bounded below by some $\varepsilon > 0$, and, then, observe that $F(\eta)$ is compact. By compactness of $F(0)$, one can find finitely many $g_1, \ldots, g_J \in \mathcal{Y}$ such that $D\Psi(t, \varphi, 0)$ is surjective for all $(t, \varphi) \in F(0)$. Lemma 11.0.2 then implies there exists a neighborhood U of $F(0) \times \{0\}$ in $(0, \infty) \times C^0 \times \mathbb{R}^J$ such that the conclusions of that lemma hold for $M = \Psi^{-1}(0) \cap U$.

Now, consider the projection $\pi : M \to \mathbb{R}^J$ given by $\pi(t, \varphi, \eta) = \eta$. Since the tangent space of M at (t, φ, η) is equal to the null space of $D\Psi(t, \varphi, \eta)$, and formula (11.5) holds for $N = 1$ at each $(t, \varphi, \eta) \in \Psi^{-1}(0)$ (replacing φ^* by φ and Λ and Γ by the corresponding derivatives computed at (t, φ, η) instead of $(t^*, \varphi^*, 0)$) we get

$$\text{null}(D\pi) = \{(s, \psi, 0) \in \mathbb{R} \times C^0 \times \mathbb{R}^J : \dot{\varphi}s + (\Lambda - I)\psi = 0\}$$

$$\text{range}(D\pi) = \{\sigma \in \mathbb{R}^J : \dot{\varphi}s + (\Lambda - I)\psi + \Gamma\sigma = 0$$

$$\text{for some } (s, \psi) \in \mathbb{R} \times C^0\}.$$

The reasoning leading to (11.6)–(11.7) is also valid in the present situation and we can compute the dimensions of $(D\pi)$ and range$(D\pi)$ by studying the finite-dimensional equation (11.6) with the use of the surjectivity of the map $(s, P\psi, \sigma) \to \dot{\varphi}_s + (\Lambda - I)P\psi + P\Lambda\sigma$. It is then possible to show that the Fredholm index of $D\pi$, dim $(D\pi) - \text{codim range}(D\pi)$, is equal to one. Since M and π are of class C^2, we can now apply Smale's version of Sard's theorem to get that the set of regular values of π is residual in $\mathbb{R}^J$. In particular, there are regular values arbitrarily near zero. On the other hand, the upper semicontinuity of $F(\eta)$, guarantees that $F(\eta) \times \{\eta\} \subset U$ for sufficiently small η. For such η which are regular values of π, we have that any solution of (11.2) of period $\leq 3A/2$, lying in K, must correspond to a point $(t, \varphi, \eta) \in M$. Since $D\Psi$ is surjective at points of M and η is a regular value of π implies $D\pi(t, \varphi, \eta) : (D\Psi(t, \varphi, \eta)) \mapsto \mathbb{R}^J$ is also surjective. Thus $(s, \psi) \to D\Psi(t, \varphi, \eta)(s, \psi, 0)$ is surjective, implying that the solution of (11.2) with initial condition φ is a nondegenerate periodic solution with period t. This finishes the proof that $\mathcal{G}_{3/2}(K, 3A/2)$ is dense in $\mathcal{G}_2(K, A)$.

5) $\mathcal{G}_2(K, A)$ is dense in $\mathcal{G}_{3/2}(K, A)$.

Fix $f \in \mathcal{G}_{3/2}(K, A)$. Each periodic solution $x^*(t)$ of (11.1) of period $t^* \leq A$ and lying in K is nondegenerate. Therefore, there exists a neighborhood of its orbit which contains no other periodic orbits of period close to t^*, and, under small perturbations of (11.1), the periodic solution and its period depend continuously on the perturbation. If N_1, N_2, N_3 are chosen as in the proof of Lemma 11.0.2 and since $x^*(t)$ is a periodic nondegenerate solution of (11.1) with any of the periods $N_j t^* > 1$, $j = 1, 2, 3$, there are unique orbits Γ_j of periods near $N_j t^*$ and changing continuously with the perturbation, for $j = 1, 2, 3$. The orbits of period near $N_j t^*$, $j = 1, 2$ also have periods near $N_1 N_2 t^* = N_3 t^*$, and, therefore, $\Gamma_1 = \Gamma_2 = \Gamma_3 \overset{\text{def}}{=} \Gamma$. Since N_1, N_2 are relatively prime we have that the period of Γ is near t^* and depends continuously on the perturbation. By compactness, there are only finitely many periodic solutions of (11.1) of periods $\leq A$ and lying in K. To prove that $\mathcal{G}_2(K, A)$ is dense in $\mathcal{G}_{3/2}(K, A)$, it is sufficient to make a small perturbation in a neighborhood of each periodic solution.

Assume $x^*(t)$ is a nondegenerate periodic solution of (11.1) with least period $t^* \leq A$. Let $y^j(t)$, $j = 1, \ldots, d$ be solutions of the variational equation of (11.1) which form a basis for the generalized eigenspace corresponding to all characteristic multipliers of $x^*(t)$ having $|\mu| = 1$. Without loss of generality we take $y1(t) = \dot{x}^*(t)$. Letting $Y(t) = \big(y^1(t), \ldots, y^d(t)\big)$, there exists a $d \times d$ matrix M with all eigenvalues in the unit circle such that $Y(t + t^*) = Y(t)M$. After changing the basis so that M is in Jordan canonical form, it is not difficult to perturb $Y(t)$ and M to a differentiable function $Y^\varepsilon(t)$ and M^ε, for ε small, so that $Y^0 = Y$, $M^0 = M$, and the eigenvalues of M^ε are all off the unit circle except for the eigenvalue 1 which is simple, and $Y^\varepsilon(t + t^*) = Y^\varepsilon(t)M^\varepsilon$. In order to perturb (11.1) as

$$\dot{x}(t) = f(x_t) + g(x_t) \tag{11.8}$$

and have the periodic solution $x^*(t)$ of (11.1) transformed to an hyperbolic periodic solution of (11.8) we can try to choose g so that $x^*(t)$ is still a solution of (11.8) and $Y^\varepsilon(t)$ is a solution of the linear variational equation of (11.8) around $x^*(t)$. It is not difficult to show that this can be accomplished choosing $g \in \mathcal{Y}$. If we denote

$$\delta_N \varphi = \big(\varphi(0), \varphi(-r/N), \varphi(-2r/N), \ldots, \varphi(-r)\big),$$

then the appropriate functions g are of the form

$$g(\varphi) = G(\delta_N \varphi),$$

with the function $G(x_1, \ldots, x_{N+1})$ satisfying

$$\dot{Y}^\varepsilon(t) = f'(x_t^*)Y_t^\varepsilon + \left[\frac{\partial G}{\partial x_1}, \ldots, \frac{\partial G}{\partial x_{N+1}} \right]_{\delta_N x_t^*} \delta_N Y_t^\varepsilon. \tag{11.9}$$

It is, therefore, enough to find a function G satisfying this equation. Assume there exist sequences $t_N \subset [0, T]$, v_N of d vectors with norm smaller or equal to 1, and $\varepsilon_N \to 0$ as $N \to \infty$ such that $Y^{\varepsilon_N}(t_N - k/N)v_N = 0$ for all $0 \leq k \leq N$. Given an arbitrary $\theta \in [-1, 0]$, there exists a sequence k_N such that $0 \leq k_N \leq N$ and $k_N/N \to \theta$ as $N \to \infty$. Taking subsequences if necessary, we get $t_N \to \tau$, $v_N \to \omega$, $\varepsilon_N \to 0$, $Y(\tau + \theta)\omega = 0$. Therefore $Y_\tau \omega = 0$ since the columns of Y_τ are linearly independent, we must have $\omega = 0$. Consequently, for N sufficiently large and ε close to zero, the columns of $\delta_N Y_t^\varepsilon$ are linearly independent for all $t \in [0, T]$. Since M^ε is nonsingular and $\delta_N Y_{t+T}^\varepsilon = (\delta_N Y_t^\varepsilon) M^\varepsilon$, the columns of $\delta_N Y_t^\varepsilon$ are linearly independent for all $t \in \mathbb{R}$, whenever N is large and ε is close to zero. As $\delta_N Y^\varepsilon$ is a matrix of dimension $n(N+1) \times d$, the equation (11.9) for the unknown $\left[\frac{\partial G}{\partial x_1}, \ldots, \frac{\partial G}{\partial x_{N+1}} \right]_{\delta_N x_t^*}$ is underdetermined for N large and ε close to zero, and we can get one particular solution by multiplying (11.9) by the Moore-Penrose generalized inverse of $\delta_N Y_t^\varepsilon$

$$(\delta_N Y_t^\varepsilon)^+ = \left[(\delta_N Y_t^\varepsilon)^T (\delta_N Y_t^\varepsilon) \right]^{-1} (\delta_N Y_t^\varepsilon)^T$$

where the superscript T denotes transpose. We get

$$\left[\frac{\partial G}{\partial x_1}, \ldots, \frac{\partial G}{\partial x_{N+1}} \right]_{\delta_N x_t^*} = \left(\dot{Y}^\varepsilon(t) - f'(x_t^*) Y_t^\varepsilon \right) (\delta_N Y_t^\varepsilon)^+ . \qquad (11.10)$$

Since we want equation (11.8) to be a local perturbation of (11.1) around x_t^*, we look for a function G of compact support, small as $\varepsilon \to 0$, vanishing over $\gamma = \{ (x^*(t), x^*(t - r/N), \ldots, x^*(-r)) : 0 \le t \le t^* \}$ and satisfying (11.10). Choosing a local tubular coordinate system $(u_1, u_2, \ldots, u_{n(N+1)})$ around γ with γ corresponding to $\{u_2 = \ldots = u_{n(N+1)} = 0\}$ and u_1 of period t^*, we must then have

$$G(u_1, 0, \ldots, 0) = 0, \quad \frac{\partial G}{\partial u_j}(u_1, 0, \ldots, 0) = \gamma_j(u_1),$$

$$j = 1, \ldots, n(N+1),$$

where $\gamma_j(u_1)$ are given by the right-hand side of (11.10). Since f is of class C^k, the γ_j are of class C^{k-1}. But, as we want $g \in \mathcal{X} = \mathcal{X}^k$, we need G to be a C^k function. We can achieve this by integral averaging, in order to recover the missing degree of smoothness, as

$$G(u_1, \ldots, u_{n(N+1)}) = \sum_{j=2}^{m} u_j \int_0^\infty \gamma_j(u_1 + v u_j) \rho(v) dv$$

where $\rho : [0, \infty) \to \mathbb{R}$ is C^∞, has compact support and satisfies $\rho(v) = 1$ for $v \in [0, 1]$. After multiplication by a C^∞ bump function of compact support and equal to 1 near γ, we get G such that the perturbation of (11.1) defined by (11.9) is a small local perturbation of (11.1) around x_t^*, with x_t^* being an hyperbolic solution of the perturbed equation (11.8). By adding such local perturbations around each one of the (finitely many) nonhyperbolic solutions of (11.1) lying in K and having periods in $[0, A]$, we get small perturbations $(f + g) \in \mathcal{G}_2(K, A)$ of f. This finishes the proof that $\mathcal{G}_2(K, A)$ is dense in $\mathcal{G}_{3/2}(K, A)$.

We are now in the situation of being able to use the induction procedure introduced by Peixoto for ordinary differential equations. Since $\mathcal{G}_2(K, 3A/2)$ is dense in $\mathcal{G}_2(K, A)$ (by 4)) and $\mathcal{G}_2(KA)$ is dense in $\mathcal{G}_{3/2}(K, A)$ (by 5)), it follows by induction that $\mathcal{G}_2(K, A)$ is dense in $\mathcal{G}_2(K, B)$ for all $B < A$. It was mentioned before that, for any $f \in \mathcal{G}_0(K)$, there exists a neighborhood $\mathbb{N}$ of f in $\mathcal{X}$ such that the periods of nonconstant periodic solutions of $\dot{x}(t) = g(x_t)$, $g \in \mathbb{N}$ lying in K are bounded below by some $\varepsilon > 0$. Thus $\mathbb{N} \subset \mathcal{G}_2(K, \varepsilon)$, implying that $\mathcal{G}_2(K, \varepsilon)$ is dense in $\mathcal{X}$ and, thus, also $\mathcal{G}_2(K, A)$ is dense in $\mathcal{X}$ for all A. Since $\mathcal{G}_0(K)$ is dense in $\mathcal{X}$ (by 2)), it follows that $\mathcal{G}_2(K, A)$ is dense in $\mathcal{X}$ for all A. The set

$$\mathcal{G}_2 = \{f \in \mathcal{X} : \text{all critical points and all periodic solutions}$$
$$\text{of (3.3) are hyperbolic}\}$$

can be expressed as a countable intersection of sets of the form $\mathcal{G}_2(K, A)$ with K compact and $A > 0$. Consequently, $\mathcal{G}_2$ is residual in $\mathcal{X}$, finishing the proof of the theorem. ∎

It remains to prove Lemma 11.0.3. For this proof, we use the following result:

Lemma 11.0.4. *Let $x^*(t)$ be a periodic solution of (11.1) of least period $t^* > 0$. Then, for sufficiently large N, the map*

$$t \mapsto \left(x^*(t), x^*(t - r/N), x^*(t - 2r/N), \ldots, x^*(t - r)\right)$$

is a one-to-one regular (that is the derivative $\neq 0$ everywhere) mapping of the reals mod t^* into $\mathbb{R}^{n(N+1)}$.

Proof: If the statement is not true, there would exist arbitrarily large N such that either: 1) there are $t1 \not\equiv t_2 (\mod t^*)$ with $x^*(t_1 - kr/N) = x^*(t_2 - kr/N)$ for all $0 \leq k \leq N$, or 2) there is t_3 with $\dot{x}^*(t_3 - kr/N) = 0$ for all $0 \leq k \leq N$. Consequently, one could find a sequence of integers $N_m \to \infty$ as $m \to \infty$ with either $t_1(N_m)$ and $t_2(N_m)$ or $t_3(N_m)$ as above, and take convergent subsequences $t_j(N_m) \to \tau_j$ as $m \to \infty$. On the other hand, for any $\theta \in [-r, 0]$, there exists a sequence $0 \leq k_m \leq N_m$ such that $-rk_m/N_m \to \theta$ as $m \to \infty$. If 1) holds and $\tau_1 \not\equiv \tau_2 (\mod t^*)$, then $x^*(\tau_1 + \theta) = x^*(\tau_2 + \theta)$ for all $\theta \in [-r, 0]$, contradicting that t^* is the least period of x^*. If 1) holds and $\tau_1 \equiv \tau_2 (\mod t^*)$, then

$$(t_1 - t_2)^{-1}\left[x^*(t_1 - kr/N) - x^*(t_2 - kr/N)\right] \to \dot{x}(\tau_1 + \theta)$$

as $m \to \infty$, and $\dot{x}(\tau_1 + \theta) = 0$ since each term in the sequence vanishes. This would imply $\dot{x}^*_{\tau_1} = 0$, a contradiction since x^* is nonconstant. Finally, if 2) holds, then $\dot{x}^*_{\tau_3} = 0$, also a contradiction. ∎

Proof: (of Lemma 11.0.3) Let $\delta_N : C^0 \to \mathbb{R}^{n(N+1)}$ denote the map

$$\delta_N \varphi = \big(\varphi(0), \varphi(-r/N), \varphi(-2r/N), \ldots, \varphi(-r)\big).$$

Then $g \in \mathcal{Y}$ is equivalent to $g(\varphi) = G(\delta_N \varphi)$ for some N and some $G \in C^k(\mathbb{R}^{n(N+1)}, \mathbb{R}^n)$ of compact support.

Suppose first that $t^* > r$. Given any $\xi \in C^0$, there is a $z \in C^0([-r, t^*], \mathbb{R}^n)$ such that $z_0 = 0$, $z_{t^*} = \xi$, and there is a $y^* \in C^1([-r, t^*], \mathbb{R}^n)$ with $y_0^* = 0$ and arbitrarily close to z, uniformly on $[-r, t^*]$. Defining $\gamma^*(t) = \dot{y}^*(t) - f'(x_t^*)y_t^*$ and applying the variation of constants formula, we can get the solution of $\dot{y}(t) = f'(x_t^*)y_t + \gamma(t)$, $y_0 = 0$ arbitrarily close to z, uniformly on $[-r, t^*]$, by choosing γ sufficiently close to γ^* in $L^1(0, t^*)$. By Lemma 11.0.4, taking N sufficiently large one may define a function G on $\{\delta_N x_t^*; t \in \mathbb{R}\}$ by $G(\delta_N x_t^*) = \gamma'(t)$ and then extend G to the whole of $\mathbb{R}^{n(N+1)}$ as a C^k function of compact support to get $\gamma(t) = g(x_t^*)$ for some $g \in \mathcal{Y}$.

Since $t^* > r$, $\varLambda = D_\eta x_{t^*}^*(\varphi^*, 0)$ is compact (see Theorem 11.0.1-4). It follows that the range of $\varLambda - I$ has finite codimension in C^0. Let $\xi_1, \ldots, \xi_J$ be a basis for a linear complement of range$(\varLambda - I)$. By the argument of the preceding paragraph, one can get the value of the solution $y^i t^*$ of $\dot{y}(t) = f'(x_t^*)y_t + g_i(x_t^*)$, $y_0 = 0$, arbitrarily close to ξ_i, by choosing $g_1, \ldots, g_J$ appropriately in $\mathcal{Y}$. The approximation of the ξ_i by the $y_{t^*}^i$ can be made so close that $y_{t^*}^1, \ldots, y_{t^*}^J$ form a basis for a linear complement of range$(\varLambda - I)$. Using the notation on the proof of Theorem 11.0.1-4 with this choice of g_j in the definition of $f_\eta(\varphi) = f(\varphi) + \sum_{j=1}^{J} \eta_j g_j(\varphi)$, it is clear that $\varGamma = D_\eta x_{t^*}^*(\varphi^*, 0)$ can be defined in terms of the solution of

$$\dot{y}(t) = f'(x_t^*)y_t + g(x_t^*), \quad y_0 = 0 \tag{11.11}$$

as

$$\varGamma \sigma = y_{t^*} \quad \text{with} \quad g = \sum_{j=1}^{J} \sigma_j g_j. \tag{11.12}$$

Thus, for $\sigma^i = (\delta_1^i, \ldots, \delta_J^i)$ with δ_j^i the Kronecker delta, we have $\varGamma \sigma^i = y_{t^*}^i$, $i = 1, \ldots, J$. It follows that the map $(\psi, \sigma) \mapsto (\varLambda - I)\psi + \varGamma \sigma$ is surjective, and therefore, by equation (11.5), $D\varPsi(t^*, \varphi^*, 0)$ is also surjective.

Now suppose $0 < t^* \le r$ and consider the problem

$$(\varLambda - I)\psi, +\varGamma \sigma = \xi \tag{11.13}$$

for $\xi \in C^0$ given. Since $\varGamma \sigma$ satisfies (11.11–11.12), and $\varLambda$ satisfies (11.3–11.4), this equation is equivalent to the system

$$\psi(t^* + \theta) - \psi(\theta) = \xi(\theta), \quad -r \le \theta \le -t^* \tag{11.14}$$
$$[(\varLambda - I)\psi + \varGamma \sigma](\theta) = \xi(\theta), \quad -t^* \le \theta \le 0. \tag{11.15}$$

The general solution of (11.14) is $\psi = \psi_0 + \psi_1$ where ψ_0 is a particular solution of the equation and $\psi_1 \in C^0$ is any function of period t^*. Fixing ψ_0, (11.15) becomes

$$[(\Lambda - I)\psi_1 + \Gamma\sigma](\theta) = \xi_1(\theta), \quad -t^* \le \theta \le 0 \qquad (11.16)$$

where $\xi_1 = \xi - (\Lambda - I)\psi_0$. Let C_p be the space of the t^*-periodic continuous functions of $[-t^*, 0]$ into $\mathbb{R}^n$, and let $L: C_p \to C([-t^*, 0], \mathbb{R}^n)$ assign to each element $\psi_1 \in C_p$ the constant function with value $\psi_1(0)$. Then

$$V = \left\{(\Lambda - I)\psi_1\big|_{[-t^*, 0]} : \psi_1 \in C^0 \text{ is } t^*\text{-periodic}\right\}$$

is equal to

$$\left\{(\Lambda - L)\psi_1\big|_{[-t^*, 0]} + (L - I)\psi_1\big|_{[-t^*, 0]} : \psi_1 \in C_p\right\}.$$

Since $(\Lambda - L)|_{[-t^*, 0]}$ is compact (for the same reason that Λ is when $t^* > 1$) and $(\Lambda - I)|_{[-t^*, 0]}$ is an isomorphism identifying C_p and

$$C_0 = \left\{\varphi \in C([-t^*, 0], \mathbb{R}^n); \varphi(-t^*) = 0\right\},$$

it follows that V has finite codimension in C_0. Noting that $\xi_1(-t^*) = 0$, we can proceed as for $t^* > 1$ to get $g_1, \ldots, g_J \in \mathcal{Y}$ such that the map $(\psi, \sigma) \mapsto (\Lambda - I)\psi + \Gamma\sigma$ is surjective, implying that $D\Psi(t^*, \varphi^*, 0)$ is also surjective and finishing the proof of the lemma. ∎

It is interesting to restrict the class of functions $\mathcal{X}$; for example, to consider only differential difference equations of the form

$$\dot{x}(t) = F\big(x(t), x(t-1)\big). \qquad (11.17)$$

To obtain a generic theorem about this restricted class of equations is more difficult since there is less freedom to construct perturbations. For example, the functions $g \in \mathcal{Y}$ used in the proof of Theorem 11.0.1 cannot be used in the present case. Nevertheless, Theorem 11.0.1 still holds for these equations. The proof of this fact follows the sane general scheme as the proof of Theorem 11.0.1, but the proofs of denseness of $\mathcal{G}_{3/2}(K, 3A/2)$ in $\mathcal{G}_2(K, A)$ and of $\mathcal{G}_2(K, A)$ in $\mathcal{G}_{3/2}(K, A)$ are very different. The role played by Lemma 11.0.4 in the construction of the perturbations of (11.1) used in the proof of the denseness of $\mathcal{G}_{3/2}(K, 3A/2)$ in $\mathcal{G}_2(K, A)$ is now played by the following lemma after approximating F by an analytic function.

Lemma 11.0.5. *If $x(t)$ is a periodic solution of Equation (11.16) of least period $t^* > 0$, and F is analytic, then the map*

$$y(t) = \big(x(t), x(t-1)\big)$$

is one-to-one and regular except at a finite number of t values in the reals mod t^*.

Proof: It can be proved that x is analytic. Thus, any self-intersection of $y(t)$ is either isolated or forms an analytic arc. In the latter case, there exists an analytic function σ defined in an interval I with $\dot{\sigma}(t) \neq 0$ and $\sigma(t) \neq t$ such that $y(t) = y(\sigma(t))$, $x(t) = x(\sigma(t))$. Thus,

$$\dot{x}(t) = F(y(t)) = F\Big(y(\sigma(t))\Big) = \dot{x}(\sigma(t)).$$

By differentiation, we get $\dot{x}(t) = x(\sigma(t))\dot{\sigma}(t)$, implying that $\dot{\sigma}(t) \equiv I$. Hence for some τ, $x(t) = x(t + \tau)$ for $t \in I$ and thus, by analyticity, for all t. Therefore A is a multiple of t^* and the lemma is proved. ∎

One may consider an even more restrictive class of equations of the form

$$\dot{x}(t) = F(x(t - 1)).$$

The analogue of Theorem 11.0.1 for this class is still an open question, since the generic properties of periodic solutions of these equations have not been established.

Properties of local stable and unstable manifolds of critical points and periodic orbits can be found in Hale [72]. The first proof that G_0^k, G_1^k, $G_{3/2}^k(T)$, $G_2^k(T)$ are open in $\mathcal{X}^k(I, M)$, $k \geq 1$ was given by Oliva [147], [148]. Mallet-Paret [135] proved Theorem 11.0.1 even for the more general case when the Whitney topology is used. Although the proof follows the pattern that was developed in Peixoto [165] (see also Abraham and Robbin [2]), Lemmas 11.0.3, 11.0.4, 11.0.5 contain essential new ideas. The analyticity used in the proof of Lemma 11.0.5 is due to Nussbaum [146]. For Smale's version of Sard's theorem, see [190] or [2]. Camargo in [19] extended the main results of this Chapter to second order retarded functional differential equations in $\mathbb{R}^n$, that is, generically, all critical points and all periodic solutions of the equation $\ddot{x}(t) = f(x_t, \dot{x}_t)$ are hyperbolic. Using ideas from generic theory, Chow and Mallet-Paret [33] introduced Fuller's index to obtain a new class of periodic solutions of certain delay equations, for example, the equation

$$\dot{x}(t) = -[\alpha x(t - 1) + \beta x(t - 2)]f(x(t)).$$

The generic theory for $NFDE$ also received some attention. We mention the paper of de Oliveira [40] in which he proved that the sets G_0^k and G_1^k are generic.

A An Introduction to the Conley Index Theory in Noncompact Spaces

by Krzysztof P. Rybakowski

This appendix serves to introduce the reader to the main aspects of the Conley index theory.

In its original form for (two-sided) flows on compact or locally compact spaces the theory is due mainly to Conley, although people like R. Easton, R. Churchill, J. Montgomery and H. Kurland should also be mentioned. The interested reader is referred to the monograph [34] for an account of the original version of the theory.

Conley's theory, in its original form, was developed primarily for ODE. By means of some special constructions, certain parabolic PDE and RFDE can also be treated in this original version of the theory. However, this imposes severe restrictions on the equations like, for example, the existence and knowledge of a bounded positively invariant set.

In papers [177]–[182], [185], Conley's theory was extended to large classes of semiflows on noncompact spaces. In particular, not only RFDE and parabolic PDE, but also certain classes of NFDE and hyperbolic equations can be treated quite naturally by this extended theory. In the above cited papers, some applications to all these classes of equations are given.

We may consider Conley's original version of the Conley index to be a generalization of the classical Morse index theory on compact manifolds: Morse assigns an index to every nondegenerate equilibrium of a gradient system, Conley assigns an index to every compact isolated invariant set of a not necessarily gradient ODE.

The extended Conley index theory is, in a sense, analogous to the Palais-Smale extension of the classical Morse index to noncompact spaces.

Although our only application will be to RFDE on $\mathbb{R}^m$, we will present the theory for general semiflows. This will clarify the main ideas. We begin with a well-known concept:

Definition A.0.1. *Given a pair* (X, π), π *is called a local semiflow (on X) if the following properties hold:*

1) *X is a topological space, $\pi : D \to X$ is a continuous mapping, D being an open subset of $\mathbb{R}^+ \times X$. (We write $x\pi t$ for $\pi(t, x)$).*
2) *For every $x \in X$ there is an ω_x, $0 < \omega_x \leq \infty$, such that $(t, x) \in D$ if and only if $0 \leq t < \omega_x$.*
3) *$x\pi 0 = x$ for $x \in X$.*

4) If $(t, x) \in D$ and $(s, x\pi t) \in D$, then $(t+s, x) \in D$ and $x\pi(t+s) = (x\pi t)\pi s$.

Remark A.0.2. If $\omega_x = \infty$ for all $x \in X$, then π is called a *(global) semiflow (on X)*.

(Local) semiflows are also called (local) dynamical (or, more appropriately, (local) semidynamical) systems.

Example A.0.3. Let F be a locally Lipschitzian RFDE on an m-dimensional manifold M and let Φ be the corresponding solution map. Write $\varphi \pi_F t = \Phi_t \varphi$ whenever the right-hand side is defined. Then π_F is a local semiflow on C^0. We call π_F the local semiflow generated by the solutions of F. We omit the subscript $_F$ and write $\pi = \pi_F$ if no confusion can arise. If $M = \mathbb{R}^m$, then $F(\varphi) = (\varphi(0), f(\varphi))$, where $f : C^0 \to \mathbb{R}^m$ is locally Lipschitzian. In this case, we will write π_f instead of π_F.

In the previous chapters of this book, several important concepts were defined relative to the local semiflow π_F, like that of a solution and of an invariant set. It is useful to extend these concepts to general local semiflows π on a topological space X. In particular, let $\mathcal{J}$ be an interval in $\mathbb{R}$ and $\sigma : \mathcal{J} \to X$ be a mapping. σ is called a solution (of π) if for all $t \in \mathcal{J}$, $s \in \mathbb{R}^+$ for which $t + s \in \mathcal{J}$, it follows that $\sigma(t)\pi s$ is defined and $\sigma(t)\pi s = \sigma(t + s)$. If $0 \in \mathcal{J}$ and $\sigma(0) = x$ then we may say that σ is a solution through x. If $\mathcal{J} = (-\infty, \infty)$, then σ is called a global (or full) solution.

If Y is a subset of X, then set:

$$I^+(Y) = \{x \in X \mid x\pi[0, \omega_x) \subset Y\},$$
$$I^-(Y) = \{x \in X \mid \text{there is a solution } \sigma : (-\infty, 0] \to X$$
$$\text{through } x \text{ with } \sigma(-\infty, 0] \subset Y\}.$$
$$I(Y) = I^+(Y) \cap I^-(Y).$$

Y is called positively invariant if $Y = I^+(Y)$,

Y is called negatively invariant if $Y = I^-(Y)$,

Y is called invariant if $Y = I(Y)$.

In particular, if $\omega_x = \infty$ for every $x \in Y$, then Y is invariant if and only if for every $x \in Y$ there exists a full solution σ through x for which $\sigma(\mathbb{R}) \subset Y$.

For a general subset Y of X, $I^+(Y)$ (resp. $I^-(Y)$, resp. $I(Y)$) is easily seen to be the largest positively invariant (resp., negatively invariant, resp. invariant subset of Y). $I^+(Y)$ (resp. $I^-(Y)$) is often called the stable (resp. unstable) manifold of $K = I(Y)$, relative to Y.

To illustrate these concepts with an example, suppose that f is an RFDE on $\mathbb{R}^m$ of class C^1 and 0 is a hyperbolic equilibrium of f (cf. [72], Chapter 10). Then the well-known saddle-point property implies that there is a direct sum decomposition $C^0 = U \oplus S$ and a closed neighborhood Y of 0 such that $K = \{0\}$ is the largest invariant set in Y, i.e., $\{0\} = I(Y)$. Moreover, the sets

$I^+(Y)$ and $I^-(Y)$ are tangent to S, resp. to U, at zero. There is a small ball $B_\delta \subset Y$ such that $I^+(Y) \cap B_\delta$ (resp. $I^-(Y) \cap B_\delta$) are diffeomorphic to $S \cap B_\delta$ (resp. $U \cap B_\delta$). Finally, the ω-limit set of every solution starting in $I^+(Y)$ (resp. the α-limit set of every solution defined on $(-\infty, 0]$ and remaining in $I^-(Y)$) is equal to $\{0\}$. Therefore, the qualitative picture near the equilibrium looks as in Fig. A.1.

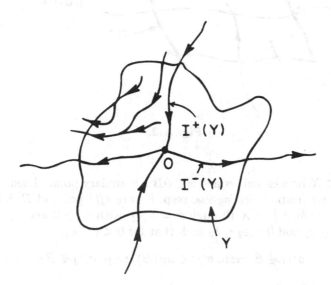

Fig. A.1.

The set $K = \{0\}$ has the important property of being isolated by Y. More generally, if K is a closed invariant set and there is a neighborhood U of K such that K is the largest invariant set in U, then K is called an *isolated invariant set*. On the other hand, if N is a closed subset of X and N is a neighborhood of $K = I(N)$, i.e., if the largest invariant set in N is actually contained in the interior of N, then N is called an *isolating neighborhood of K*. Hence, in the situation of Fig. A.1, $K = \{0\}$ is an isolated invariant set and Y is an isolating neighborhood of K. Let us analyze the example a little further: The isolating neighborhood Y is rather arbitrary, i.e., its boundary ∂Y is unrelated in any way to the semiflow π. However, Fig. A.1 suggests that one should be able to choose the set Y in such a way that ∂Y is "transversal" to π, i.e., such that orbits of solutions of π cross Y in one or the other direction (Fig. A.2). In fact, this is, for example, the case for ODE, where such special sets, called isolating blocks are used in connection with the famous Ważewski principle. The transversality of ∂Y with respect to π implies that every point x is of one of the following three types: it is either a strict egress, or a strict ingress or a bounce-off point.

Let us define those three concepts for an arbitrary local semiflow π.

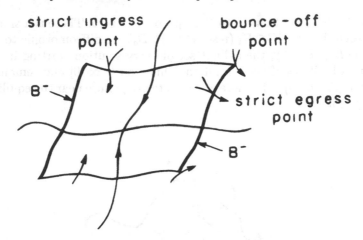

strict ingress
point

bounce-off
point

B^-

strict egress
point

B^-

Fig. A.2.

Let $B \subset X$ be a closed set and $x \in \partial B$ a boundary point. Then x is called a *strict egress* (resp. *strict ingress*, resp. *bounce-off*) point of B, if for every solution $\sigma : [-\delta_1, \delta_2] \to X$ through $x = \sigma(0)$, with $\delta_1 \geq 0$ and $\delta_2 > 0$ there are $0 \leq \varepsilon_1 \leq \delta_1$ and $0 < \varepsilon_2 \leq \delta_2$ such that for $0 < t \leq \varepsilon_2$:

$$\sigma(t) \notin B \ (\text{resp. } \sigma(t) \in \text{int}(B), \text{ resp. } \sigma(t) \notin B),$$

and for $-\varepsilon_1 \leq t < 0$:

$$\sigma(t) \in \text{int}(B) \ (\text{resp. } \sigma(t) \notin B, \text{ resp. } \sigma(t) \notin B).$$

By B^e (resp. B^i, resp. B^b) we denote the set of all strict egress (resp. strict ingress, resp. bounce-off) points of the closed set B. We finally set $B^+ = B^i \cup B^b$ and $B^- = B^e \cup B^b$.

We then have the following:

Definition A.0.4 (Isolating block). *A closed set $B \subset X$ is called an isolating block, if*

(i) $\partial B = B^e \cup B^i \cup B^b$,
(ii) B^e *and* B^i *are open in* ∂B.

Note that for general semiflows, $B^e \cap B^b$ may be nonempty, and consist of points $x \in B^e$ for which there is no solution defined for some negative times.

If B is an isolating block such that B^- is not a strong deformation retract of B, then there is a nonempty, positively invariant set in B. This is an important special case of Ważewski principle and was one of the motivations for developing the Conley index theory for ODE. (Cf. [34]).

Since ω-limit sets of compact trajectories are invariant sets, Ważewski principle plus some compactness assumptions imply that $I(B) \neq \emptyset$. Moreover, it is obvious that B is an isolating neighborhood of $I(B)$. The important converse problem arises: Given an isolated invariant set K, is there an isolating neighborhood set B of K which is an isolating block? Fig. A.2 suggests that this should be the case for hyperbolic equilibria, but we will try to give a general answer.

For two-sided flows on compact manifolds, the existence of isolating blocks was first proved by Conley and Easton [35]. The proof uses the theory of fibre bundles and it needs both the two-sidedness or the flow as well as the compactness of the underlying space in a very crucial way, i.e., it applies essentially only to ordinary differential equations in finite dimensions.

An alternative proof, still for ODE, was given by Wilson and Yorke [202]. These authors construct two special Lyapunov functions V_1 and V_2 and define $B = \{x \mid V_1(x) \leq \varepsilon, V_2(x) \leq \varepsilon\}$, for some $\varepsilon > 0$. This resembles Ważewski's original idea to use isolating blocks in the form of the so-called regular polyfacial sets, i.e., sets whose boundaries consist, piecewise, of level surfaces of special Lyapunov-like functions.

Although Wilson and Yorke still use compactness and the two-sidedness of the flow in an essential way, a portion of their proof can be utilized in generalizing the existence result for isolating blocks to semiflows on non-necessarily compact spaces.

The following elementary observation gives a first hint of how to proceed:

Proposition A.0.5. *Let π be a local semiflow on the metric space X, K be an isolated invariant set and N be an isolating neighborhood of K.*

Suppose that there exist continuous functions $V_i : N \to \mathbb{R}$, $i = 1, 2$, satisfying the following properties:

(i) *If $\sigma : \mathcal{J} \to N$ is a solution (of π) and $V_1(\sigma(t)) \neq 0$ (resp. $V_2(\sigma(t)) \neq 0$) for all $t \in \mathcal{J}$, then $t \to V_1(\sigma(t))$ is strictly increasing (resp. $t \to V_2(\sigma(t))$ is strictly decreasing).*

(ii) *If $x \in N$, then $x \in K$ if and only if $V_1(x) = 0$ and $V_2(x) = 0$.*

(iii) *If $\{x_n\} \subset N$ is a sequence such that $V_1(x_n) \to 0$ and $V_2(x_n) \to 0$ as $n \to \infty$, then $\{x_n\}$ contains a convergent subsequence.*

Under these hypotheses, there is an $\varepsilon_0 > 0$ such that whenever $0 < \varepsilon_1, \varepsilon_2 \leq \varepsilon_0$, then the set $B = \mathrm{Cl}\{x \in N \mid V_1(x) < \varepsilon_1, V_2(x) < \varepsilon_2\}$ is an isolating block for K (i.e., such that B is also an isolating neighborhood of K). ("Cl" denotes closure.)

Remark A.0.6. Property (i) means that V_1 and V_2 are Lyapunov-like functions for the semiflow, one of them increasing and the other decreasing along solutions of π. Property (iii) looks very much like the Palais-Smale condition (cf. [159] or [30]). (iii) is automatically satisfied if N is a compact metric space, and it will lead us to the concept of admissibility which will enable us to extend the Conley index theory to noncompact spaces and (one-sided) semiflows.

Let us sketch the proof: first observe that there is an $\varepsilon_0 > 0$ such that

$$B_{\varepsilon_0} := \mathrm{Cl}\{x \in N \mid V_1(x) < \varepsilon_0 \quad \text{and} \quad V_2(x) < \varepsilon_0\} \subset \mathrm{Int}\, N. \qquad (A.1)$$

In fact, if this is not true, then there exists a sequence $\{x_n\} \subset N$ such that $V_1(x_n) \to 0$ and $V_2(x_n) \to 0$ as $n \to \infty$, but $x_n \in \partial N$ for all n. Hence, by property (iii), we may assume that $\{x_n\}$ converges to some $x \in N$. By continuity, $V_1(x) = 0 = V_2(x)$. Hence, by (ii), $x \in K$. However, $x \in \partial N$ which is a contradiction (since N isolates K and therefore $\partial N \cap K = \emptyset$). Let $0 < \varepsilon_1$, $\varepsilon_2 \leq \varepsilon_0$ be arbitrary and set

$$B = \mathrm{Cl}\{x \in N \mid V_1(x) < \varepsilon_1,\ V_2(x) < \varepsilon_2\}.$$

To prove that B is an isolating block for K, note first that $K \subset \mathrm{Int}\, B$ (by (ii)). Moreover, by (A.1),

$$\partial B = \{x \in B \mid V_1(x) = \varepsilon_1 \quad \text{or} \quad V_2(x) = \varepsilon_2\}.$$

Now, using property (i), it is easily proved that

$$B^e = \{x \in \partial B \mid V_1(x) = \varepsilon_1 \quad \text{and} \quad V_2(x) < \varepsilon_2\},$$
$$B^i = \{x \in \partial B \mid V_2(x) = \varepsilon_2 \quad \text{and} \quad V_1(x) < \varepsilon_1\},$$
$$B^b \supset \{x \in \partial B \mid V_1(x) = \varepsilon_1 \quad \text{and} \quad V_2(x) = \varepsilon_2\}.$$

This implies that B is an isolating block and completes the proof.

Using Proposition A.0.5, let us now prove the existence of an isolating block in the simplest case of a hyperbolic equilibrium of a linear RFDE. This will illustrate some of the ideas of the general case without introducing any technicalities:

Proposition A.0.7. *If 0 is a hyperbolic equilibrium of the linear RFDE*

$$\dot{x} = Lx_t \qquad (A.2)$$

on $\mathbb{R}^m$, then there exist arbitrarily small isolating blocks for $K = \{0\}$.

Proof: Let π be the semiflow generated by (A.2). It is a global semiflow. By results in [72] there is a direct sum decomposition $C^0 = U \oplus S$, $\dim U < \infty$, such that $\Phi(t)U \subset U$ and $\Phi(t)S \subset S$, for $t \geq 0$, $\Phi(t)|_U$ can be uniquely extended to a group of operators, and there are constants $M, \alpha > 0$ such that

$$\|\Phi(t)\varphi\| \leq Me^{-\alpha t}\|\varphi\| \qquad \text{for} \quad \varphi \in S, t \geq 0$$
$$\|\Phi(t)\varphi\| \leq Me^{+\alpha t}\|\varphi\| \qquad \text{for} \quad \varphi \in U, t \leq 0. \qquad (A.3)$$

Let $k = \dim U$ and $\Psi : \mathbb{R}^k \to U$ be a linear isomorphism. If $A_1 : U \to U$ is the infinitesimal generator of the group $\Phi(t)|_U$, $t \in \mathbb{R}$, then there exists a $k \times k$-matrix B such that $\Psi^{-1}A_1\Psi = B$. It follows that $\mathrm{re}\,\sigma(B) > 0$. Hence there exists a positive definite matrix D such that $B^T D + DB = I$, where I is the identity matrix.

Now choose $\tau > 0$ such that $M \cdot (t+1)e^{-\alpha t} < 1/2$ for $t \geq \tau$. For $\varphi \in C^0$ define

$$V_1(\varphi) = \left(\Psi^{-1} P_U \varphi\right)^T D \left(\Psi^{-1} P_U \varphi\right), \tag{A.4}$$

$$V_2(\varphi) = \sup_{0 \leq t \leq \tau} \left[(t+1)\|\Phi(t) P_S \varphi\|\right], \tag{A.5}$$

where P_U and P_S are the projections onto U, S resp., corresponding to the above direct sum decomposition. An easy computation shows that

$$\liminf_{t \to 0^+} \frac{1}{t}\left(V_1\big(\Phi(t)\varphi\big) - V_1(\varphi)\right) > 0 \quad \text{if} \quad \varphi \notin S$$

and

$$\limsup_{t \to 0^+} \frac{1}{t}\left(V_2\big(\Phi(t)\varphi\big) - V_2(\varphi)\right) < 0 \quad \text{if} \quad \varphi \notin U.$$

Therefore V_1 and V_2 are easily seen to satisfy all assumptions of Proposition A.0.5 (property (iii) follows from the fact that U is finite-dimensional). The proposition is proved. ∎

If we try to prove the existence of isolating blocks for general semiflows by using Proposition A.0.5, we have to find an hypothesis which implies property (iii) of that Proposition. Such an hypothesis can be formulated by means of the following fundamental concept:

Definition A.0.8. *Let π be a local semiflow on the metric space X, and N be a closed subset of X. N is called π-admissible (or simply admissible, if no confusion can arise) if for every sequence $\{x_n\} \subset X$ and every sequence $\{t_n\} \subset \mathbb{R}^+$ the following property is satisfied: if $t_n \to \infty$ as $n \to \infty$, $x_n \pi t_n$ is defined and $x_n \pi[0, t_n] \subset N$ for all n, then the sequence of end points $\{x_n \pi t_n\}$ has a convergent subsequence.*

N is called strongly π-admissible *(or strongly admissible) if N is π-admissible and π does not explode in N, i.e., if whenever $x \in N$ is such that $\omega_x < \infty$, then $x\pi t \notin N$ for some $t < \omega_x$.*

Remark A.0.9. Admissibility is an asymptotic compactness hypothesis: if the solution through x_n stays in N long enough ($t_n \to \infty$!) then $\{x_n \pi t_n\}$ is a relatively compact set. Obviously, every compact set N in X is admissible, hence the concept is trivial for ODE. However, bounded sets N are π-admissible for many semiflows in infinite dimensions, like the semiflows generated by RFDE (see below) as in the Example A.0.3 or those generated by certain neutral equations and many classes of parabolic and even hyperbolic PDE. (See [177], [181], [180], [73].)

The assumption that π does not explode in N is quite natural and it implies that as long as we stay in N, we can treat π like a global semiflow. However, it is useful for the applications not to assume a priori that π is a global semiflow on X, since many local semiflows cannot be modified outside a given set N without destroying their character (e.g., the fact that they are generated by a specific equation).

Example A.0.10. (cont.) Suppose that $M = \mathbb{R}^m$ and $f : C^0 \to \mathbb{R}^m$ is locally Lipschitzian. Let $N \subset C^0$ be closed and bounded and $f(N)$ be bounded. Then the Arzelà-Ascoli Theorem easily implies that N is a π_f-admissible (cf. the proof of Theorem 3.6.1 in [72]). Moreover, Theorem 2.3.2 in [72] (or rather its proof) implies that π_f does not explode in N. It follows that N is strongly π_f-admissible.

Similar statements are of course true for general manifolds M. They are related to the fact that under quite natural assumptions, the solution operators $\Phi(t)$, $t \geq r$ are conditionally compact.

More generally, if π is a (local) semiflow on a complete metric space X and its solution operator $T(t_0)$ is, for some $t_0 > 0$, a conditional α-contraction, then every bounded set $N \subset X$ is π-admissible.

Let us note the following simple

Lemma A.0.11. *([177]) If $N \subset X$ is closed and strongly admissible, then $I^-(N)$ and $I(N)$ are compact.*

In other words, the largest invariant set $K = I(N)$ in N and its unstable manifold relative to N are both compact.

Proof: If $\{y_n\} \subset I^-(N)$, then there are solutions $\sigma_n : (-\infty, 0] \to N$ such that $\sigma_n(0) = y_n$, $n \geq 1$. Let $x_n = \sigma_n(-n)$. Obviously $x_n \pi[0, t_n] \subset N$ and $t_n \to \infty$, where $t_n = n$. Hence admissibility implies that $x_n \pi t_n = y_n$, $n \geq 1$, is a relatively compact sequence. Since the diagonalization procedure easily implies that both $I^-(N)$ and $I(N)$ are closed, the result follows. ∎

Let us also note that for N as in Lemma A.0.11, $\omega_x = \infty$ whenever $x \in I^+(N)$.

We are now in a position to state a main result on the existence of isolating blocks:

Theorem A.0.12. *([177]) If K is an isolated invariant set and N is a strongly admissible isolating neighborhood of K, then there exists an isolating block B such that $K \subset B \subset N$.*

Hence we assert the existence of arbitrarily small isolating blocks for K as long as K admits a strongly admissible, but otherwise arbitrary, isolating neighborhood. The proof of Theorem A.0.12, given in [177], is rather technical, but we should at least try to indicate its main ideas.

Let U be an open set such that $K \subset U$ and $\mathrm{Cl}\, U \subset N$, e.g., $U = \mathrm{Int}\, N$. Replacing N by $\mathrm{Cl}\, U$, if necessary, we may assume without loss of generality that $N = \mathrm{Cl}\, U$. Also, for the sake of simplicity, assume that π is a global semiflow. Finally, we may assume that $K \neq \emptyset$, since otherwise $B = \emptyset$ is an isolating block for K. Define the following mappings:

$$s_N^+ : N \to \mathbb{R}^+ \cup \{\infty\}, \quad s_N^+(x) = \sup\{t \mid x\pi[0,t] \subset N\},$$
$$t_U^+ : U \to \mathbb{R}^+ \cup \{\infty\}, \quad t_U^+(x) = \sup\{t \mid x\pi[0,t] \subset U\},$$
$$F : X \to [0,1], \quad F(x) = \min\{1, \mathrm{dist}(x, I^-(N))\},$$
$$G : X \to [0,1], \quad G(x) = \mathrm{dist}(x,K)/\big(\mathrm{dist}(x,K) + \mathrm{dist}(x, X \setminus N)\big);$$
$$g_U^+(x) := \inf\{(1+t)^{-1} G(x\pi t) \mid 0 \le t < t_U^+(x)\},$$
$$g_N^-(x) := \sup\{\alpha(t)F(x\pi t) \mid 0 \le t \le s_N^+(x), \text{ if } s_N^+(x) < \infty,$$
$$\text{and } 0 \le t < \infty, \text{ if } s_N^+(x) = \infty\},$$

g_U^+ is defined on U, g_N^- is defined on N, $\alpha : [0, \infty) \to [1,2)$ is a fixed monotone C^∞-diffeomorphism.

Then the following lemma holds (see [177]):

Lemma A.0.13.

(i) s_N^+ is *upper-semicontinuous*, t_U^+ is *lower-semicontinuous*.

(ii) g_U^+ is *upper-semicontinuous, and g_U^+ is continuous in a neighborhood of K. Moreover, if $g_U^+(x) \neq 0$, then*

$$\dot{g}_U^+(x) = \liminf_{t \to 0^+} (1/t)\big(g_U^+(x\pi t) - g_U^+(x)\big) > 0.$$

If $g_U^+(x) = 0$, then for every $t \in \mathbb{R}^+$, $x\pi t \in U$ and $g_U^+(x\pi t) = 0$.

(iii) g_N^- is *upper-semicontinuous. If $t_U^+(x) = s_N^+(x)$ on U, then g_N^- is continuous on U. Moreover, if $g_N^-(x) \neq 0$ then*

$$\dot{g}_N^-(x) = \limsup_{t \to 0^+} (1/t)\big(g_N^-(x\pi t) - g_N^-(x)\big) < 0.$$

If $g_N^-(x) = 0$, then for every $t \le s_N^+(x)$, $g_N^-(x\pi t) = 0$.

Therefore, taking N_1 to be an appropriate isolating neighborhood of K such that $N_1 \subset U$ and defining V_1 and V_2 to be the restrictions to N_1 of g_U^+ and g_N^- resp., we see that all assumptions of Proposition A.0.5 are satisfied except that maybe V_2 is not continuous. In particular, property (iii) of that proposition is a consequence of the fact that $I^-(N)$ is compact (see Lemma A.0.11 above). Therefore, the set

$$\tilde{N} = \mathrm{Cl}\,\tilde{U}, \quad \text{where} \quad \tilde{U} = \{x \in N_1 \mid V_1(x) < \varepsilon,\ V_2(x) < \varepsilon\}, \quad \varepsilon \text{ small},$$

is not an isolating block, in general. However, this set has some properties of an isolating block, e.g., that $t_{\tilde{U}}^+(x) = s_{\tilde{N}}^+(x)$ on $\tilde{U}$. Therefore, we can repeat the same process by taking $\tilde{N}$ to be a new isolating neighborhood of K, and defining $g_{\tilde{U}}^+, g_{\tilde{N}}^-$ as above. Now Lemma A.0.13, (iii) implies that $g_{\tilde{N}}^-$ is continuous on U. Hence taking $\tilde{N}_1$ to be an isolating neighborhood of K with $\tilde{N}_1 \subset \tilde{U}$ and letting $\tilde{V}_1$ and $\tilde{V}_2$ to be the restrictions to $\tilde{N}_1$ of $g_{\tilde{U}}^+$ and

$g_{\tilde{N}}^-$, we can satisfy all the hypotheses of Proposition A.0.5, thus proving the theorem.

If a set $K \neq \emptyset$ satisfies the assumptions of Theorem A.0.12 then, of course, there are infinitely many isolating blocks for K. Moreover, if we perturb the semiflow π a little (for instance, by perturbing the righthand side of an RFDE) then an isolating block with respect to the unperturbed semiflow, in general, is no longer an isolating block with respect to the perturbed semiflow. However, all isolating blocks for a given set K have a common property which may roughly be described as follows: take an admissible isolating block B for K, and collapse the subset B^- of B to one point. Then the resulting quotient space B/B^- is independent of the choice of B, modulo homeomorphisms or deformations preserving the base points [B]. Therefore the homotopy type of B/B^- is independent of the choice of B and this homotopy type is what we call the Conley index of K.

Before giving a precise definition of the Conley index, let us recall a few concepts from algebraic topology.

Definition A.0.14. *Let Y be a topological space and Z be a closed set in Y. If $Z \neq \emptyset$, then define Y/Z to be the set of all equivalence classes of the following equivalence relation $x \sim y$ if and only if $x = y$ or $x, y \in Z$. Y/Z is endowed with the quotient space topology. If $Z = \emptyset$, choose any point $p \notin Y$, give the union $Y \cup \{p\}$ the sum topology and set $Y/Z = Y/\emptyset := (Y \cup \{p\})/\{p\}$.*

Let $[Z]$ denote either the equivalence class of Z in Y/Z (if $Z \neq \emptyset$), or else the equivalence class of $\{p\}$.

Then in each case the space Y/Z is regarded as a pointed space with the distinguished base point $[Z]$.

Definition A.0.15. *Let (X, x_0) and (Y, y_0) be two pointed spaces. We say that (X, x_0) and (Y, y_0) are homotopy equivalent if there exist continuous base point preserving maps $f : X \to Y$, $g : Y \to X$ such that $f \circ g$ and $g \circ f$ are homotopic (with base point preserving homotopies) to the respective identity maps. The homotopy type of (X, x_0), denoted by $h(X, x_0)$ is the class of all pointed spaces which are homotopy equivalent to (X, x_0).*

Definition A.0.16. *Let X be a metric space. Then $\mathcal{I}$ (or $\mathcal{I}(X)$, if confusion may arise) is the set of all pairs (π, K), where π is a local semiflow on X, and K is an isolated, π-invariant set admitting a strongly π-admissible isolating neighborhood.*

Remark A.0.17. $\mathcal{I}$ *is the class of all (π, K) satisfying the hypotheses of Theorem A.0.12. Therefore, there exists a strongly admissible isolating block B for K.*

Now we have the following uniqueness result:

Theorem A.0.18. *([177], [179]) Let $(\pi, K) \in \mathcal{I}$, and $B, \tilde{B}$ be two strongly admissible isolating blocks for K (relative to the semiflow π). Then*

$(B/B^-, [B^-])$ and $(\tilde{B}/\tilde{B}^-, [\tilde{B}^-])$ are homotopy equivalent. Consequently, the homotopy type $h(B/B^-, [B^-])$ only depends on the pair $(\pi, K) \in \mathcal{I}$ and we write

$$h(\pi, K) = h(B/B^-, [B^-]).$$

$h(\pi, K)$ is called the Conley index of (π, K).

Remark A.0.19. If π is clear from the context, we write $h(K) = h(\pi, K)$ and speak of the Conley index of K.

Before giving a few hints about the proof of Theorem A.0.18, let us compute the index of $K = \{0\}$ in Proposition A.0.7.

In fact, by Proposition A.0.5, the set

$$B = \mathrm{Cl}\{\varphi \in C^0 \mid V_1(\varphi) < 1, \ V_2(\varphi) < 1\}$$

is a isolating block for $K = \{0\}$. But B is easily seen to be equal to $\{\varphi \in C^0 \mid V_1(\varphi) \leq 1, \ V_2(\varphi) \leq 1\}$. Moreover,

$$B^- = \{\varphi \in B \mid V_1(\varphi) = 1\}.$$

Let $H : B \times [0,1] \to B$ be defined as $H(\varphi, s) = P_U\varphi + (1-s)P_S\varphi$.

Since $H(B^- \times [0,1]) \subset B^-$, H induces a continuous, base point preserving homotopy $\tilde{H} : B/B^- \times [0,1] \to B/B^-$. $\tilde{H}$ is a strong deformation retraction of $(B/B^-, [B^-])$ onto $(B_1/\partial B_1, [\partial B_1])$ where

$$B_1 = \{\varphi \in U \mid V_1(\varphi) \leq 1\}.$$

Now B_1 is an ellipsoid in $U \cong \mathbb{R}^k$, hence the pair $(B_1, \partial B_1)$ is homeomorphic to (E^k, S^{k-1}), where E^k is the unit ball, and S^{k-1} is the unit sphere in $\mathbb{R}^k$.

Now E^k/S^{k-1} is homeomorphic to the pointed k-dimensional sphere (S^k, s_0) (with a base-point preserving homeomorphism). Altogether we obtain that

$$h(B/B^-, [B^-]) = h(B_1/\partial B_1, [\partial B_1])$$
$$= h(E^k/S^{k-1}, [S^{k-1}]) = h(S^k, s_0) =: \Sigma^k.$$

We obtain the following corollary:

Corollary A.0.20. *(cf. [181]) Under the assumptions of Proposition A.0.7*

$$h(\pi, \{0\}) = \Sigma^k$$

where $k = \dim U$ and Σ^k is the homotopy type of a pointed k-sphere (S^k, s_0).

Hence the Conley index of a hyperbolic equilibrium of a linear RFDE is determined by the dimension of its unstable manifold. This result is of crucial importance in the applications to be discussed later.

Let us now indicate a few ideas involved in the proof of Theorem A.0.18.

Let B be a strongly admissible, isolating block for K. Write $N_1 = B$, $N_2 = B^-$. For $t \geq 0$, let N_2^{-t}, called the *t-exit ramp* of (N_1, N_2), be the set of all $x \in N_1$ such that $x\pi s \in N_2$ for some $0 \leq s \leq t$ (see Fig. A.3). Moreover, let $N_1^t = \{x \in N_1 \mid$ there is a y such that $y\pi[0, t] \subset N_1$ and $y\pi t = x\}$. (See Fig. A.4.) Figures A.3 and A.4 suggest that $N_1^t \setminus N_2^{-t}$ can be made arbitrarily small in the sense that whenever V is an arbitrary neighborhood of K, then $N_1^t \setminus N_2^{-t} \subset V$ for large t. This may be described as squeezing the block B. Of course, the result of the squeezing is not a block. However, by using the semiflow π as a natural homotopy mapping, we can prove that $B/B^- = N_1/N_2$ is homotopy equivalent to N_1^t/N_2^{-t}. If $\tilde{B}$ is another block, then, by what we said above, $N_1^t \setminus N_2^{-t} \subset \tilde{B}$ for large t. Therefore we obtain a mapping $f : B/B^- \to \tilde{B}/\tilde{B}^-$ roughly as a composition of a "squeezing" followed by an inclusion. Similarly, a mapping $g : \tilde{B}/\tilde{B}^- \to B/B^-$ is defined. The deformation nature of the squeezing implies that $f \circ g$ and $g \circ f$ are homotopic to the corresponding identity maps. This proves Theorem A.0.18.

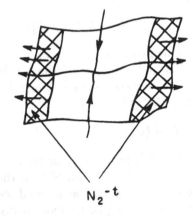

$$N_2^{-t}$$

Fig. A.3.

Let us note that the pairs (N_1, N_2^{-t}) and (N_1^s, N_2^{-t}) generated by the isolating block B inherit certain properties of the pair (B, B^-). Such pairs are called index or quasi-index pairs (for K). Hence (B, B^-) is a special index pair for K. A precise definition of index and quasi-index pair is given in [179]. One can show that whenever (N_1, N_2) is an index or quasi-index pair for K, then N_1/N_2 is homotopy equivalent to B/B^-, where B is an isolating block for B (of course, the usual admissibility assumption has to be imposed). Hence the Conley index can be defined by general index or quasi-index pairs. However, the special index pairs (B, B^-) induced by isolating blocks have several advantages: e.g., they permit the use of arbitrary homology and cohomology modules, whereas only the Čech cohomology groups can be meaningfully used with general index pairs. This is due to the fact that

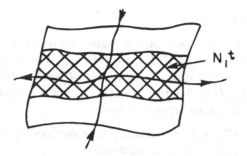

Fig. A.4.

the inclusion $B^- \subset B$ is a cofibration, a result which is not true for general index pairs.

The Conley index as defined in Theorem A.0.18 has an important property of being invariant under continuations of the semiflow. What is meant by this expression is that under certain admissible changes of the pair (π, K), the index $h(\pi, K)$ remains invariant. The changes of (π, K) are described by introducing a parameter $\lambda \mapsto (\pi(\lambda), K(\lambda))$ where λ varies over elements in some metric space Λ and $(\pi(\lambda), K(\lambda)) \in \mathcal{I}$. Call the resulting map α. When is α "admissible" in the sense that it leaves the index invariant?

A plausible condition is that 1) the map $\lambda \mapsto \pi(\lambda)$ is continuous in some sense (e.g., that $\pi(\lambda)$ represent RFDE $F(\lambda)$ with continuously varying $\lambda \to F(\lambda)$), and 2) that there is a set N such that no bifurcations of invariant sets occur at the boundary of N as λ is varied.

This situation is analogous to that of the Leray-Schauder Fixed Point Index, which remains constant under homotopies as long as no fixed-points appear on the boundary of the set considered. We need a third, technical assumption, which is, in a sense, a collective admissibility condition on N. More precisely, we have

Definition A.0.21. *Let X be a metric space and N be a closed set in X. Let $\{\pi_n\}$ be a sequence of local semiflows on X. N is called $\{\pi_n\}$-admissible if for every choice of sequences $\{x_n\} \subset X$, $\{t_n\} \subset \mathbb{R}^+$ satisfying (i) $t_n \to \infty$, as $n \to \infty$, (ii) $x_n \pi_n t_n$ is defined and $x_n \pi_n [0, t_n] \subset N$ for every n, it follows that the sequence $\{x_n \pi_n t_n\}$ of endpoints has a convergent subsequence.*

Of course, if $\pi_n \equiv \pi$ for all n, then this definition reduces to the admissibility condition given previously.

Example A.0.22. (cont.) Let $M = \mathbb{R}^m$ and $f_n : C^0 \to \mathbb{R}^m$, $n \geq 1$, be a sequence of locally Lipschitzian maps. If $N \subset C^0$ is a closed bounded set such that the set $\bigcup_{n=1}^{\infty} f_n(N)$ is bounded, then N is $\{\pi_n\}$-admissible where $\pi_n = \pi_{f_n}$.

This follows, as before in the case of one semiflow, by an application of the Arzelà-Ascoli theorem.

We can now formulate

Definition A.0.23. *Let Λ be a metric space and $\alpha : \Lambda \to \mathcal{I}$ be a mapping. Since for $\lambda \in \Lambda$, $\alpha(\lambda) = \big(\pi(\lambda), K(\lambda)\big)$, we write $\alpha_1(\lambda) = \pi(\lambda)$, $\alpha_2(\lambda) = K(\lambda)$. Let $\lambda_0 \in \Lambda$. We say that α is $\mathcal{I}$-continuous at λ_0 if there is an isolating neighborhood N of $\alpha_2(\lambda_0)$ (relative to $\alpha_1(\lambda_0)$), and a neighborhood W of λ_0 in Λ such that the following properties hold:*

1) *For every $\lambda \in W$, N is an isolating neighborhood of $\alpha_2(\lambda)$, relative to $\alpha_1(\lambda)$, and N is strongly $\alpha_1(\lambda)$-admissible.*
2) *For every sequence $\{\lambda_n\} \subset W$ converging to λ_0:*
   (i) N is $\{\alpha_1(\lambda_n)\}$-admissible.
   (ii) the sequence $\{\alpha_1(\lambda_n)\}$ of local semiflows converges to the local semiflow $\alpha_1(\lambda_0)$, as $n \to \infty$.

We remark that if $\{\pi_n\}$ is a sequence of local semiflows on X, then we say that π_n converges to the local semiflow π as $n \to \infty$ $(\pi_n \to \pi$, as $n \to \infty)$ if whenever $x_n \to x$ in X, $t_n \to t$ in $\mathbb{R}^+$ and $x\pi t$ is defined, then $x_n\pi_n t_n$ is defined for n sufficiently large, and $x_n\pi_n t_n \to x\pi t$.

This, in fact, is a very weak type of convergence, e.g., we have

Example A.0.24. (cont.) Let $f_n : C^0 \to \mathbb{R}^m$, $n \geq 0$, be a sequence of locally Lipschitzian mappings such that $f_n(\varphi) \to f_0(\varphi)$, as $n \to \infty$, uniformly on compact subsets of C^0. Then $\pi_n \to \pi_0$ as $n \to \infty$, where $\pi_n = \pi_{f_n}$, $n \geq 0$. This is an easy exercise left to the reader.

Definition A.0.23 gives precise conditions on the map α to be "admissible". In fact we have

Theorem A.0.25. *([177], [179]) If $\alpha : \Lambda \to \mathcal{I}$ is $\mathcal{I}$-continuous (i.e., if α is $\mathcal{I}$-continuous at λ_0, for every $\lambda_0 \in \Lambda$), then the index $h(\alpha(\lambda))$ is constant on connected components of Λ. In other words, if λ_1, λ_2 belong to the same connected component of Λ, then $h\big(\alpha(\lambda_1)\big) = h\big(\alpha(\lambda_2)\big)$. (Note $\alpha(\lambda) = (\pi(\lambda), K(\lambda))$.)*

In particular, if $\Lambda = [0,1]$ then $h\big(\alpha(0)\big) = h\big(\alpha(1)\big)$. This latter relation is basic in the applications of the index. The idea is, of course, to "deform" (or "continue") a given equation to a simpler equation for which the index is known. This will yield the index with respect to the original system.

Not even an intuitive description of the proof of Theorem A.0.25 can be given here.

Before turning to some applications of the index, let us state a result which shows that, in a certain sense, the Conley index is a finite-dimensional concept:

Theorem A.0.26. *([177]) Let $(\pi, K) \in \mathcal{I}$ and B be a strongly admissible isolating block. Then the natural inclusion and projection mappings include the following isomorphisms of the Čech cohomology:*

$$H^*(B/B^-, \{[B^-]\}) \cong H^*(B, B^-)$$
$$\cong H^*(B \cap I^-(B), B^- \cap I^-(B))$$
$$\cong H^*\big((B \cap I^-(B))/(B^- \cap I^-(B)), \{[B^- \cap I^-(B)]\}\big).$$

Recall that $I^-(B)$ is the unstable manifold of K relative to B.

Theorem A.0.26 is also valid for arbitrary index pairs (N_1, N_2). The proof follows by an application of the tautness and continuity properties of the Čech cohomology.

To see the significance of Theorem A.0.26 suppose that X is an open subset of a Banach space E. Moreover, assume that the solution operator $T(t_0)$ of π, is, for some $t_0 > 0$, a C^1-map whose derivative can be decomposed as a sum of a contraction and a compact map. (This is the case for semiflows generated by many RFDE and NFDE, but also by semilinear parabolic and even some hyperbolic PDE.) Then $I^-(B)$ has finite Hausdorff dimension. Consequently, under these hypotheses, the Čech cohomology of the Conley index is that of a finite-dimensional space. In particular, only finitely many of the groups $H^{*q}(B/B^-, \{[B^-]\})$ are nontrivial. This latter result also gives a heuristic explanation of why Ważewski's principle is applicable to many infinite-dimensional problems despite the fact that, e.g. the infinite dimensional unit sphere is a strong deformation retract of the closed unit ball.

We will now give a few applications of the Conley index to RFDE on $M = \mathbb{R}^m$. In previous chapters, the union $A(F)$ of all global bounded orbits of the RFDE(F) was studied. Conditions were given to assure that $A(F)$ is bounded (hence compact), connected and attracts all compact sets. In this case $A(F)$ is a maximal (hence isolated) compact invariant set which has an attractor nature. In the next few pages, we will exhibit a class of RFDE on $M = \mathbb{R}^m$, for which the set $A(F)$ is bounded, i.e. it is a maximal compact invariant set, but $A(F)$ is not necessarily an attractor. The condition roughly is asymptotic linearity of f and "non-criticality" at infinity. We will also compute the index of $A(F)$ and make some statement about the structure of $A(F)$. In particular, $A(F)$ will, with one exception, have a nonempty unstable manifold. Furthermore, although $A(F)$ need not be connected (we give an example of that) it is irreducible (index-connected), i.e. $A(F)$ cannot be decomposed as a disjoint union of two sets with nonzero Conley index. Note that we will write $A(f)$ for $A(F)$, where $F(\varphi) = (\varphi(0), f(\varphi))$.

We begin with the following result:

Theorem A.0.27. *Consider a sequence f_n, $n = 1, 2, \ldots$ of continuous mappings from C^0 to $\mathbb{R}^m$ such that every f_n is locally Lipschitzian, and let $\pi_n = \pi_{f_n}$ be the corresponding sequence of local semiflows on C^0. Assume the following hypotheses:*

(H1) There is a closed set $G \subset C^0$ such that for all $a \geq 0$, $a \cdot G \subset G$, and there is a continuous mapping $L : C^0 \to \mathbb{R}^m$ which is locally Lipschitzian and such that $L(a\varphi) = aL(\varphi)$ for $a \geq 0$, $\varphi \in G$.

(H2) For every $K > 0$, there is an $M > 0$ such that, for all n and every $\varphi \in G$ for which $\|\varphi\| \leq K$, it follows that $\|f_n(\varphi)\| \leq M$ and $\|L(\varphi)\| \leq M$.
(H3) If $\varphi_n \in G$ and $\|\varphi_n\| \to \infty$ as $n \to \infty$ then

$$\frac{\|f_n(\varphi_n) - L(\varphi_n)\|}{\|\varphi_n\|} \to 0 \quad \text{as } n \to \infty.$$

(H4) If $t \to \sigma(t)$ is a bounded solution of π_L on $(-\infty, \infty)$, then $\sigma(t) \equiv 0$ for all $t \in \mathbb{R}$.

Under these hypotheses, there is an $M_0 > 0$ and an n_0 such that, for all $n \geq n_0$, and every global bounded solution $t \to \sigma(t)$ of π_n such that $\sigma[\mathbb{R}] \subset G$, it follows that $\sup_{t \in \mathbb{R}} \|\sigma(t)\| \leq M_0$.

Remark A.0.28. In the applications of Theorem A.0.27 in this section, $G = C^0$ and L is a linear mapping, hence (H1) is automatically satisfied. However, we give the more general version of Theorem A.0.27 with the view of possible applications to "nonnegative" mappings L. In these cases, G would be the "nonnegative cone" of C^0. In [181], this general version has been applied to nonnegative solutions of parabolic PDE.

Notice that in the statement of Theorem A.0.27 as well as in its proof we use "$\|\ \|$" to denote both the Euclidean norm in $\mathbb{R}^m$ and the induced sup-norm in C^0. Confusion should not arise.

Proof: (of Theorem A.0.27)   Notice first that if $t \to \sigma(t)$ is a global bounded solution of the semiflow π_f, then there is a unique continuous mapping $x : (-\infty, \infty) \to \mathbb{R}^m$ such that $\sigma(t) = x_t$ for every $t \in \mathbb{R}$. Moreover $t \to x(t)$ is a solution of the RFDE(f) on $(-\infty, \infty)$. Now suppose that the theorem is not true. Then, taking subsequences if necessary, we may assume that there is a sequence of global bounded solutions $t \to \sigma_n(t) \in G$ of π_n such that $\alpha_n = \sup_{t \in \mathbb{R}} \|\sigma_n(t)\| \to \infty$, $\alpha_n \neq 0$ and $\|\sigma_n(0)\| > \alpha_n - 1$. Let $x^n : (-\infty, \infty) \to \mathbb{R}^m$ be the corresponding sequence of mappings such that $x_t^n = \sigma_n(t)$ for $t \in \mathbb{R}$.

Let $y^n(t) = \frac{x^n(t)}{\alpha_n}$, and $\tilde{\sigma}_n(t) = y_t^n$, $t \in \mathbb{R}$. By (H1), $\tilde{\sigma}_n(t) \in G$ for all n, and all $t \in \mathbb{R}$. Let $\tilde{f}_n(\varphi) = \frac{f_n(\alpha_n\varphi)}{\alpha_n}$. Then $\tilde{f}_n : C^0 \to \mathbb{R}^m$ is locally Lipschitzian. We will show that for every $\rho \geq 0$, $\sup_{\substack{\varphi \in G \\ \|\varphi\| \leq \rho}} \|\tilde{f}_n(\varphi) - L(\varphi)\| \to 0$ as $n \to \infty$.

In fact, let $\varepsilon > 0$ be arbitrary. Then by hypothesis (H3) there are $K > 0$ and n_1 such that whenever $\varphi \in G$, $\|\varphi\| > K$, $n \geq n_1$, then $(\|f_n(\varphi) - L(\varphi)\|)/\|\varphi\| < \varepsilon/\rho$.

Moreover, by hypothesis (H2), there is an $M > 0$ such that for all n and all $\varphi \in G$, $\|\varphi\| \leq K : \|f_n(\varphi)\| \leq M$ and $\|L(\varphi)\| \leq M$. Choose $n_2 \geq n_1$ such that $(2M)/\alpha_n < \varepsilon$ for $n \geq n_2$. Then we have for every $\varphi \in G$, $\|\varphi\| \leq \rho$ and every $n \geq n_2$ (using hypothesis (H1)):

$$\|\tilde{f}_n(\varphi) - L(\varphi)\| = \alpha_n^{-1} \cdot (\|f_n(\alpha_n \cdot \varphi) - L(\alpha_n \cdot \varphi)\|) \leq (2M)/\alpha_n < \varepsilon,$$

if $\alpha_n \cdot \|\varphi\| \leq K$,

$$\|\tilde{f}_n(\varphi) - L(\varphi)\| =$$

$$= \|\varphi\| \cdot (\alpha_n \cdot \|\varphi\|)^{-1} \cdot \|f_n(\alpha_n \cdot \varphi) - L(\alpha_n \cdot \varphi)\| < \rho \cdot \varepsilon / \rho = \varepsilon,$$

if $\alpha_n \cdot \|\varphi\| > K$. Hence our claim is proved.

Let $\tilde{\pi}_n = \pi_{\tilde{f}_n}$. Then it follows that $t \to \tilde{\sigma}_n(t)$ is a global bounded solution of $\tilde{\pi}_n$.

It follows from what we have just proved and from hypothesis (H2) (using the example above) that N is $\{\pi_n\}$-admissible, for every closed bounded set $N \subset C^0$. Since $\|\tilde{\sigma}_n(t)\| \leq 1$ for every $t \in \mathbb{R}$ and every n, it follows that for every $t \in \mathbb{R}$, the sequence $\{\tilde{\sigma}_n(t)\}$ is precompact in C^0. Hence $\{y^n\}$ is precompact on $[-r + t, t]$.

If $r > 0$, this means that $\{y^n\}$ is equicontinuous at every $t \in \mathbb{R}$, hence, using an obvious diagonalization procedure, we conclude that there exists a subsequence $\{y^{n_k}\}$ of $\{y^n\}$ and a continuous map $y : (-\infty, \infty) \to \mathbb{R}^m$ such that $y^{n_k}(t) \to y(t)$ as $k \to \infty$, uniformly on compact intervals.

If $r = 0$, then the RFDE involved are in fact, ODE and we obtain

$$\|y^n(t) - y^n(t_0)\| \leq \int_{t_0}^{t} \left\| \tilde{f}_n\left(y_n(s)\right) \right\| ds.$$

Hence, again $\{y^n\}$ is equicontinuous at every $t \in \mathbb{R}$ and we obtain a y and a subsequence $\{y^{n_k}\}$ as above.

A simple limit argument now shows that $t \to y(t)$ is a global bounded solution of the RFDE(L), which in view of hypothesis (H4), implies that $y(t) \equiv 0$. However,

$$\|\tilde{\sigma}_{n_k}(0)\| = \alpha_{n_k}^{-1} \cdot \|\sigma(0)\| > \alpha_{n_k}^{-1}(\alpha_{n_k} - 1) \to 1 \quad \text{as} \quad k \to \infty.$$

Hence $\|y_0^{n_k}\| \to 1$ as $k \to \infty$, a contradiction which proves the theorem. ∎

Using results in [72] we see that if $G = C^0$ and L is linear and bounded, then hypothesis (H4) is equivalent to the requirement that $\varphi = 0$ be a hyperbolic equilibrium of $\dot{x} = Lx_t$.

We thus have the following

Theorem A.0.29. *Let $f : C^0 \to \mathbb{R}^m$ be a locally Lipschitzian and completely continuous mapping. Furthermore, let $L : C \to \mathbb{R}^m$ be a bounded linear mapping. Suppose that*

$$\lim_{\|\varphi\| \to \infty} \frac{f(\varphi) - L(\varphi)}{\|\varphi\|} = 0.$$

If zero is a hyperbolic equilibrium of L and d is the dimension of the unstable manifold U of L, then $A(f)$ is bounded, hence compact. Moreover, $h(\pi_f, A(f)) = \Sigma^d$.

Proof: Let $f_\sigma = (1-\sigma)f + \sigma L$, $\sigma \in [0,1]$, and let $\pi_\sigma = \sigma_{f_\sigma}$. Write $\mathcal{J}_\sigma := A(f_\sigma)$. Using Theorem A.0.27 and a simple compactness argument, it is easily seen that there is a closed bounded set $N \subset C^0$ such that $\mathcal{J}_\sigma \subset \mathrm{Int}\, N$ and N is strongly π_σ-admissible for every $\sigma \in [0,1]$. Hence $(\pi_\sigma, \mathcal{J}_\sigma) \in \mathcal{I}$ and it is easily seen that the mapping $\alpha : \sigma \to (\pi_\sigma, \mathcal{J}_\sigma)$ is $\mathcal{I}$-continuous. It follows, by Theorem A.0.25, that $h(\pi_0, \mathcal{J}_0) = h(\pi_1, \mathcal{J}_1)$. However, $(\pi_0, \mathcal{J}_0) = (\pi_f, A(f))$, $(\pi_1, \mathcal{J}_1) = (\pi_L, \{0\})$. Hence, by Corollary A.0.20

$$h(\pi_f, A(f)) = \Sigma^d$$

and the proof is complete.   ∎

We will now draw a few conclusions from Theorem A.0.29.

First, let us define the following concept.

Definition A.0.30. *A pair $(\pi, K) \in \mathcal{I}$ is called* irreducible, *if K cannot be decomposed as a disjoint union $K = K_1 \cup K_2$ of two compact sets (both these sets would necessarily be invariant) such that*

$$h(\pi, K_1) \neq \bar{0} \quad and \quad h(\pi, K_2) \neq \bar{0}.$$

Let us remark that $\bar{0}$ is the homotopy type of a *one-point* pointed space. It is clear that e.g., $h(\pi, \emptyset) = \bar{0}$.

Definition A.0.30 generalizes the concept of connectedness; in fact, if K is connected, then (π, K) is irreducible, of course.

Moreover, we have the following:

Proposition A.0.31. *(cf. [180]) If $(\pi, K) \in \mathcal{I}$ and $h(\pi, K) = \bar{0}$ or $h(\pi, K) = \Sigma^k$ for some $k \geq 0$, then (π, K) is irreducible.*

The purely algebraic-topological proof of Proposition A.0.31 is omitted.

As a consequence of Proposition A.0.31, we see that $(\pi_f, A(f))$ is irreducible. Later on we will see that $A(f)$ does not have to be connected. Still irreducibility implies the following

Proposition A.0.32. *(cf. [180]) Assume all hypotheses of Theorem A.0.29 are satisfied. Let $K \subset A(f)$ be an isolated π_f-invariant set and suppose that*

$$h(\pi_f, K) \neq \bar{0} \quad and \quad h(\pi_f, K) \neq \Sigma^d.$$

Then there exists a global bounded solution $t \to x(t)$ of the RFDE(f) such that for some t_0, $x_{t_0} \notin K$ but either the α- or the ω-limit set of $t \to x_t$ is contained in K (or maybe both).

In other words, although the orbit of $t \to x_t$ is not fully contained in K, it either emanates from K or tends to K, or both. If π_f is gradient-like, this means that there is a heteroclinic orbit joining a set of equilibria $L_1 \subset K$ with some other set of equilibria. In the special case that $K = \{0\}$, this also gives us existence of nontrivial equilibria of π_f. Incidentally, this procedure, applied to semilinear parabolic equations, proves the existence of nontrivial solutions of elliptic equations ([181], [182]).

The proof of Proposition A.0.32 is obtained by noticing that if the proposition is not true, then, there exists a compact set K' disjoint from K and such that $K \cup K' = A(f)$. However, the irreducibility of $(\pi_f, A(f))$ then leads to a contradiction, since $h(\pi_f, K) \neq \bar{0}$ and $h(\pi_f, K') \neq \bar{0}$.

Proposition A.0.32 gives some (rather crude) information about the inner structure of the set $A(f)$. Of course, $A(f) \neq \emptyset$, since otherwise $h(\pi_f, A(f)) = \bar{0} \neq \Sigma^d$, a contradiction.

We will now give some more information about $A(f)$. First we consider the case $d = 0$:

Proposition A.0.33. *If the assumptions of Theorem A.0.29 are satisfied and if $d = 0$, then the RFDE(f) is point-dissipative. Consequently, the set $A(f)$ is a connected global attractor for the semiflow π_f.*

Proof: Let τ be as in the proof of Proposition A.0.7 and $V = V_2$, where V_2 is given by (A.5). Since $k = d = 0$, it follows that $U = \{0\}$, i.e., $S = C^0$ and hence, noticing that $\Phi(t)\varphi = \varphi\pi_L t$, we have

$$V(\varphi) = \sup_{0 \leq t \leq \tau} [(1+t)\|\varphi\pi t\|].$$

From the definition of V it follows easily that

$$\|\varphi\| \leq V(\varphi) \quad \text{and}$$
$$|V(\varphi) - V(\psi)| \leq (1+\tau)\|\varphi - \psi\| \tag{A.6}$$

for all $\varphi, \psi \in C^0$.

Let $\tilde{f}_n(\varphi) = n^{-1} f(n\varphi)$. As in the proof of Theorem A.0.27 we obtain that for every $\rho \geq 0$, $\sup_{\|\varphi\| \leq \rho} \|\tilde{f}_n(\varphi) - L\varphi\| \to 0$ as $n \to \infty$.

Let $\pi_n = \pi_{\tilde{f}_n}$. A simple estimate implies that

$$\limsup_{t \to 0^+} \frac{1}{t}(V(\varphi\pi_n t) - V(\varphi)) \leq -\frac{1}{1+\tau}\|\varphi\| + (1+t)\|\tilde{f}_n(\varphi) - L\varphi\| \tag{A.7}$$

for every $\varphi \in C^0$.

By Theorem A.0.29, $A(f)$ is compact. Hence, in order to show that the RFDE(f) is point-dissipative it suffices to prove that every solution of the RFDE(f) is bounded. Suppose this is not true, and let $t \to x(t)$ be a solution of the RFDE(f) such that $\sup_{t \geq 0} \|x_t\| = \infty$.

There is an n_0 such that for every $n \geq n_0$ and every φ with $\|\varphi\| \leq 2(1+\tau)$, $(1+\tau)\|\tilde{f}_n(\varphi) - L\varphi\| \leq \frac{1}{2(1+\tau)}$.

There is an $n \geq n_0$ such that $\|n^{-1}x_0\| < 1$. Hence, by (A.6), $V(n^{-1}x_0) < (1+\tau)$. Consequently, there exists a first time $t_1 > 0$ such that $V(n^{-1}x_{t_1}) = (1+\tau)$. Since $t \to n^{-1}x(t)$ is a solution of the RFDE$(\tilde{f}_n)$, it follows from (A.6), (A.7) that for t in a neighborhood of t_1

$$\limsup_{h \to 0^+} \frac{1}{h}\left(V(n^{-1}x_{t+h}) - V(n^{-1}x_t)\right) \le$$

$$\le -\frac{1}{1+\tau}\|n^{-1}x_t\| + (1+\tau)\|\tilde{f}_n(n^{-1}x_t) - L(n^{-1}x_t)\|$$

$$\le -\frac{1}{1+\tau}(1-\varepsilon) + \frac{1}{2(1+\tau)} < 0.$$

Here $\varepsilon > 0$ is a small number. This implies that $t \to V(n^{-1}x_t)$ is strictly decreasing in a neighborhood of $t = t_1$ and, in particular, that

$$V(n^{-1}x_{t_2}) > V(n^{-1}x_{t_1}) = (1+\tau) \quad \text{for some} \quad t_2 < t_1,$$

a contradiction to our choice of t_1. This contradiction proves that the RFDE(f) is point-dissipative and this, in turn, implies the remaining assertions of the proposition.   ∎

We will now prove, that $A(f)$ has a non-empty unstable manifold, provided $d \ge 1$. Hence in this case, the RFDE(f) is not point-dissipative, and we may expect $A(f)$ to satisfy a saddle-point property, at least generically. However, no proof of the latter conjecture is available.

Proposition A.0.34. *If the assumptions of Theorem A.0.29 are satisfied and if $d \ge 1$, then there is a global solution $t \to x(t)$ of the RFDE(f) such that*

$$\sup_{t \le 0} \|x(t)\| < \infty \quad \text{and} \quad \sup_{t \ge 0} \|x(t)\| = \infty.$$

Remark A.0.35. Hence $x_t \to A(f)$ as $t \to -\infty$ but x_t is unbounded as $t \to +\infty$.

Proposition A.0.34 is a special case of Theorem 3.4 in [178]. The proof is obtained as follows: if the proposition is not true, then every global solution of π_f, bounded on $(-\infty, 0]$, is also bounded on $[0, \infty)$. Take a bounded neighborhood N of $A(f)$. It follows that $I^-(N) = I(N) = A(f)$. Using this and the arguments from the proof of Theorem A.0.12, one shows the existence of an isolating block $B \ne \emptyset$ for $A(f)$ such that $B^- = \emptyset$. Hence $h(\pi_f, A(f))$ is the homotopy type of the disjoint union of the set B with a one point set $\{p\}$. Now an algebraic-topological argument implies that the d-sphere (S^d, s_0), $d \ge 1$, is *not* homotopy equivalent to such a disjoint union of sets. This is a contradiction and proves the proposition.

Using ideas from the proofs of Theorems A.0.27 and A.0.29 we also obtain the following result:

Theorem A.0.36. *Let $f : C^0 \to \mathbb{R}^m$ be a locally Lipschitzian mapping. Suppose φ_0 is an equilibrium of the RFDE(f), i.e a constant function such that $f(\varphi_0) = 0$. If f is Fréchet-differentiable at φ_0 and if 0 is a hyperbolic equilibrium of the linear RFDE(L), $L = f'(\varphi_0)$, with a d-dimensional unstable manifold, then $K = \{\varphi_0\}$ is an isolated π_f-invariant set, $h(\pi_f, \{\varphi_0\})$ is defined and*

$$h(\pi_f, \{\varphi_0\}) = \Sigma^d.$$

Sketch of Proof: We may assume w.l.o.g. that $\varphi_0 = 0$. Let $f_\sigma = (1 - \sigma)f + \sigma L$, $\pi_\sigma = \pi_{f_\sigma}$, and $K_\sigma = \{0\}$. We claim that there is a closed set $N \subset C^0$ such that N is a strongly π_σ-admissible isolating neighborhood of K_σ, for every $\sigma \in [0, 1]$. Assuming this for the moment, we easily see that the map $\sigma \to (\pi_\sigma, K_\sigma)$ is well-defined and $\mathcal{I}$-continuous. Hence Theorem A.0.25 and Corollary A.0.20 imply the result. Now, if our claim is not true, there is a sequence $\sigma_n \in [0, 1]$ converging to some $\sigma \in [0, 1]$ and a sequence of bounded solutions $t \to x^n(t)$ of $\dot{x} = f_{\sigma_n}(x_t)$ on $(-\infty, \infty)$ such that $0 \neq \sigma = \sup_{t \in \mathbb{R}} \|x^n(t)\| \to 0$ and $\|x^n(0)\| > \alpha_n - 1$. Let $\tilde{f}_n(\varphi) = (\alpha_n^{-1}) \cdot f_{\sigma_n}(\alpha_n \varphi)$. Then $\tilde{f}_n$ is locally Lipschitzian. Now it is easily seen that $\tilde{f}_n \to L$ uniformly in a bounded neighborhood of zero. Therefore, the arguments from the proof of Theorem A.0.27 lead to a contradiction and complete the proof. ∎

We will now apply our previous results to vector-valued Levin-Nohel equations. The relevant facts are contained in the following well-known proposition (see Theorem 5.0.3 for the scalar case).

Proposition A.0.37. *([72]) Let $r > 0$ and $b : [-r, 0] \to \mathbb{R}$ be a C^2-function such that $b(-r) = 0$, $b'(\theta) \geq 0$, $b''(\theta) \geq 0$, for $-r \leq \theta \leq 0$, and there is a $-r \leq \theta_0 \leq 0$, such that $b''(\theta_0) > 0$. Moreover, let $G : \mathbb{R}^m \to \mathbb{R}$ be a C^1-function, and $g = \nabla G$ be locally Lipschitzian. Consider the following RFDE:*

$$\dot{x}(t) = -\int_{-r}^0 b(\theta)g(x(t + \theta))d\theta. \tag{$8_{b,G}$}$$

Then the local semiflow $\pi = \pi_{b,G}$ generated by solutions of $(8_{b,G})$ is gradient-like with respect to the following function:

$$V(\varphi) = G(\varphi(0)) + \frac{1}{2}\int_{-r}^0 b'(\theta)\left[\int_\theta^0 g(\varphi(s))ds\right]^2 d\theta.$$

Moreover, every equilibrium φ_0 of $(8_{b,G})$ is constant, $\varphi_0(\theta) \equiv a$, $\theta \in [-r, 0]$ and $g(a) = 0$.

For the analysis of equilibria of $(8_{b,G})$ we need the following lemma:

Lemma A.0.38. *Let b be as in Proposition A.0.37 and let A be a symmetric, real-valued $m \times m$ matrix. Consider the following linear RFDE*

$$\dot{x}(t) = -\int_{-r}^0 b(\theta)Ax(t + \theta)d\theta. \tag{$8_{b,A}$}$$

Then the following properties hold:

1) $\varphi = 0$ is a hyperbolic equilibrium of $(8_{b,A})$ if and only if A is nonsingular.

2) If $\varphi = 0$ is a hyperbolic equilibrium of $(8_{b,A})$, then $h(\pi, \{0\}) = \Sigma^d$, where π is the semiflow generated by $(8_{b,A})$ and $d = d^-(A)$.

Here, $d^-(A)$ is the total algebraic multiplicity of all negative eigenvalues of A.

Proof: If $m = 1$, the result is well-known and follows by a simple analysis of the characteristic equation of $(8_{b,A})$. If $m > 1$, let us diagonalize A, using the fact that A is symmetric. We thus obtain that $(8_{b,A})$ is equivalent to a system of m uncoupled one dimensional equations

$$\dot{x}^i = -\lambda_i \int_{-r}^0 b(\theta) x^i(t + \theta) d\theta, \qquad (A.8)$$

where λ_i, $i = 1, \ldots, m$ are the (possibly multiple) eigenvalues of A. Therefore, the unstable manifold of $\varphi = 0$ with respect to $(8_{b,A})$ is easily seen to be $d = d^-(A)$-dimensional. Now Corollary A.0.20 implies the result. ∎

We are now ready to state our main result about equation $(8_{b,G})$.

Theorem A.0.39. *Let b, G, and g be as in Proposition A.0.37. Moreover assume the following hypotheses:*

1) *G is a Morse function, i.e., $G \in C^2(\mathbb{R}^m)$ and whenever $\nabla G(x_0) = 0$, then the Hessian $\left(\frac{\partial^2 G(x_0)}{\partial x_i \partial x_j}\right)_{i,j}$ is nonsingular.*
2) *There is a symmetric, nonsingular $m \times m$-matrix A_∞ such that*

$$\frac{g(x) - A_\infty x}{\|x\|} \to 0 \quad as \quad \|x\| \to \infty.$$

Then the following statements hold:

(i) *If x_0 is a zero of g such that $d^-(A_0) \neq d^-(A_\infty)$, where $A_0 = \left(\frac{\partial^2 G(x_0)}{\partial x_i \partial x_j}\right)_{i,j}$ is the Hessian of G at x_0, then there is another zero x_1 of g and a bounded solution $t \to x(t)$ of $(8_{b,G})$ defined for $t \in (-\infty, \infty)$ such that*

either 1° $\lim_{t \to -\infty} x(t) = x_0$ and $\lim_{t \to +\infty} x(t) = x_1$,
or 2° $\lim_{t \to +\infty} x(t) = x_0$ and $\lim_{t \to -\infty} x(t) = x_1$.

(ii) *If $d^-(A_\infty) = 0$, then $(8_{b,G})$ is point-dissipative, hence the union $A(b, G)$ of all global bounded orbits of $(8_{b,G})$ is a connected global attractor.*
(iii) *If $d^-(A_\infty) \geq 1$, then there is a zero $\tilde{x}_0$ of g and a solution $t \to \tilde{x}(t)$ of $(8_{b,G})$ defined for $t \in (-\infty, \infty)$ such that $\lim_{t \to -\infty} x(t) = x_0$ but $\sup_{t \geq 0} \|\tilde{x}(t)\| = \infty$.*

The proof of Theorem A.0.39 is an easy consequence of the preceding results.

In the situation of Theorem A.0.39, the union $A(b, G)$ of all full bounded orbits of $\pi = \pi_{b,G}$ is itself bounded, hence compact. Now if $A(b, G)$ is connected, part (i) of Theorem A.0.39 is trivial. Hence in order to show the significance of our results it is necessary to prove that $A(b, G)$ is not connected, in general.

In fact, we have the following

Proposition A.0.40. *For every $b : [-r, 0] \to \mathbb{R}$, $r > 0$ satisfying the assumptions of Proposition A.0.37 and every positive number $c > 0$, there is an analytic function $G : \mathbb{R} \to \mathbb{R}$ such that $g := G'$ has exactly three zeros $a_1 < a_2 < a_3$, all of them simple, such that $\lim_{|s| \to \infty} \frac{g(s) + cs}{|s|} = 0$, and such that the set $A(b, G)$ is disconnected and consists of the three equilibria $\varphi_i(\theta) \equiv a_i$, $\theta \in [-r, 0]$, $i = 1, 2, 3$, and an orbit joining φ_2 with φ_3.*

Proof: Choose $\tilde{a}_1 = 0$. Let $f(s) = -cs$. Then, there is a unique $\lambda > 0$ such that $x(t) = e^{\lambda t}$, $t \in \mathbb{R}$, is a solution of $(8_{b,F})$ where $F(x) = \int_0^x f(s)ds$. Let $t_0 > 0$ be arbitrary and let $y(h)$, $h \in [0, r]$, be defined as

$$y(h) = x(t_0) - \int_0^h \left(\int_{-r}^{-s} b(\theta) f\big(x(s + t_0 + \theta)\big) d\theta \right) ds.$$

Hence, there is a $0 < h_1 < r$ such that $y(h) > 0$ for $h \in [0, h_1]$, hence $y(h_1) > x(t_0)$. Define a continuous function $\tilde{f}$ such that $\tilde{f} = f$ on $(-\infty, x(t_0)]$, $\tilde{f}$ is affine on $[x(t_0), y(h_1)]$, $\tilde{f}(y(h_1)) = 0$ and $\tilde{f}$ is affine on $[y(h_1), \infty)$ with negative slope (see Fig. A.5) equal $-c$.

If $\varphi(\theta) = x(t_0 + \theta)$, $\theta \in [-r, 0]$, let $\tilde{y}$ be the solution of $(8_{b,\tilde{F}})$ through φ, where $\tilde{F}(x) = \int_0^x \tilde{f}(s)ds$. Then obviously $\tilde{y}(h) > y(h)$ for $h \in (0, h_1]$. If we define $\tilde{x}(t) = x(t)$, $t \leq t_0$, $\tilde{x}(t) = \tilde{y}(t - t_0)$, $t > t_0$, then $\tilde{x}$ is a solution of $(8_{b,\tilde{F}})$ on $\mathbb{R}$. Moreover, $\tilde{x}(t) \to \infty$ as $t \to \infty$, for otherwise $x(t)$ would go to a zero β of $\tilde{f}$, $\beta > y(h_1)$, a contradiction. Consequently, there is an $s_1 > t_0$ such that $\tilde{x}(s_1 + \theta) > y(h_1)$ for $\theta \in [-r, 0]$. Now perturb $\tilde{f}$ a little on an interval $[x(t_0) - \varepsilon, y(h_1)]$, where $\varepsilon > 0$ is a small number, to obtain a C^1-function $\tilde{g}$ which has exactly three simple zeros $\tilde{a}_1 = 0 < \tilde{a}_2 < \tilde{a}_3 = y(h_1)$ (see Fig. A.6).

Let $\tilde{G}(x) = \int_0^x \tilde{g}(s)ds$. If the perturbation is small, then the unique solution $t \to \tilde{\tilde{x}}(t)$ of $(8_{b,\tilde{G}})$ which emanates from $\tilde{a}_1 = 0$ (i.e. $\tilde{\tilde{x}}(t) \to \tilde{a}_1$ as $t \to -\infty$) and staying to the right of $\tilde{a}_1$, such that $\tilde{\tilde{x}}(s_1 + \theta) > y(h_1)$ for $\theta \in [-r, 0]$. Hence $\tilde{\tilde{x}}(t) \to \infty$ as $t \to \infty$ and this implies that there is no bounded orbit of $(8_{b,\tilde{G}})$ emanating from $\tilde{a}_1 = 0$.

Furthermore the unique orbit emanating from $\tilde{a}_3 = y(t_1)$ and staying to the left of $\tilde{a}_3$ must run to $\tilde{a}_2$. In fact it cannot hit $\tilde{a}_1$ as is easily seen by examining the Lyapunov function V of Proposition A.0.37: as $t \to \tilde{\tilde{x}}(t)$ emanates from $\tilde{a}_1$ and hits $\tilde{a}_3$ it follows that $\tilde{G}(\tilde{a}_3) < \tilde{G}(\tilde{a}_1)$. Hence there is no orbit emanating from a_3 and hitting a_1, because otherwise $\tilde{G}(\tilde{a}_3) > \tilde{G}(\tilde{a}_1)$. It follows that the set $A(b, \tilde{G})$ defined above consists of the three equilibria $\tilde{\varphi}_i = \tilde{a}_i$, $i = 1, 2, 3$ and an orbit running from $\tilde{a}_3$ to $\tilde{a}_1$. Hence the proposition is proved except for the fact that $\tilde{g}$ is not analytic. Now one can approximate $\tilde{g}$ by analytic functions g using Whitney's Lemma. Alternatively, one may

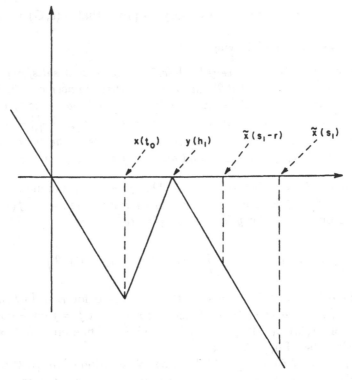

Fig. A.5.

use the proof of Lemma 2.5 in [91] to conclude that if g is sufficiently close to $\tilde{g}$ and $G(x) = \int_0^x g(s)ds$, then G, g and $A(b, G)$ satisfy all the statements of our proposition. The proof is complete. ∎

We can compute the index $h(\pi_{b,G}, A(b, G))$ as follows: Let $g_\sigma(s) = -\sigma \cdot c \cdot s + (1 - \sigma)g(s)$, $\sigma \in [0, 1]$, and $G_\sigma(x) = \int_0^x g_\sigma(s)ds$. Then it easily follows that the map $\alpha : \sigma \to (\sigma_{b,G_\sigma}, A(b, G_\sigma))$ is well-defined and $\mathcal{I}$-continuous. Consequently, by Theorem A.0.25 and Lemma A.0.38

$$h(\pi_{b,G}, A(b, G)) = h(\pi_{b,G}, \{0\}) = \Sigma^1.$$

Moreover, by Theorem A.0.36,

$$h(\pi_{b,G}, \{\varphi_i\}) = \begin{cases} \Sigma^1 & \text{if } i = 1, 3, \\ \Sigma^0 & \text{if } i = 2. \end{cases}$$

This illustrates very clearly the concept of irreducibility: there is no heteroclinic orbit running from or to φ_1. By Proposition A.0.32, this is only possible if the index of $(\pi_{b,G}, \{\varphi_1\})$ is either $\bar{0}$ or equal to the index of $(\pi_{b,G}, A(b, G))$, and this is indeed the case. By the same token, since

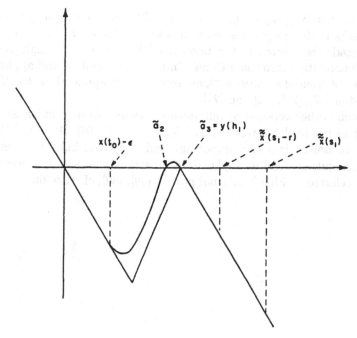

Fig. A.6.

$h(\pi_{b,G}, \{\varphi_2\}) = \Sigma^0 \neq h(\pi_{b,G}, A(b,G))$, it follows from Proposition A.0.32 that φ_2 is the "target" or the "source" of a heteroclinic orbit, the former being the case here.

Note that Proposition A.0.31 gives no criterion to detect heteroclinic orbits emanating from or tending to an equilibrium φ_i, if the index of this equilibrium is $\bar{0}$ or equal $h(\pi_{b,G}, A(b,G))$. In fact, in our example, $\{\varphi_1\}$ is isolated in $A(b,G)$ and $\{\varphi_3\}$ is the source of a heteroclinic orbit, although both equilibria have the same index equal $h(\pi_{b,G}, A(b,G)) = \Sigma^1$.

Concluding Remarks

In this appendix, we have only presented the simplest aspects of the Conley index theory on noncompact spaces and some applications to functional differential equations. In particular, we have entirely omitted the discussion of the Morse index as a category (see [34] and [179]). In many cases, invariant sets K admit a so-called Morse decomposition. Classical examples include finite sets of equilibria in K. A question arises as to the existence of heteroclinic orbits connecting such equilibria. We discussed this question above in a very simple setting, but much more can be said leading to the notion of index triples, the connection-index, generalized Morse inequalities and the connection matrix ([34], [179], [185], [61], [62]). Interesting applications of these results to delay equations have appeared in [136], [137] and [141].

The reader is referred to the book [183] for a more thorough presentation of the infinite-dimensional Conley index theory and several applications to partial differential equations. Some more recent applications to PDE are contained in [57], [20], [22] and [21].

For some other versions of infinite dimensional Conley index and their applications the reader is referred to [15], [143], [42], [67], [105] and [106].

Floer homology is a very important and nontrivial infinite dimensional Conley-like index, useful in applications to strongly indefinite systems. The reader is referred to Floer's original papers [59], [60], cf. also [6].

References

1. V. E. Abolina and A. D. Myshkis. Mixed problems for quasi-linear hyperbolic systems in the plane. *Mat. Sb. (N.S.)*, 50(92):423–442, 1960.
2. R. Abraham and J. Robbin. *Transversal Mappings and Flows*. Benjamin, 1967.
3. N. Alikakos and G. Fusco. A dynamical systems proof of the Krein-Rutman theorem. *Proc. Royal Soc. Edinburgh*, 117A:209–214, 1991.
4. S. Angenent. The Morse-Smale property for a semi-linear parabolic equation. *J. Differential equations*, 62:427–442, 1986.
5. S. Angenent. The zero set of a solution of a parabolic equation. *J. Reine Angew. Math.*, 390:79–96, 1988.
6. S. Angenent and R. van de Vorst. A superquadratic indefinite system and its Morse-Conley-Floer homology. *Math. Z.*, 390:243–248.
7. J. Arrieta, A. Carvalho, and Hale J. K. A damped hyperbolic equation with critical exponent. *Comm. PDE*, 17:841–866, 1992.
8. A. V. Babin and M. I. Vishik. Attractors of evolution equations. In *Studies in Math. and Applications 25*. Elsevier, 1992.
9. J. Ball. Continuity properties and global attractors of generalized semiflows and the Navier-Stokes equations. *J. Nonlinear Science*, 7:475–502, 1997.
10. J. Ball. Erratum-Continuity properties and global attractors of generalized semiflows and the Navier-Stokes equations. *J. Nonlinear Science*, 8:233, 1998.
11. C. Bardos, G. Lebeau, and J. Rauch. Sharp sufficient conditions for the observation, control and stabilization of waves from the boundary. *SIAM J. Control*, 1991.
12. L. Barreira, A. Katok, and Ya. Pesin. Nonuniformly hyperbolic dynamical systems. In preparation.
13. L. Barreira and Ya. Pesin. *Lyapunov exponents and smooth ergodic theory*. University Lecture Series 23. Amer. Math. Soc., 2002.
14. L. Barreira and J. Schmeling. Sets of "non-typical" points have full topological entropy and full hausdorff dimension. *Israel J. Math.*, 116:29–70, 2000.
15. V. Benci. A new approach to the Morse-Conley theory and some applications. *Ann. Mat. Pura Appl.*, 4-158:231–305, 1991.
16. E. Beretta and Y. Kuang. Convergence results in a well-known delayed predator-prey system. *J. Math. Anl.Appl.*, 204:840–853, 1996.
17. J. E. Billotti and J. P. LaSalle. Periodic dissipative processes. *Bull. AMS (N.S.)*, 6:1082–1089, 1971.
18. R. K. Brayton. Bifurcation of periodic solutions in a nonlinear difference-differential equation of neutral type. *Quar. Appl. Marh.*, 24:215–244, 1966.

19. I. Camargo. Second-order retarded functional differential equations: generic properties. In *Advanced Topics in the Theory of Dynamical Systems. Eds G. Fusco, M. Iannelli and L. Salvadori - Notes and Reports in Mathematics in Science and Engineering*, number 6, pages 51–73. Academic Press, 1989.

20. M. C. Carbinatto and K. P. Rybakowski. On a general Conley index continuation principle for singular perturbation problems. *Ergodic Th. Dyn. Systems*, to appear.

21. M. C. Carbinatto and K. P. Rybakowski. On convergence, admissibility and attractors for damped wave equations on squeezed domains. *Proc. Royal Soc. Edinburgh*, to appear.

22. M. C. Carbinatto and K. P. Rybakowski. Conley index continuation and thin domain problems. *Topological Methods in Nonlinear Analysis*, 16:201–251, 2000.

23. J. Carr. *Applications of Center Manifold Theory*. Springer-Verlag, New York, 1981.

24. M. L. Cartwright. Almost periodic differential equations. *J. Differential equations*, (5):167–181, 1962.

25. N. Chafee. Asymptotic behavior for solutions of a one dimensional parabolic equation with Newmann boundary conditions. *J. Differential equations*, 18:111–135, 1975.

26. N. Chafee and E. F. Infante. A bifurcation problem for a nonlinear partial differential equation of parabolic type. *Aplicable Analysis*, 4:17–37, 1974.

27. M. Chen, X. Chen, and J. K. Hale. Structural stability of time-periodic one-dimensional parabolic equations. *J. Differential equations*, 96:355–418, 1992.

28. X.-Y. Chen and P. Poláčik. Gradient-like structure and Morse decompositions for time-periodic one-dimensional parabolic equations. *J. Differential equations*, 7:73–107, 1995.

29. J. W. Cholewa and J. K. Hale. Some couterexamples in dissipative systems. *Dyn. of Cont. Discrete and Impulsive Systems*, 1998.

30. S. N. Chow and J. K. Hale. *Methods of Bifurcation Theory*. Springer-Verlag, New York, 1982.

31. S. N. Chow, J. K. Hale, and W. Huang. From sine waves to square waves in delay equations. *Proc. Roy. Soc. Edinburgh*, 120-A:223–229, 1992.

32. S. N. Chow and J. Mallet-Paret. Integral averaging and bifurcation. *J. Differential equations*, 26:112–159, 1977.

33. S. N. Chow and J. Mallet-Paret. The Fuller index and global Hopf bifurcation. *J. Differential equations*, 29:66–85, 1978.

34. C. C. Conley. Isolated invariant sets and the Morse index. In *CBMS*. Providence, R. I., 1978.

35. C. C. Conley and R. Easton. Isolated invariant sets and isolating blocks. *Trans. AMS*, 158:35–61, 1971.

36. K. Cooke and J. A. Yorke. Some equations modelling growth processes and gonorrhea epidemics. *Mathematical Biosciences*, (16):75–101, 1973.

37. K. L. Cooke and D. W. Krumme. Differential-difference equations and nonlinear initial-boundary value problems for linear hyperbolic differential equations. *J. Math. Anal. Appl.*, 24:372–387, 1968.

38. G. Cooperman. α-condensing maps and dissipative systems. In *Ph. D.Thesis*. Brown University, Providence, 1978.

39. C. Dafermos. Asymptotic behavior of solutions of evolutionary equations. In *Nonlinear Evolution Equations, Ed. M. Crandall, p.103-123*. Academic Press, Providence.

40. J. C. F. de Oliveira. The generic G_1-property for a class of NFDE's. *J. Differential equations*, 31(3):329-336, 1979.

41. J. C. F. de Oliveira and L. Fichmann. Discontinuous solutions of neutral functional differential equations. *Publ. Matem.*, 37:369-386, 1993.

42. M. Degiovanni and M. Mrozek. The Conley index for maps in the absence of compactness. *Proc. Royal Soc. Edinburgh*, 123A:75-94, 1993.

43. T. Faria. Normal forms for periodic retarded functional equations. *Proc. Royal Soc. Edinburgh*, 120-A:21-46, 1997.

44. T. Faria. Bifurcation aspects for some delayed population models with diffusion. In *Differential Equations with Applications to Biology (ed. S. Ruan, G. Wolkowicz and J. Wu)*, number 21, pages 143-158. Fields Institute Communications, 1999.

45. T. Faria. Normal forms and Hopf bifurcation for partial differential equations with delays. *Trans. AMS*, 352:2217-2238, 2000.

46. T. Faria. Normal forms for semilinear functional differential equations in Banach spaces and applications. part ii. *Disc. Cont. Dyn. Systems*, 7:155-176, 2001.

47. T. Faria. Stability and bifurcation for a delayed predator-prey model and the effect of diffusion. *J. Math. Anal. Appl.*, 254:433-463, 2001.

48. T. Faria, W. Huang, and J. Wu. Smoothness of center manifolds for maps and formal adjoints for semilinear functional differential equations in general Banach spaces. preprint.

49. T. Faria and L. T. Magalhães. Normal forms for retarded functional differential equations and applications to Bogdanov-Takens singularity. *J. Differential equations*, 122:201-224, 1995.

50. T. Faria and L. T. Magalhães. Normal forms for retarded functional differential equations with parameters and applications to Hopf singularity. *J. Differential equations*, 122:181-200, 1995.

51. T. Faria and L. T. Magalhães. Realization of ordinary differential equations by retarded functional differential equations in neighborhoods of equilibrium points. *Proc. Royal Soc. Edinburgh*, 125-A:759-776, 1995.

52. T. Faria and L. T. Magalhães. Restrictions on the possible flows of scalar retarded functional differential equations in neighborhoods of singularities. *J. Dynam. Diff. Eqs.*, 8(1):35-70, 1996.

53. A. Fathi, M. Herman, and J.-C. Yoccoz. A proof of Pesin's stable manifold theorem. In *Geometric Dynamics- Lecture Notes in Math.*, pages 177-215. Springer, 1983.

54. L. Fichmann. A compact dissipative dynamical system for a difference equation with diffusion. *J. Differential equations*, 138:1-18, 1997.

55. L. Fichmann and W. M. Oliva. Collision of global orbits in C^∞ retarded functional differential equations. In *Topics in Functional Differential and Difference Equations*. Fields Inst. Commun., 29, 2001.

56. L. Fichmann and W. M. Oliva. One-to-oneness and hyperbolicity. In *Topics in Functional Differential and Difference Equations*, pages 113-131. Fields Inst. Commun., 29, 2001.

57. B. Fiedler and C. Rocha. Heteroclinic orbits of semilinear parabolic equations. *J. Differential equations*, 125:239-281, 1996.

58. B. Fiedler and C. Rocha. Orbit equivalence of global attractors of semilinear parabolic differential equations. *Trans. AMS*, 352,n.1:257–284, 1999.

59. A. Floer. Morse theory for Lagrangian intersections. *J. Diff. Geometry*, 28:513–547, 1988.

60. A. Floer. Symplectic fixed points and holomorphic spheres. *Comm. Math. Phys.*, 120:575–611, 1989.

61. R. Franzosa. The connection matrix theory for Morse decompositions. *Trans. AMS*, 311:561–592, 1989.

62. R. Franzosa and K. Mischaikow. The connection matrix theory for semiflows on (not necessarily compact) metric spaces. *J. Differential equations*, 71:270–287, 1988.

63. G. Fusco and W. M. Oliva. Jacobi matrices and transversality. *Proc. Royal Soc. Edinburgh*, 117A:231–243, 1988.

64. G. Fusco and W. M. Oliva. Transversality between invariant manifolds of periodic orbits for a class of monotone dynamical systems. *J. Dynam. Diff.Eqs.*, 2(1):1–17, 1990.

65. G. Fusco and W. M. Oliva. A Perron theorem for the existence of invariant subspaces. *Ann. Mat. Pura ed Appl.*, CLX(IV):63–76, 1991.

66. F. R. Gantmacher and M. Krein. Sur les matrices compltement non ngatives et oscillatoires. *Compositio Math.*, (vol.4):445–476, 1937.

67. M. Gęba, M. Izydorek, and A. Pruszko. The Conley index in Hilbert spaces and its applications. *Studia Math.*, 134(3):217–233, 1999.

68. M. Gobbino and M. Sardella. On the connectedness of attractors for dynamical systems. *J.Differential equations*, 133:1–14, 1997.

69. J. Gukhenheimer and P. Holmes. *Nonlinear Oscilations, Dynamical Systems and Bifurcations*. Springer-Verlag, 1983.

70. J. K. Hale. Sufficient conditions for stability and instability of functional differential equations. *J. Differential equations*, (1):452–482, 1965.

71. J. K. Hale. Smoothing properties of neutral equations. *An. Acad. Brasil. Ci.*, 45:49–50, 1973.

72. J. K. Hale. *Theory of Functional Differential Equations*. Springer-Verlag, 1977.

73. J. K. Hale. Topics in Dynamic Bifurcation Theory. In *CBMS*, number 47. AMS, Providence, 1981.

74. J. K. Hale. Infinite dimensional dynamical systems. In *Geometric Dynamics-Ed. J. Palis. Lect. Notes in Math*, number 1007, pages 379–400. World Scientific, 1983.

75. J. K. Hale. Asymptotic Behavior and dynamics in infinite dimensions. In *Research Notes in Math.*, number 132, pages 1–41. Pitman, Boston, 1985.

76. J. K. Hale. Flows on centre manifolds for scalar functional differential equations. *Proc.Royal Soc. Edinburgh*, 101A:193–201, 1985.

77. J. K. Hale. Local flows for functional differential equations. *Contemporary Math.*, 56:185–192, 1986.

78. J. K. Hale. Asymptotic Behavior of Dissipative Systems. In *Mathematical Surveys and Monographs*, number 25. AMS, Providence, 1988.

79. J. K. Hale. Effects of delays on dynamics. In *Topological Methods in Differential Equations and Inclusions (ed. Granas and Frigon)*, pages 191–238. Kluwer, 1995.

80. J. K. Hale. Dissipation and attractors. In *Int. Conf. on Differential Equations*, pages 622–637. World Scientific, 1999.

81. J. K. Hale and W. Huang. Periodic doubling in singularly perturbed delay equations. *J. Differential equations*, (114):1–23, 1994.

82. J. K. Hale and W. Huang. Periodic solutions of singularly perturbed delay equations. *Z. Angew. Math. Phys.*, (47):57–88, 1996.

83. J. K. Hale, J. P. LaSalle, and M. Slemrod. Theory of a general class of dissipative systems. *J. Math. Anal. Appl.*, 39:177–191, 1972.

84. J. K. Hale and X.-B. Lin. Symbolic dynamics and nonlinear semiflows. *Ann. Mat. Pura ed Appl.(IV)*, CXLIV:229–260, 1986.

85. J. K. Hale and O. Lopes. Fixed point theorems and dissipative processes. *J. Differential equations*, 13:391–402, 1973.

86. J. K. Hale and S. M. V. Lunel. *Introduction to Functional Differential Equations*. Springer-Verlag, 1993.

87. J. K. Hale, L. Magalhães, and W. M. Oliva. An Introduction to Infinite Dimensional Dynamical Systems-Geometric Theory. In *Applied Mathematical Sciences*, volume 47. Springer-Verlag, New York, 1984.

88. J. K. Hale and W. M. Oliva. One-to oneness for linear retarded functional differential equations. *J. Differential equations*, 20:28–36, 1976.

89. J. K. Hale and G. Raugel. Lower semicontinuity of attractors of gradient systems and applications. *Ann. Mat. Pura ed Appl.(IV)*, CLIV:281–326, 1989.

90. J. K. Hale and G. Raugel. Convergence in gradient like systems. *ZAMP*, 43:63–124, 1992.

91. J. K. Hale and K. P. Rybakowski. On a gradient-like integro-differential equation. *Proc. Royal Soc. Edinburgh*, 92A:77–85, 1982.

92. J. K. Hale and J. Scheurle. Smoothness of bounded solutions of nonlinear evolution equations. *J. Differential equations*, 56:142–163, 1985.

93. J. K. Hale and S. M. Tanaka. Square and pulse waves with two delays. *J. Dynam. Diff. Eq.*, 12:1–30, 2000.

94. J. K. Hale and M. Weedermann. On perturbations of delay-differential equations with periodic orbits. preprint, 2001.

95. R. S. Hamilton. The inverse function theorem of Nash and Moser. *Bull. AMS*, 7(1):65–222, 1982.

96. A. Haraux. Two remarks on dissipative hyperbolic systems. Sem. College de France 7. In *Research Notes in Math.*, number 122, pages 161–179. Pitman, Providence, 1985.

97. X. Z. He. Stability and delays in a predator-prey system. *J. Math. Anal. Appl.*, 198:335–370, 1996.

98. D. Henry. Small solutions of linear autonomous functional differential equations. *J. Differential equations*, 8:494–501, 1970.

99. D. Henry. Geometric Theory of Semilinear Parabolic Equations. In *Lect. Notes in Math.-Springer Verlag*, number 840. Providence, 1981.

100. D. Henry. Some infinite dimensional Morse-Smale systems defined by parabolic differential equations. *J. Differential equations*, 59:165–205, 1985.

101. Y. Hino, S. Murakami, and T. Naito. Functional Differential Equations with Infinite Delay. In *Lecture Notes in Math*. Springer-Verlag, New York, 1991.

102. M. W. Hirsh, C. C. Pugh, and M. Shub. Invariant Manifolds. In *Lecture Notes in Math.*, volume 583. Springer-Verlag, 1977.

103. W. Hurewicz and H. Wallman. *Dimension Theory*. Princeton Un. Press, 1948.

104. N. Iwasaki. Local decay of solutions for symmetric hyperbolic systems and coercive boundary conditions in exterior domains. *Pub. Res. Inst. Math. Sci.,Kyoto Univ.*, 5:193–218, 1969.

105. M. Izydorek and K. P. Rybakowski. On the Conley index in Hilbert spaces in the absence of uniqueness. *Fund. Math.*, 171:31–52, 2002.
106. M. Izydorek and K. P. Rybakowski. The Conley index in Hilbert spaces and a problem of Angenent and van der Vorst. *Fund. Math.*, to appear.
107. J. P. Kahane. Mesures et dimensions. In *Lecture Notes in Math.*, volume 565. Springer-Verlag, New York, 1976.
108. J. K. Kaplan and J. A. Yorke. Ordinary differential equations which yield periodic solutions of differential delay equations. *J. Math. Anal. Appl.*, (48):317–325, 1974.
109. T. Kato. *Perturbation theory of linear operators.* Springer-Verlag, 1980.
110. A. Katok and L. Mendoza. Dynamical systems with non-uniformly hyperbolic behavior. In *supplement to Introduction to the Modern Theory of Dynamical Systems, by A. Katok and B. Hasselblatt.* Cambrigge University Press, 1995.
111. S. Kobayashi and K. Nomizu. *Foundations of Differential Geometry, Vol I.* Interscience Publishers, 1963.
112. V. Kolmanovskii and K.M. Myshkis. *Introduction to the theory and applications of functional-differential equations.* Kluweer Academic Publishers, Dordrecht, 1999.
113. M.A. Krasnoselski, Je.A. Lifshits, and A.V. Sobolev. *Positive Linear Systems-The method of Positive Operators.* Heldermann Verlag, 1989.
114. Y. Kuang. *Delay Differential Equations with Applications in Population Dynamics.* Academic Press, Boston, 1993.
115. J. Kurzweil. Invariant manifolds for flows. In *Differential Equations and Dynamical Systems*, pages 431–468. Academic Press, 1967.
116. J. Kurzweil. Global solutions of functional differential equations. In *Lecture Notes in Math.*, volume 144. Springer-Verlag, New York, 1970.
117. J. Kurzweil. Invariant manifolds I. *Comm. Math. Univ. Carolinae*, 11:336–390, 1970.
118. J. Kurzweil. Small delays dont matter. In *Lecture Notes in Math.*, volume 206, pages 47–49. Springer-Verlag, New York, 1971.
119. O.A. Ladyzenskaya. A dynamical system generated by the Navier Stokes equation. *J. Soviet Math.*, 3:458–479, 1975.
120. O.A. Ladyzenskaya. On the determination of minimal global attractors for the Navier-Stokes and other partial differential equations. In *Russian Mathematical Surveys*, number 42:6, pages 27–73. 1987.
121. S. Lang. *Differential manifolds.* Addison Wesley, 1962.
122. LaSalle. The Stability of Dynamical Systems. In *CBMS Regional Conference Series*, pages 27–73. SIAM 25, 1976.
123. J.J. Levin and J. Nohel. On a nonlinear delay equation. *J. Math. Anal. Appl.*, (8):31–44, 1964.
124. J. Lewowicz. Stability properties of a class of attractors. *Trans. AMS*, (185):183–198, 1973.
125. X. Lin, J.W.-H So, and J. Wu. Centre manifolds for partial differential equations with delays. *Proc. Royal Soc. Edinburgh*, 122A:237–254, 1992.
126. O. Lopes. *Asymptotic fixed point theorems and forced oscillations in neutral equations.* Ph.D. Thesis, Brown University, Providence RI, 1973.
127. K. Lu. A Hartman-Grobman theorem for scalar reaction-diffusion equations. *J. Differential equations*, 93:364–394, 1991.
128. W. Ma and Y. Takeushi. Stability analysis on a predator-prey system with distributed delays. *J. Computational and Appl. Math.*, 88:79–94, 1998.

129. R. Mañé. On the dimension of the compact invariant sets of certain nonlinear maps. In *Lecture Notes in Math.*, volume 898, pages 230–242. Springer-Verlag, New York, 1981.

130. R. Mañé. Lyapunov exponents and stable manifolds for compact transformations. In *Geometric Dynamics - Lecture Notes in Math.*, volume 1007, pages 522–577. Springer-Verlag, New York, 1983.

131. Luis T. Magalhães. Invariant manifolds for functional differential equations close to ordinary differential equations. *Funkcial. Ekvac.*, (28):57–82, 1985.

132. Luis T. Magalhães. Persistence and smoothness of hyperbolic invariant manifolds for functional differential equations. *SIAM J. Math. Anal.*, (18):670–693, 1987.

133. J. Mallet-Paret. Negatively invariant sets of compact maps and an extension of a theorem of Cartwright. *J. Differential equations*, (22):331–348, 1976.

134. J. Mallet-Paret. Generic and qualitative properties of retarded functional differential equations. In *Symposium of Functional Differential Equations, São Carlos, Aug. 1975, Atas da SBM*. Sociedade Brasileira de Matematica, Sao Carlos-SP-Brasil, 1977.

135. J. Mallet-Paret. Generic periodic solutions of functional differential equations. *J. Differential equations*, (25):163–183, 1977.

136. J. Mallet-Paret. Morse decompositions and global continuation of periodic solutions for singularly perturbed delay equations. In *Systems of nonlinear partial differential equations*. Oxford, 1982.

137. J. Mallet-Paret. Morse decompositions for delay-differential equations. *J. Differential equations*, (72):270–315, 1988.

138. P. Massatt. Attractivity properties of α-contractions. *J. Differential equations*, 48:326–333, 1983.

139. H. Matano. Convergence of solutions of one-dimensional semilinear parabolic equations. *J.Math.Kyoto Univ.*, 18:221–227, 1978.

140. H. Matano. Nonincrease of the lap-number of a solution for a one-dimensional semilinear parabolic equation. *J. Fac. Sci.Univ. Tokyo*, 29:401–441, 1982.

141. C. McCord and K. Mischaikow. On the global dynamics of attractors for scalar delay equations. *J. Amer. Math. Soc.*, 9:1095–1133, 1996.

142. M. C. Memory. Stable and unstable manifolds for partial functional differential equations. *Nonlinear Anal.*, TMA 16:131–142, 1991.

143. M. Mrozek and K.P. Rybakowski. A cohomological Conley index for maps on metric spaces. *J. Differential equations*, 90:143–171, 1991.

144. J. Nagumo and M. Shimura. Self-oscillation in a transmission line with a tunnel diode. *Proc. IRE*, 49:1281–1291, 1961.

145. R. Nussbaum. Some asymptotic fixed point theorems. *Trans. AMS*, 171:349–375, 1972.

146. R. Nussbaum. Periodic solutions of analytic functional differential equations are analytic. *Michigan Math. J.*, 20:249–255, 1973.

147. W. M. Oliva. Functional differential equations on compact manifolds and an approximation theorem. *J. Differential equations*, (5):483–496, 1969.

148. W. M. Oliva. Functional differential equations on manifolds. pages 103–116. Soc. Brasileira de Matematica - Colecao Atas n. 1, 1971.

149. W. M. Oliva. Some open questions in the geometric theory of retarded functional differential equations. In *Proc. 10th Brazilian Colloq. Math.*, Pocos de Caldas- MG- Brasil, 1975. Impa.

150. W. M. Oliva. Functional Differential Equations- generic theory. In *Dynamical Systems-An International Symposium, vol.I, 195–208 eds. L. Cesari, J.K. Hale and J.P. LaSalle*, New York, 1976. Academic Press.

151. W. M. Oliva. The behavior at infinity and the set of global solutions of retarded functional differential equations. In *Proc of the Symposium on Functional Differential Equations p.103–126*, Sao Carlos(S)- Brasil, 1977. Sociedade Brasileira de Matematica- Colecao Atas.

152. W. M. Oliva. Stability of Morse-Smale maps. pages 1–49. D.Mat.Aplicada-IME- Un. Sao Paulo-Brasil, 1982.

153. W. M. Oliva. Retarded equations on the sphere induced by linear equations. *J. Differential equations*, (49):453–472, 1983.

154. W. M. Oliva. Morse–Smale semiflows. Openness and A-stability. In *Proc. of the Conference on Differential Equations and Dynamical Systems -to appear*, volume 31. Fields Institute Communications, 2002.

155. W. M. Oliva, J. C. F. de Oliveira, and J. Sola-Morales. An infinite-dimensional Morse-Smale map. *NoDEA*, 1:365–387, 1994.

156. W. M. Oliva, M. N. Kuhl, and L. T. Magalhães. Diffeomorphisms of R^n with oscillatory jacobians. *Publ. Matem.*, 37:255–269, 1993.

157. W. M. Oliva and P. Z. Taboas. Existence of periodic orbits, set of global solutions and behavior near equilibrium for volterra equations of retarded type. *Portugaliae Mathematica*, 54:165–186, 1997.

158. V. Oseledets. A multiplicative ergodic theorem. liapunov characteristic numbers for dynamical systems. *Trans. Moscow Math. Soc.*, 19:197–221, 1968.

159. R. Palais and S. Smale. Morse theory on Hilbert manifolds. *Bull. Amer. Math. Soc.*, 70:165–171, 1964.

160. J. Palis. On Morse-Smale dynamical systems. *Topology*, 3:385–404, 1968.

161. J. Palis and S. Smale. Structural stability theorems. In *Global Analysis, Proc. Symp. Pure Math.*, number 14, pages 223–231, Providence, 1970. AMS.

162. J. Palis and F. Takens. Stability of parametrized families of gradient vector-fields. *Ann. of Math.*, 118:383–421, 1983.

163. J. Palis and F. Takens. *Hyperbolicity and Sensitive Chaotic Dynamics at Homoclinic Bifurcations*. Cambridge University Press, 1993.

164. L. Pandolfi. Feedback stabilization of functional differential equations. *Bolletino della Unione Matematica Italiana*, 11:626–635, 1975.

165. M. M. Peixoto. On an approximation theorem of Kupka and Smale. *J. Differential Equations*, 3:214–227, 1966.

166. Ya. Pesin. Families of invariant manifolds corresponding to nonzero characteristic exponents. *Math. USSR-Izv.*, 10:1261–1305, 1976.

167. P. Poláčik and K. P. Rybakowski. Imbedding vector fields in scalar parabolic equations with dirichlet bvps. *Ann. Scula Norm. Sup. Pisa*, XXII:737–749, 1995.

168. V. M. Popov. Pointwise degeneracy of linear, time invariant, delay differential equations. *J. Differential equations*, 11:541–561, 1972.

169. M. Prizzi. Realizing vector fields without loss of derivatives. *Ann. Scuola Norm. Sup. Pisa Cl. Sci. (4)*, XXVII:289–307, 1998.

170. C. Pugh and M. Shub. Ergodic attractors. *Trans. Amer. Math. Soc.*, 312:1–54, 1989.

171. M. Raghunathan. A proof of Oseledec's multiplicative ergodic theorem. *Israel J. Math.*, 32:356–362, 1979.

172. G. Raugel. Global attractors in partial differential equations. 2001. to appear.
173. H. L. Royden. *Real Analysis*. The Macmillan Co., New York, 1963.
174. D. Ruelle. Ergodic theory of differentiable dynamical systems. *Inst. Hautes tudes Sci. Publ. Math.*, 50:27–58, 1979.
175. D. Ruelle. Characteristic exponents and invariant manifolds in Hilbert spaces. *Ann. of Math. (2)*, 115:243–290, 1982.
176. D. Ruelle. *Elements of Differentiable Dynamics and Bifurcation Theory*. Academic Press Inc., 1989.
177. K. P. Rybakowski. On the homotopy index for infinite dimensional systems. *Trans. AMS*, 269:351–383, 1982.
178. K. P. Rybakowski. On the Morse index for infinite-dimensional semiflows. In *Dynamical Systems II (Bednarek /Cesari eds.)*. Academic Press, 1982.
179. K. P. Rybakowski. The Morse index, repeller-attractor pairs and the connection index for semiflows on non compact spaces. *J. Differential equations*, 47:66–98, 1983.
180. K. P. Rybakowski. Irreducible invariant sets and asymptotically linear functional differential equations. *Boll. Unione Mat. Ital.*, 6-3B:245–271, 1984.
181. K. P. Rybakowski. Trajectories joining critical points of nonlinear parabolic and hyperbolic partial differential equations. *J. Differential equations*, 51:182–212, 1984.
182. K. P. Rybakowski. Nontrivial solutions of elliptic boundary value problems with resonance at zero. *Annali Mat. Pura ed Appl. IV*, CXXXIX:237–278, 1985.
183. K. P. Rybakowski. *The Homotopy Index and Partial Differential Equations*. Springer Verlag, 1987.
184. K. P. Rybakowski. Realization of arbitrary vector fields on invariant manifolds of delay equations. *J. Differential equations*, 114:222–231, 1994.
185. K. P. Rybakowski and E. Zehnder. A Morse equation in Conley's index theory for semiflows on metric spaces. *Ergodic Th. Dyn. Systems*, 5:123–143, 1985.
186. R. J. Sacker and G. R. Sell. A spectral theory for linear differential systems. *J. Differential equations*, 27:320–358, 1978.
187. R. J. Sacker and G. R. Sell. Dynamics of Evolutionary Equations. *J. Differential equations*, 38:135–160, 1980.
188. G. R. Sell and Y. You. Dynamics of Evolutionary Equations. In *Applied Mathematical Sciences*, volume 143. Springer Verlag, 2002.
189. M. Shub. *Global Stability of Dynamical Systems*. Springer-Verlag, New York, 1987.
190. S. Smale. An infinite dimensional version of Sard's Theorem. *Amer. J. Math.*, 87:861–866, 1965.
191. S. Smale. Differentible dynamical systems. *Bull.Amer.Math. Soc.*, 73:747–817, 1967.
192. H. L. Smith. *Monotone dynamical systems. An introduction to the theory of competitive and cooperative systems*, volume 41 of *Mathematical Surveys and Monographs*. American Mathematical Society, 1995.
193. N. Sternberg. One-to oneness of the solution map in retarded functional differential equations. *J. Differential equations*, 85:201–213, 1990.
194. N. Sternberg. A Hartmann-Grobman theorem for maps. *Partial and Functional Differential Equations*, 1992.
195. N. Sternberg. A Hartman-Grobman theorem for a class of retarded functional differential equations. *J. Math. Anal. Appl.*, 176:156–163, 1993.

196. M. Taboada and Y. C. You. Invariant manifolds for retarded semilinear wave equations. *J. Differential equations*, 114:337–369, 1994.

197. F. Takens. Singularities of vector fields. *Pub. Math. IHES*, 43:47–100, 1974.

198. R. Temam. *Infinite Dimensional Dynamical Systems in Mechanics and Physics*. Springer-Verlag, New York, 1988.

199. P. Thieullen. Fibrés dynamiques asymptotiquement compacts. Exposants de Lyapunov. Entropie. Dimension. *Ann. Inst. H. Poincare Anal. Non Linéaire*, 4:49–97, 1987.

200. C. C. Travis and G. F. Webb. Existence and stability for partial functional differential equations. *Trans. AMS*, 200:395–418, 1974.

201. M. Weedermann. Normal forms for neutral functional differential equations. In *Topics in Functional Differential and Difference Equations (ed. T. Faria and P. Freitas)*, volume 29, pages 361–368. Fields Institute Communications, 2001.

202. S. W. Wilson and J. A. Yorke. Lyapunov functions and isolating blocks. *J. Differential equations*, 13:106–123, 1973.

203. M. Wojtkowski. Invariant families of cones and Liapunov exponents. *Ergodic Th. Dyn. Systems*, 5:145–161, 1985.

204. E. M. Wright. A nonlinear differential difference equation. *J. Reine Angew. Math.*, 194:66–87, 1955.

205. J. Wu. *Theory and Applications of Partial Functional Differential Equations*. Springer-Verlag, New York, 1996.

206. D. Xiao and S. Ruan. Multiple bifurcations in a delayed predator-prey system with nonmonotonic functional response. *J. Differential equations*, 176:495–510, 2001.

207. J. Yorke. Noncontinuable solutions of differential-delay equations. *Proc. AMS*, 21:648–657, 1969.

208. K. Yoshida. The Hopf bifurcation and its stability for semilinear diffusion equations with time delay arising in Ecology. *Hiroshima Math. J.*, 12:321–348, 1982.

209. T. J. Zelenyak. Stabilization of solutions of boundary value problems for a second order parabolic equation with one space variable. *J. Differential equations*, 4:17–22, 1968.

210. T. Zhao, Y. Kuang, and H. L. Smith. Global existence of periodic solutions in a class of delayed Gause-type predator-prey systems. *Nonlinear Anal.*, TMA 28:1373–1394, 1997.

Index

Applied Mathematical Sciences

(continued from page ii)

(continued on next page)

Applied Mathematical Sciences

(continued from previous page)